Offshore Energy and Marine Spatial Planning

The generation of offshore energy is a rapidly growing sector, competing for space in an already busy seascape. This book brings together the ecological, economic, and social implications of the spatial conflict this growth entails. Covering all energy-generation types (wind, wave, tidal, oil, and gas), it explores the direct and indirect impacts the growth of offshore energy generation has on both the marine environment and the existing uses of marine space.

Chapters explore main issues associated with offshore energy, such as the displacement of existing activities and the negative impacts it can have on marine species and ecosystems. Chapters also discuss how the growth of offshore energy generation presents new opportunities for collaboration and co-location with other sectors, for example, the co-location of wild-capture fisheries and wind farms.

The book integrates these issues and opportunities, and demonstrates the importance of holistic marine spatial planning for optimising the location of offshore energy-generation sites. It highlights the importance of stakeholder engagement in these planning processes and the role of integrated governance, with illustrative case studies from the United States, United Kingdom, northern Europe, and the Mediterranean. It also discusses trade-off analysis and decision theory and provides a range of tools and best practices to inform future planning processes.

Katherine L. Yates is a Lecturer at The University of Salford, United Kingdom, specialising in spatial planning, distribution modelling, and stakeholder engagement. She is also a National Environmental Research Council Knowledge Exchange Fellow working with the United Kingdom Marine Management Organisation.

Corey J. A. Bradshaw is Matthew Flinders Fellow in Global Ecology and Professor in the College of Science and Engineering at Flinders University in Adelaide, Australia. His research is mainly in the area of global-change ecology—how human endeavour and climate fluctuations have altered past, present, and future ecosystems.

Earthscan Oceans

Transboundary Marine Spatial Planning and International Law
Edited by S.M. Daud Hassan, Tuomas Kuokkanen, Niko Soininen

Marine Transboundary Conservation and Protected Areas
Edited by Peter Mackelworth

Marine and Coastal Resource Management
Principles and Practice
Edited by David R. Green and Jefferey Payne

Citizen Science for Coastal and Marine Conservation
Edited by John A. Cigliano and Heidi L. Ballard

Ocean Energy
Governance Challenges for Wave and Tidal Stream Technologies
Edited by Glen Wright, Sandy Kerr and Kate Johnson

Offshore Energy and Marine Spatial Planning
Edited by Katherine L. Yates and Corey J. A. Bradshaw

For further details please visit the series page on the Routledge website:
http://www.routledge.com/books/series/ECOCE

Offshore Energy and Marine Spatial Planning

Edited by Katherine L. Yates
and Corey J. A. Bradshaw

LONDON AND NEW YORK

from Routledge

First published 2018 by Routledge

2 Park Square, Milton Park, Abingdon, Oxon OX14 4RN
605 Third Avenue, New York, NY 10017

Routledge is an imprint of the Taylor & Francis Group, an informa business

First issued in paperback 2021

British Library Cataloguing-in-Publication Data
A catalogue record for this book is available from the British Library

Library of Congress Cataloging-in-Publication Data
A catalog record for this book has been requested.

ISBN: 978-1-138-95453-3 (hbk)
ISBN: 978-0-367-50850-0 (pbk)

Typeset in Bembo
by codeMantra

Contents

List of figures and tables

Figures

Tables

Acknowledgements

This book combined the efforts of many people, to all of whom we are grateful. All authors were selected for both their academic expertise and real-world experiences in marine spatial planning, and we thank them for their substantial contributions. For many of them, their chapters represented work outside the 'day job', and their added investment in sharing their knowledge is much appreciated. We also thank our colleagues for their support, suggestions, and critical comments. In particular, we thank our reviewers—peer review is an essential aspect of academic rigour and one that helped to shape the book's content, ensuring that it provides readers with as complete and critically reflective overview of marine spatial planning and offshore energy as possible. As well as extensive internal review, this book benefited from many external reviewers to whom we are grateful. Our thanks to David Schoeman, Matthias Kloppmann, François Bastardie, Heather Ritchie, Joao Ferriera, Sylvaine Giakoumi, Gaynor Bagnall, Katie Smyth, Priscilla Lopes, Katrina Davis, Bryce Beukers-Stewart, Hannah Worsely, Linus Hammer, Mags Adams, Rachelle Gould, Kieren Reilly, Stelios Katsanevakis, Wesley Flannery, Marta Coll, Pablo Sanchez-Jerez, Prue Addison, Michael Bell, Ramon Filgueira, Tilen Genov, Stuart Robertson, Sue Kidd, Thomas Smyth, Vanessa Stelzenmuller, Maria Beger, Hugh Possingham, Jacqueline Tweddle, and Sue Kidd.

List of contributors

Karen A. Alexander is an interdisciplinary Research Fellow in the Centre for Marine Socioecology at the University of Tasmania. She has wide-ranging interests, centring on marine governance. Karen specialises in issues around the transition to a green (blue) economy, and more recently, her research has focused on stakeholder engagement and social license for sectors such as offshore renewable energy and aquaculture.

Matthew Ashley is an interdisciplinary scientist with a background in marine ecology and social sciences. His research applies methods from natural and social sciences to investigate the effect of policy, management, and anthropogenic activities on marine ecology, and the economic and social effects on resource users and local communities. He is currently a Research Fellow at the University of Plymouth.

Melanie Austen is Head of Science at Plymouth Marine Laboratory leading the Sea and Society science area. Mel was originally a benthic ecologist specialising in meiofauna, but in the last 15 years, she has been developing and leading United Kingdom- and European Union-funded collaborative marine research that directly interfaces marine ecology and ecosystem modelling with environmental economics to support policy development and management for sustainable ecosystems. This research encompasses the environmental benefits and costs of a diversity of human activity in the seas such as renewable energy extraction, shipping, tourism and recreation, fishing and conservation.

Silvana N.R. Birchenough is Senior Marine Ecologist and Group Manager for Advice & Assessment at the Centre for Environment Fisheries and Aquaculture Science (Cefas), United Kingdom, with 15 years of postdoctoral experience in marine ecology. She is a researcher and scientific advisor on issues related to ecology and human activities. Her main focus is understanding long-term benthic changes resulting from climate change and ocean acidification. More recently, her work has explored the potential impacts of multiple stressors (e.g., temperature, pH changes and metals) on commercial species, generating science evidence of expected effects under a changing climate. She also does field studies looking at processes and function of benthic infauna following disturbance events.

Corey J. A. Bradshaw is Matthew Flinders Fellow in Global Ecology and Professor in the College of Science and Engineering at Flinders University in Adelaide, Australia. His research is mainly in the area of global-change ecology—how human endeavour and climate fluctuations have altered past, present, and future ecosystems.

Eran Brokovich is a marine ecologist at The Ministry of Energy in Israel. His research background includes exploring and understanding anthropogenic threats and their impacts on coral reefs, as well as a variety of other marine ecosystems. His current work focuses on examining the implications of hydrocarbon operations in the marine realm and how best to manage and mitigate effects on marine biodiversity while considering energy demands.

Priscilla M. Brooks is Vice President and Director of Ocean Conservation at the Conservation Law Foundation in Boston, Massachusetts, USA. She holds Master's and Ph.D. degrees in environmental and resource economics from the University of Rhode Island. Her work focuses on protecting New England's ocean wildlife and habitat and advancing sustainable use of ocean resources.

Benjamin Burkhard is Professor for Physical Geography at Leibniz Universität in Hannover, Germany. His main fields of research and teaching are landscape ecology, human-environmental interactions, and ecosystem services.

Ian Davies is the Renewables and Energy Programme Manager at Marine Scotland Science in Aberdeen. Originally trained as a geologist and geochemist at Cambridge and Edinburgh universities, he came to Aberdeen in 1974 to work on marine environmental science, executing and managing projects on marine pollution and the effects of chemicals on marine organisms. Links with several universities have led to successful supervision of M.Sc. and Ph.D. studentships. He currently leads scientific aspects of marine renewable energy and offshore oil and gas for the Scottish Government through Marine Scotland, using science to inform policy, planning, and licensing for sustainable exploitation of energy in Scottish waters.

Penelope Donohue has an Honours degree in Business from Coventry University (2000–2003) and an Honours degree in Marine Biology from Plymouth University (2006–2010). Penelope was awarded a Ph.D. in biology from Glasgow University in 2015 during which she investigated the physiological response of marine calcifying biota to global climate change. During her Ph.D., she did a placement with Marine Scotland (Scottish

Government) reviewing the displacement of commercial sea fishing as a result of policy and administrative activities in the marine environment. Penelope has a range of experience in business, academia and research and, most recently, the civil service. Penelope joined Forest Research as Assistant Statistician in IFOS-Statistics in 2015. Her current roles involved contributions to policy development, monitoring, and evaluation of sustainable forest management by providing statistical information, analysis and advice.

Erik van Doorn is Research Associate at the Walther Schücking Institute for International Law at Kiel University in Germany. His field of expertise is international law of the sea, with a main interest in the international regulation of marine resources, but also new uses of the ocean and the effects of climate change. His research has focused on fisheries, mineral resources of the deep sea, and marine planning. Since 2017, Erik has been member of the Scientific Steering Committee of the international project Surface Ocean – Lower Atmosphere Study (SOLAS).

Charles N. Ehler has been a consultant to the Marine Spatial Planning Programme of the Intergovernmental Oceanographic Commission (IOC) of UNESCO in Paris, France, since 2006. He has worked internationally on coastal and marine spatial planning for the past 45 years, including 27 years as a senior executive in the National Oceanic and Atmospheric Administration (NOAA) in Washington, DC. He is the co-author of the 2009 IOC guide to "Marine spatial planning: a step-by-step approach toward ecosystem-based management".

Sarah Fiona Gahlen is a lawyer and practices at the law firm Lebuhn & Puchta in Hamburg, Germany. She is specialised in international maritime law and questions of ocean governance and the law of the sea. Prior to joining the bar, Sarah did her doctoral studies on the liability for accidents at sea at the Max Planck Institute for Comparative and International Private Law, and obtained her Ph.D. in 2015. She then was a postdoctoral fellow at the Walther Schücking Institute for International Law in Kiel, where her research focused on the legal framework governing marine spatial planning.

Kira Gee is Research Associate at Helmholtz Zentrum Geesthacht, Germany. She has many years of experience in marine spatial planning through a range of projects in the North Sea and Baltic Sea. Apart from marine spatial planning, her main fields of research are (energy) landscapes, perceptions of the sea, and cultural ecosystem services.

Andrew B. Gill is the Director of the Cape Eleuthera Institute, Bahamas. He is also a Visiting Research Fellow at Cranfield University, United Kingdom, affiliated with the Offshore Renewable Energy Engineering Centre, Cranfield Energy and Power. He has over 20 years of experience as an applied aquatic ecologist with a particular interest in the interactions between aquatic organisms (freshwater, coastal, and marine) and marine and offshore renewable energy developments.

Lucy Greenhill is a Research Fellow in Marine Policy and Governance at the Scottish Association for Marine Science. Her research focuses on changes in governance, and the role that marine spatial planning can play in enabling greater participation and integration across interests, disciplines and knowledge systems, to support better decision-making in terms of social and ecological objectives. Lucy has over 10 years of experience in planning and licensing of marine industries, including in particular marine renewable energy, aquaculture, and oil and gas. She comes from an applied public policy background, having worked for the Joint Nature Conservation Committee, the statutory advisor to United Kingdom government departments on offshore activities in relation to nature conservation.

Tara Hooper is an interdisciplinary researcher at Plymouth Marine Laboratory. Her work focuses on interpreting ecological information in terms of ecosystem services and in exploring ecosystem service approaches to impact assessment, which draws on both her background in marine ecology and her Ph.D. in environmental economics. She has a particular interest in the environmental and social impacts of marine renewable energy, and her research has included tidal range, offshore wind, and most recently, wave and tidal current technologies.

Ron Janssen is an Associate Professor in the Department of Spatial Economics/SPINlab at the Vrije Universiteit, Amsterdam. He specialises in decision support for environmental and spatial management. Ron's main topics of research are decision analysis, spatial analysis/evaluation, and spatial planning and design.

Tricia K. Jedele has worked as an attorney on environmental law and policy issues for 17 years. She worked for the Rhode Island Attorney General's office and later for Conservation Law Foundation. She spent more than a decade of her legal career advocating for ocean-use planning and was an instrumental voice in the development of the State of Rhode Island's Ocean Special Area Management Plan.

Alice R. Jones works as a postdoctoral researcher in the marine biology group at the University of Adelaide (Australia). She did her Ph.D. at the National Oceanography Centre, Southampton (United Kingdom), where she researched the distributions and environmental preferences of cetaceans, seabirds and basking sharks in southwest United Kingdom. Alice is a quantitative ecologist with a keen interest in marine systems. Her research is focused on the effect that human activities and environmental change have on the distribution of species, populations and ecosystems, and the impact that this has for their survival and ability to provide ecosystem services.

Adrian Judd is a Principal Marine Advisor within the Centre for Environment, Fisheries and Aquaculture Science (Cefas) Evidence and Interpretation Group. Adrian has over 20 years' experience of providing expert scientific advice to regulators and policy makers, nationally and internationally, on the assessment and management of the marine environmental impacts arising from human activities. Adrian is leading work under the blue growth theme of the Cefas Science and Evidence Strategy. Adrian is part of the United Kingdom delegation to the OSPAR Environmental Impacts of Human Activities Committee, and is leading the work programme to deliver a cumulative-effects assessment for the North East Atlantic as part of the OSPAR Quality Status Report. Adrian is also working for the Welsh Government as a member of the drafting team for the Welsh National Marine Plan—developing, interpreting, and communicating ecological, economic, social, and cultural policies, evidence, and implementation guidance.

Simon Jude is a Senior Lecturer in Environmental Decision-Making based at Cranfield University where his research focuses on the development and application of interdisciplinary approaches to investigate and improve environmental decision-making. Marine environmental risks and decision-making form a major component of this work, with recent research projects investigating issues ranging from marine vertebrate interactions with wave and tidal developments, to cumulative effects assessment, and risk frameworks for use in marine licensing processes. He has direct experience of the offshore renewables industry, where he developed spatial risk analysis and visualisation tools to mitigate project risks, and contributed to the production of environmental impact assessments for European offshore wind farm and interconnector projects, while working for an offshore renewables contractor and specialist marine-risk consultancy.

Andronikos Kafas joined Marine Scotland Science (Scottish Government) in 2012 as a research scientist in Offshore Renewable Energy based in Aberdeen. His role is to provide scientific advice to the Licensing Operations Team by assessing environmental impact assessment documentation submitted in support of applications for offshore renewable energy developments. He also provides specialist advice to Marine Planning and Policy Division on interactions between the marine environment and the marine renewable energy industries. Andronikos has been involved in a variety of research projects including commercial fisheries, marine mammals, sea birds, bathymetric surveys, ecosystem services evaluation, and other geospatial projects. His research focuses on the potential impacts of offshore renewable energy developments on commercial fisheries.

Salit Kark is an Associate Professor at the School of Biological Sciences, University of Queensland (Australia) and Head of the Biodiversity Research Group. She is an ecologist and conservation scientist, with interest and expertise in the processes shaping biodiversity and their implications for conservation prioritisation, environmental decisions, and cross-boundary collaboration across ecosystems and spatial scales (from global to local). She strives to advance the links between science, practice and policy and in leading actions that allow us to improve science-based conservation.

Anna Kilponen is a marine scientist currently working as a fisheries consultant. Anna has worked with with non-governmental organisations (World Wildlife Fund, International Union for the Conservation of Nature, Conservation International) on research, conservation, science communication, and environmental education projects. Her primary interests include fisheries, marine protected area management, community-based conservation, and sustainable development. She holds a Master of Science degree in marine biology and conservation from Stockholm University, Sweden.

Sarah E. Lester is an Assistant Professor in the Geography Department at Florida State University. She received her Ph.D. in marine ecology from the University of California Santa Barbara. Her research interests include marine conservation and protected areas, natural resource management and policy, marine spatial planning and ocean zoning, sustainable seafood, fisheries management, and biogeography and macroecology. Her recent research focuses particularly on understanding the effectiveness of marine protected areas, applying trade-off analysis to marine spatial planning, designing ocean zoning and fisheries management around small island states, and understanding offshore aquaculture siting and development—she applies an interdisciplinary lens to all of these research topics.

Noam Levin is an Associate Professor at the Department of Geography at The Hebrew University of Jerusalem, and an Honorary Associate Professor at the University of Queensland. Noam studies geographical and environmental patterns and processes of land-cover changes in the face of human- and climate-induced changes using remote sensing and geographic information systems tools. He has a great interest in maps, and in exploring new methods to analyse spatial information, from historical maps, spatial data layers, aerial photographs, and satellite images.

Stephen C. Mangi is a Senior Environmental Economist at the Centre for Environment, Fisheries and Aquaculture Science (Cefas) with 19 years of practical experience in evaluating different fisheries-management tools including restrictions on gears, fishing effort, and closed areas in sustaining fisheries. He has been principal investigator for projects in the Western Indian Ocean, Mediterranean Sea, United Kingdom, and Caribbean. His main research interests are in social and economic impact assessment, marine protected areas, sustainable fisheries management, and quantification and valuation of ecosystem services.

Tessa Mazor is a Postdoctoral Fellow at the Commonwealth Scientific and Industrial Research Organisation (CSIRO, Australia) in the Oceans & Atmosphere business unit. Tessa is a spatial ecologist with research interests in finding sustainable solutions within marine systems. Her research background includes conservation planning and prioritisation, marine spatial planning, modelling and mapping anthropogenic threats to biodiversity, and integrating social and economic objectives to improve management decisions and the implementation of conservation action.

Timothy G. O'Higgins is a multidisciplinary Research Fellow at University College Cork, Ireland. He works at the interface between environmental science, policy and economics. Tim's current research focus is on mapping the social dimensions of marine planning, aquaculture, climate adaptation, and biodiversity conservation.

Bethan C. O'Leary is Research Associate at the University of York, United Kingdom. Her research interests focus on marine protected areas and fisheries management. She is currently working to inform negotiations over a new implementing agreement for the conservation and sustainable use of biodiversity in areas beyond national jurisdiction at the United Nations.

Ana Payo-Payo is a Postdoctoral Research Fellow at IMEDEA-CSIC (Spain), with a particular interest in population ecology. Ana's work has focused on understanding the complex mechanisms driving population responses and changes in life-history traits due to environmental variability, density feedback, and consecutive perturbations (extreme climatic events, predation, and invasive species) across different spatio-temporal scales. Ana works primarily to seabirds, but the mathematical models she develops can be applied to an array of animal and plant species.

Johanna Polsenberg is the Senior Director for Ocean Governance and Policy at Conservation International in Virginia, USA. Johanna is deeply committed to finding all ways forward to environmental and social sustainability, and her career has been focused on broadening the intersection of environmental protection and resource use. She has worked in Africa, Australia, Indonesia, the South Pacific, the Caribbean, and the United States in settings as varied as the United States Congress, alternative energy start-ups, Alaskan fishing boats, and even underwater on oil platforms, coral reefs, in kelp forests, and alongside humpback whales. She holds an MBA from the University of Maryland, a Ph.D. in biological sciences from Stanford University, and a B.S. in Biochemistry and Chemistry from the University of Vermont.

Lynda Rodwell is an ecological economist, an interdisciplinary scientist focusing on the interaction of humans and the environment. Her research has focused primarily on the sustainable use of the environment considering issues such as the use of marine protected areas as a tool to achieve both conservation and economic objectives and the economic valuation of ecosystem services that support human well-being in European and African contexts. She is currently an Associate Professor at the University of Plymouth, United Kingdom.

Beth E. Scott is a Reader in Marine Ecology. She does multidisciplinary research using her expertise in marine ecology, oceanography, and fisheries sciences. Her research identifies general rules in biophysical oceanographic processes that lead to the creation of hotspots of biodiversity and predator-prey activity. Specifically, her research group defines biological and physical variables that provide the limited, patchy locations and conditions where energy is transferred across trophic levels in marine food webs. Recently, her research has focused on the effects of marine renewable energy systems on multi-trophic interactions and the methods for co-developing a marine spatial planning decision-support system with a range of stakeholders to incorporate ecosystem-service knowledge and values into effective policies.

Joel Stevens graduated in 2016 from California Polytechnic State University, San Luis Obispo with a Master of Science degree in biology. He currently works in the technology industry as a data engineer in Chicago, and his research interests include using heuristic algorithms and machine learning to generate programmatic testing frameworks for enterprise-level business intelligence applications and data warehouses.

Ruth H. Thurstan is a Lecturer at the University of Exeter. Her research interests include exploring historical, social and ecological data to understand temporal dynamics in marine social-ecological systems over the last 200 years, with a view to informing contemporary management and conservation goals.

Crow White is an Assistant Professor in the Biological Sciences Department at California Polytechnic State University, San Luis Obispo. The overarching conceptual theme of his research is to understand causes and consequences of spatial population dynamics of coastal marine species, trophic processes among species in communities and ecosystems, and interactions between marine ecological communities and human users, especially fisheries. A common applied goal of his research is to identify how the implications of the above dynamics can be used to guide sustainable management of renewable natural marine resources and the conservation of marine ecosystems.

Ben Wilson works at the Scottish Association of Marine Science and the University of the Highlands and Islands. He has worked for over 20 years on the ecology of marine mammals, fish, and birds in relation to offshore industrial activities. He currently focuses on the interactions between large marine animals and energy structures (wind, wave, tidal stream, oil, and gas devices). His research aims to understand the interactions enough to develop effective monitoring and mitigation tools that will facilitate regulators to take appropriate consenting decisions and industry to reduce their impacts. As an ecologist, his research seeks to bridge the gaps between biology, oceanography, and engineering.

Katherine L. Yates is a Lecturer at The University of Salford, United Kingdom, specialising in spatial planning, distribution modelling, and stakeholder engagement. She is also a National Environmental Research Council Knowledge Exchange Fellow working with the United Kingdom Marine Management Organisation. Katherine has worked with a wide variety of stakeholders, including government institutions, industry representatives, individual fishers, non-governmental organisations and charities. She is interested in bringing together different types of knowledge to improve understanding of the multifaceted complexities of conservation issues. Ultimately, she aims to contribute to improved marine management, for combined biodiversity and social benefits.

Glossary

Cumulative environmental effects The combined effect on the environment of multiple, different effects (e.g., noise pollution, physical damage) and/or repeated effects over time, caused by past, current, and future activities.

Displacement When an existing activity/use (such as fishing) is moved from its original place (i.e., fishers have to fish somewhere else). Displacement is generally caused by the arrival of a new activity/use or by the introduction of new spatial regulations.

Ecosystem service trade-off analysis An analytical tool used to assess quantitatively the effects of alternative management actions on sector values in a system to determine the set of actions that best maximise sectoral values and minimise inter-sectoral conflicts.

Ecosystem services The goods and services provided to society by ecosystems (e.g., seafood and profits by fisheries, recreational opportunities by coral reefs and kelp forests, storm protection by mangroves).

Efficiency frontier Outer bound of points in a trade-off plot of sector values, representing the most efficient set of management actions and their resulting optimal outcomes, in which the value to one sector cannot be increased without negatively impacting the value of other sector(s) in the system.

Environmental effect Any change to an environmental component from one state to another.

Environmental impact A change great enough to degrade or enhance an environmental component.

Environmental interaction Any form of interplay between environmental components, whether natural or human.

Exclusive economic zone A stretch of water, seabed, and subsoil adjacent to the territorial sea up to a maximum breadth of 200 nautical miles, as measured from the low-water baseline, in which the coastal state can exercise sovereignty rights over the exploration, exploitation, conservation, and management of natural resources.

Geodesign   A set of techniques and technologies for planning environments in an integrated process, including project conceptualisation, analysis, design specification, stakeholder participation and collaboration, design creation, simulation, and evaluation (among other stages).

Geographic information system   A computer system for capturing, storing, checking, and displaying data related to positions on Earth's surface. This enables people to see, analyse, and understand patterns and relationships more easily.

High seas   The part of the ocean beyond the jurisdiction of any coastal state in which all states are allowed to exercise a range of freedoms, although increasingly subject to restrictions and the duty to have due regard to the interests of other states.

Marine (or maritime) spatial planning (MSP)   A public process of analysing and allocating the spatial and temporal distribution of human activities to achieve ecological, economic, and social objectives that are usually specified through a political process.

Maritime Spatial Planning Directive   A framework directive (law) of the European Union (2014/89/EU) that makes maritime or marine spatial planning mandatory for all countries of the EU with marine waters, creates a common framework for marine spatial planning in Europe, and specifies a date (2021) for its implementation when all EU countries must implement a marine spatial plan.

Marine strategy framework   A policy framework of the European Union as constituted by Directive 2008/56/EC, achieving to reach a good environmental status of European waters in 2020 without, however, qualifying or quantifying this status.

Multi-criteria analysis   A sub-discipline of operations research that explicitly considers multiple criteria in decision-making environments.

Non-renewable energy   A finite, natural source of energy.

Renewable energy   A non-exhaustive, natural source of energy.

Sector   Interest in a particular ecosystem service (e.g., wind energy production, fisheries profit, unobstructed ocean viewscape, biodiverse food web). Sectors are comprised of individuals and/or entities with common interests and/or activities (e.g., wind energy industry, fishermen, coastal property owners, conservationists), and are not mutually exclusive.

Stakeholder   Any individual or group with an interest or concern in something (in the case of this book, the marine environment).

Territorial sea   A stretch of water, seabed, subsoil, and airspace with a maximum breadth of 12 nautical miles, as measured from the low-water baseline, in which the coastal state can exercise full sovereignty with the exception of innocent passage of all ships.

Touch table   A multi-touch (the ability for a surface to recognise the presence of more than one point of contact at a time) interactive computer interface in the form of a table.

Introduction

Marine spatial planning in the age of offshore energy

Katherine L. Yates, Johanna Polsenberg, Andronikos Kafas and Corey J. A. Bradshaw

The global demand for energy continues to grow, with a projected 30% increase over the next 25 years (International Energy Agency 2016). By 2022, global spending on offshore oil and gas development and operations is estimated to be US$114 billion, which is more than 2.5 times the US$43 billion spent in 2012 (Marine Board 2013). New, highly lucrative resource discoveries are being made regularly (Chapter 14), and multibillion-dollar investments in previously unexplored areas are announced every year (Eurasia Group 2014; Mann 2016). Likewise, the Environmental and Energy Study Institute (www.eesi.org) predicts substantial growth in offshore renewable-energy generation, with more than seven times the capacity expected in 2020 compared to 2015, and potentially more than triple the 2020 capacity by 2030 (Small et al. 2016). As many countries endeavour to meet international obligations to reduce carbon emissions (EC 2015; Heard et al. 2017), the expansion of renewable energy becomes incorporated into national strategic priorities (e.g., Department for the Economy 2009), with some countries boldly committing to legally binding targets for renewable-energy production (EC 2007; International Energy Agency 2014). Much of the increase in renewable-energy generation is focussed on offshore areas: offshore wind, wave, and tidal energy (also referred to as 'ocean energy'). Indeed, offshore renewable-energy generation is expected to expand markedly in Europe and across the world (Ecofys 2014; EC 2015; International Energy Agency 2015), with predictions that as much as 7% of the total global electrical energy will be generated from marine renewables by 2050 (Esteban and Leary 2012).

The expansion of offshore energy competes for space in an already busy seascape, and it will have many potential impacts on established patterns of sea use, rights of access, and social and cultural value systems (Kerr et al. 2014). Offshore energy developments also raise concerns about competition for access to resources and can lead to conflicts between existing (e.g., fisheries) and emerging (e.g., tidal energy) uses (Jentoft and Knol 2014; Reilly et al. 2016; Ritchie and Ellis 2010).

Effective marine management will not only need to balance the often-competing demands of existing and emerging uses, but also maintain the underlying capacity of the marine environment that supports them. This challenging task is made all the more difficult by the limitations imposed by most traditional modes of governance and management, which were, and still are in many cases, sector-based (Douvere et al. 2007). To rationalise and integrate management of the marine environment, marine spatial planning has emerged as the main tool able to mediate conflict, balance multiple objectives, and move society towards more sustainable decision-making (Calado et al. 2010; Douvere 2008; Douvere and Ehler 2009).

In this volume, we incorporate offshore energy into the complex, multidisciplinary aspects of marine spatial planning, complementing 11 topical subject chapters with four in-depth case studies. The volume begins in Chapter 1 by exploring the utility and underpinning rationale of marine spatial planning and by describing how marine spatial planning can provide an integrated framework that brings together and guides sectoral management. Marine spatial planning is ultimately about optimising the way we use our marine spaces across multiple objectives. This optimisation requires making choices between different potential uses, and trade-off analysis (Chapter 2) can be a powerful tool to help to guide decision-making and increase transparency in the planning process. As part of the assessment process, the environmental implications of potential offshore developments need to be considered, and Chapter 8 provides a thorough review of that topic. Likewise, offshore energy could have a multitude of socio-economic implications (Johnson et al. 2012; Kerr et al. 2014). Chapter 6 explores the displacement of existing uses, focussing on the fishing community as the sector most directly and detrimentally affected. Chapter 7 goes on to consider the often under-appreciated cultural ecosystem service of seascapes, and how offshore energy development can affect them.

What constitutes an optimal marine spatial plan is debatable, depending on the perceptions, values, and specific circumstances of any given stakeholder (Gopnik et al. 2012). Stakeholder participation, explored in Chapter 9 and further illustrated in Chapters 12 and 15, is central to marine spatial planning and to achieving effective and 'optimal' plans (Gopnik et al. 2012; Pomeroy and Douvere 2008; Ritchie and Ellis 2010). Stakeholders can provide valuable information, filling knowledge gaps (Yates and Schoeman 2013) and improving the understanding of the socio-economic context of a plan (Middendorf and Busch 1997; Pita et al. 2010; Reed 2008; Yates 2014). Increased understanding among stakeholders can also lead to new collaborative opportunities, and Chapter 3 discusses the ways in which offshore energy companies and conservation groups can develop mutually beneficial working relationships around the shared objective of reduced risks (both environmental and economic). Collaborations and heightened appreciation of other stakeholders' situations can also lead to innovations and new opportunities, such as the potential to co-locate compatible uses in space and time. Chapter 10 considers potential co-location opportunities between offshore energy and fisheries, both wild capture and aquaculture. The possibility of offshore energy sites to act as *de facto* marine protected areas is discussed in Chapter 11, and an in-depth case study of offshore windfarms is presented in Chapter 13.

To achieve a more comprehensive and coordinated approach, and effectively incorporate the expanding offshore energy industry into all other uses, marine spatial planning should bring together the traditional, sector-oriented management into an integrated framework (Degnbol and Wilson 2008). However, the capacity to do that in reality will depend on both effective governance, the limitations of which are explored in Chapter 4, and existing legal provisions, discussed in Chapter 5.

Many marine spatial planning processes have been initiated as a result of expanding, or desired expansions in, offshore energy (Jay 2010; Johnson et al. 2016; Jones et al. 2013), giving offshore energy a unique role as both a driver of and a stakeholder in marine spatial planning. After reading this volume, you will gain an in-depth understanding of the issues involved with incorporating offshore energy into marine spatial planning, as well as the possible tools that can be used to mitigate negative impacts and maximise opportunities.

References

Calado, H., Ng, K., Johnson, D., Sousa, L., Phillips, M. and Alves, F. (2010). Marine spatial planning: lessons learned from the Portuguese debate. *Marine Policy* 34, pp. 1341–1349.

Degnbol, D. and Wilson, D.C. (2008). Spatial planning on the North Sea: a case of cross-scale linkages. *Marine Policy* 32, pp. 189–200.

Department for the Economy. (2009). *Offshore wind and marine renewable energy in Northern Ireland strategic environmental assessment (SEA) - Non Technical Summary*, Belfast, Northern Ireland.

Douvere, F. (2008). The importance of marine spatial planning in advancing ecosystem-based sea use management. *Marine Policy* 32, pp. 762–771.

Douvere, F. and Ehler, C.N. (2009). New perspectives on sea use management: initial findings from European experience with marine spatial planning. *Journal of Environmental Management* 90, pp. 77–88.

Douvere, F., Maes, F., Vanhulle, A. and Schrijvers, J. (2007). The role of marine spatial planning in sea use management: the Belgian case. *Marine Policy* 31, pp. 182–191.

Ecofys. (2014). *Assessing the EU 2030 Climate and Energy Targets: a Briefing Paper* (Project number: DESNL14683), Utrecht, The Netherlands.

Esteban, M. and Leary, D. (2012). Current developments and future prospects of offshore wind and ocean energy. *Applied Energy* 90, pp. 128–136.

Eurasia Group. (2014). *Opportunities and Challenges for Arctic Oil and Gas Development*. Eurasia Group Report to the Wilson Center. Offshore Technology Conference. OTC Arctic Technology Conference, 10–12 February, Houston, TX, pp. 1–29. www.wilsoncenter.org/sites/default/files/Artic%20Report_F2.pdf.

EC. (2007). *Renewable Energy Roadmap, Renewable Energies in the 21st Century: Building a More Sustainable Future.* COM (2006) 848, Brussels, Belgium.

EC. (2015). The Paris Protocol—a blueprint for tackling global climate change beyond 2020. *Journal of the European Commission* 81, p. 17.

Gopnik, M. Fieseler, C., Cantral, L., McClellan, K., Pendleton, L. and Crowder, L. (2012). Coming to the table: early stakeholder engagement in marine spatial planning. *Marine Policy* 36, pp. 1139–1149.

International Energy Agency. (2014). World Energy Outlook 2014, International Energy Agency, Paris, France.

International Energy Agency. (2015). World Energy Outlook 2015, International Energy Agency, Paris, France.

International Energy Agency. (2016). World Energy Outlook 2016: Executive Summary. Paris, France, www.iea.org/publications/freepublications/publication/World EnergyOutlook2016ExecutiveSummaryEnglish.pdf.

Jay, S. (2010). Planners to the rescue: spatial planning facilitating the development of offshore wind energy. *Marine Pollution Bulletin* 60, pp. 493–499.

Jentoft, S. and Knol, M. (2014). Marine spatial planning: risk or opportunity for fisheries in the North Sea? *Maritime Studies*, 13, p. 1.

Johnson, K., Kerr, S. and Side, J. (2012). Accommodating wave and tidal energy—control and decision in Scotland. *Ocean and Coastal Management* 65, pp. 26–33.

Johnson, K., Kerr, S. and Side, J. (2016). The Pentland Firth and Orkney Waters and Scotland—Planning Europe's Atlantic gateway. *Marine Policy* 71, pp. 285–292.

Jones, P., Qiu, W. and Lieberknecht, L. (2013). MESMA Work Package 6 (Governance). Deliverable 6.1. Typology of Conflicts in MESMA case studies, Department of Geography, University College London, United Kingdom.

Kerr, S., Watts, L., Colton, J., Conway, F., Hull, A., Johnson, K., Jude, S., Kannen, A., MacDougall, S., McLachlan, C., Potts, T. and Vergunst, J. (2014). Establishing an agenda for social studies research in marine renewable energy. *Energy Policy* 67, pp. 694–702.

Mann, J. (2016). BP approves investment for new offshore oil platform. *Houston Business Journal.* www.bizjournals.com/houston/news/2016/12/01/bp-approves-investment-for-new-offshore-oil.html.

Marine Board. (2013). *Best Available and Safest Technologies for Offshore Oil and Gas Operations*, National Academy of Engineering, National Research Council, London, United Kingdom. www.nae.edu/Publications/Reports/89354.aspx.

Middendorf, G. and Busch, L. (1997). Inquiry for the public good: democratic participation in agricultural research. *Agriculture and Human Values* 14, pp. 45–57.

Pita, C., Pierce, G.J. and Theodossiou, I. (2010). Stakeholders' participation in the fisheries management decision-making process: fishers' perceptions of participation. *Marine Policy* 34, pp. 1093–1102.

Pomeroy, R. and Douvere, F. (2008). The engagement of stakeholders in the marine spatial planning process. *Marine Policy* 32, pp. 816–822.

Reed, M.S. (2008). Stakeholder participation for environmental management: a literature review. *Biological Conservation* 141, pp. 2417–2431.

Reilly, K., Hagan, A.M.O. and Dalton, G. (2016). Moving from consultation to participation: a case study of the involvement of fishermen in decisions relating to marine renewable energy projects on the island of Ireland. *Ocean and Coastal Management* 134, pp. 30–40.

Ritchie, H. and Ellis, G. (2010). 'A system that works for the sea'? Exploring stakeholder engagement in marine spatial planning. *Journal of Environmental Planning and Management* 53, pp. 701–723.

Small, L., Beirne, S. and Gutin, O. (2016). Fact Sheet: Offshore Wind—Can the United States Catch up with Europe? Environmental and Energy Study Institute, Washington, DC, USA. www.eesi.org/papers.

Yates, K.L. (2014). View from the wheelhouse: perceptions on marine management from the fishing community and suggestions for improvement. *Marine Policy 48*, pp. 39–50.

Yates, K.L. and Schoeman, D.S. (2013). Spatial access priority mapping (SAPM) with fishers: a quantitative GIS method for participatory planning. *PLoS One, 8*(7), p. e68424.

Chapter 1

Marine spatial planning
An idea whose time has come

Charles N. Ehler

Introduction

Before the last century, ocean space was used mainly for two purposes: marine transport and fishing. Conflicts between these uses were infrequent, except around a few ports, with little need for integrated, comprehensive marine spatial planning (Smith 2001). But today, the human uses of ocean space are growing more intense and varied, yet fisheries are still mostly managed separately from oil and gas development, which in turn is managed separately from marine navigation and offshore renewable energy, and so on, despite real conflicts between and among these multiple uses.

This single-sector planning and management approach (one sector at a time) often fails to resolve conflicts among users of marine space, rarely dealing explicitly with trade-offs among uses (Chapter 2), and even more rarely dealing with conflicts between the cumulative effects of multiple uses on the marine environment (Chapter 8). New uses of marine space, such as ocean energy extraction (e.g., fossil fuels) and generation (e.g., wind), increase the pressure on limited marine space and the potential conflict between different uses. Single-sector management has also tended to reduce and dissipate the effect of enforcement at sea because of the scope and geographic coverage involved and the environmental conditions in which monitoring and enforcement have to operate. Relative to terrestrial environments, little 'public policing' of human activities takes place at sea.

As a consequence of this lack of coordination, integration, and enforcement, in addition to many other stressors such as pollution, invasive species, climate change, ocean acidification, and illegal fishing, marine ecosystems around the world are increasingly imperilled (Halpern et al. 2015). But awareness is growing that the ongoing degradation in marine ecosystems is, in large part, a failure issue of ineffective governance (Crowder et al. 2006, Chapter 4). Many scientists and policy analysts have advocated reforms centred on the idea of 'ecosystem-based management' (Arkema et al. 2006). However, to date, a practical method for translating this concept into operational management practice has not emerged. One step toward ecosystem-based management is the increasing worldwide interest in 'marine spatial planning'. Marine spatial planning is a process now employed in more than 60 countries around the world to identify and resolve conflicts among competing uses of ocean space, and to resolve conflicts between human uses and the natural marine environment (UNESCO-IOC 2017).

Marine spatial planning

Societal demand for marine goods and services, such as food, energy, and habitats, is rising and often exceeds the capacity of marine areas to meet all demands simultaneously. In many cases, users have free access to marine resources (including ocean space), which often leads to over-use, conflicts, and eventual degradation of some marine resources. Since many marine goods and services are not priced in the market—e.g., ecosystem services such as climate regulation and storm protection—conflicts often cannot be resolved by trade-offs made through economic analysis alone (Chapter 2). A public process must be used to decide what mix of outputs or goods and services from the marine area should be produced over time and space. That process is marine spatial planning.

Marine spatial planning (otherwise known as 'maritime spatial planning', or simply 'marine planning') is a practical way to create and organise the uses of marine space and the interactions among them. The process ideally balances the demands for development with ecosystem conservation to achieve social and economic objectives for marine regions in an open and planned way. Marine spatial planning is therefore a public process of analysing and allocating the spatial and temporal distribution of human activities in marine areas to achieve ecological, economic, and social goals and objectives that are usually specified through a political process (Ehler and Douvere 2007). More recently, marine spatial planning has been characterized as "ecosystem-based management at sea" (Katona et al. 2017).

The main characteristics of effective marine spatial planning processes include the following:

1 *Integrated*: across and among sectors and governmental agencies, and among different levels of government;
2 *Strategic and future-oriented*: focused on the long term;
3 *Participatory*: engaging stakeholders actively throughout the entire process;
4 *Adaptive*: capable of learning by doing;
5 *Ecosystem-based*: balancing ecological, economic, social, and cultural goals and objectives toward sustainable development and the maintenance of ecosystem services; and
6 *Place-* or *area-based*: focused on marine spaces that people can understand, relate to, and care about (Ehler and Douvere 2007).

Planning and managing human activities in marine areas is clear and achievable (as we do on land), whereas managing or restoring marine ecosystem functions and processes in the same area is far more difficult, and often impossible. The focus of marine spatial planning is thus on the management of human activities and not on the management of marine ecosystems.

Why marine spatial planning is needed

Most countries already designate or zone marine space for many different human activities, such as maritime transportation, oil and gas development, offshore energy, offshore aquaculture, and waste disposal. However, the problem is that usually this is done sector by sector, or case by case, without much consideration of the effects on either other human activities or the marine environment (Ehler and Douvere 2007). Consequently, this situation has led to two major types of conflicts: (1) conflicts among human uses (user–user conflicts) and (2) conflicts between human uses and the natural environment (user–nature conflicts) (Ehler and Douvere 2007). Resolving these conflicts involves trade-offs among uses and trade-offs between economic development and nature conservation (Chapter 3).

These conflicts can compromise the marine ecosystem services upon which humans and all other life on Earth depend (Chapter 8). Furthermore, without an overarching vision for the marine space that encompasses all users, decision makers usually end up only being able to react to conflict events, often when it is already too late to avoid problems. In contrast, marine spatial planning is a future oriented process that provides the opportunity to plan and shape actions that could lead to a more desirable future for the marine environment. Marine spatial planning offers a way to address both types of conflicts (user–user, user–nature) and select appropriate management actions to maintain and safeguard necessary ecosystem services. Marine spatial planning involves considering how different activities can impact users and ecosystems (Chapters 7 and 8), the impact of displacement of activities (Chapter 6), and the extent to which activities are incompatible or can coexist/co-locate (Chapters 10 and 11).

When effectively put into practice, marine spatial planning includes the following steps and outcomes:

1   *Set priorities* to increase the probability of meeting the development and conservation objectives of marine areas sustainably and equitably; it is necessary to provide a rational basis for setting priorities, and to manage and direct resources to where and when they are most needed;
2   *Create and stimulate opportunities* for new users of marine areas, including ocean energy;
3   *Co-ordinate actions and investments in space and time* to ensure positive returns from those investments, both public and private, and to facilitate complementarity among jurisdictions and institutions;

4   *Provide a spatial vision and consistent direction*, not only of what is desirable, but also of what is possible in marine areas;

5   *Protect nature*, which has its own requirements that should be respected if long-term, sustainable development is to be achieved and if large-scale environmental degradation is to be avoided or minimized;

6   *Reduce fragmentation of marine habitats*, i.e., when ecosystems are isolated into increasingly smaller fragments due to human activities and prevented from functioning properly;

7   *Avoid duplication of effort* by different public agencies and levels of government in marine spatial planning-related activities, including planning, monitoring, and permitting;

8   *Achieve a higher quality of service* at all levels of government by ensuring that permitting of human activities is streamlined when proposed development is consistent with a comprehensive spatial plan for the marine area (Ehler and Douvere 2007).

Space and time are important in marine spatial planning

Some areas of the ocean are more important than others from both ecological and socio-economic perspectives (Crowder and Norse 2008). Species, habitats, populations of animals, oil and gas deposits, sand and gravel deposits, and sustained winds or waves are all distributed unequally in space and over time. Successful marine management needs planners and managers that understand how to work with the spatial and temporal diversity of marine resources. Understanding these spatial and temporal distributions and mapping them are therefore essential components of effective marine spatial planning. Important outcomes of marine spatial planning are enhancing compatible uses and reducing conflicts among uses, as well as reducing conflicts between human activities and nature. Examining how these distributions might change due to climate change and other long-term pressures (e.g., overfishing and/ or habitat loss) on marine systems is another important challenge for marine spatial planning (Molinos et al. 2016).

Offshore energy as a principal 'driver' of marine spatial planning

Pressures from human activities have often led to initiatives to improve the management of marine areas. For example, in the 1970s, the threat of offshore oil and gas development and phosphate mining led to protection of the Great Barrier Reef (Lawrence et al. 2002). More recently, marine spatial planning has been driven by national policies to develop offshore wind energy. These policies are a result of many countries having ambitious renewable energy-generation targets associated with international climate change-mitigation agreements.

In Western Europe, Belgium, the Netherlands, and Germany have all developed and implemented marine plans that incorporated substantial consideration of offshore wind energy. In the United Kingdom, Scotland produced its marine spatial plan, and England is working on 11 regional marine spatial plans, two of which have been approved and the rest are to be completed by 2020, and all of which consider offshore energy potential (United Kingdom 2017). The recent directive on maritime spatial planning now instructs all 23 Member States of the European Union that have marine waters to prepare a legally binding maritime spatial plan by 2021, regardless of the presence of offshore energy potential as a driver (European Parliament 2014).

Offshore wind energy has also been a driver for marine spatial planning in the U.S. states of Massachusetts and Rhode Island, both of which have completed plans for their state waters that identify 'appropriate' areas for wind energy development (Commonwealth of Massachusetts 2009; State of Rhode Island 2010). Again this focus is driven by commitments to generate energy from renewable sources. The State of Rhode Island, for example, is committed to meet 15 percent of its energy needs from renewable energy resources, primarily from offshore wind farms. More recently, regional marine spatial plans, driven by plans to lease large areas of the exclusive economic zone in the Northeast and Mid-Atlantic regions of the USA for offshore wind development, have been completed and approved by the national government (Mid-Atlantic Regional Planning Body 2016; Northeast Regional Planning Body 2016).

While interests in the development of an ocean energy sector (e.g., wave, tidal) are high, large-scale commercial development and economic viability are still only intentions. As such, ocean energy (other than wind) remains in the research and development stage today, and it has not yet been a principal driver of marine spatial planning in any country. Nevertheless, it is a growing consideration. The state of Oregon, for example, completed and approved a plan for its marine waters that considered the potential of ocean energy (State of Oregon 2013). Oregon has an ideal combination of high-energy waves and available infrastructure that has led many companies to stake a claim to the State's potentially lucrative waters. Oregon's Ocean Policy Advisory Committee has gathered data to identify possible wave energy sites, including important fishing areas, important wildlife areas, and other competing uses. The resulting plan amendment identifies four "Renewable Energy Suitability Study Areas" along the Oregon coast where initial development of wave energy will be encouraged and pose the least conflict with existing ocean uses and natural resources. In 2015, Scotland approved its national marine spatial plan and produced a consultation draft for marine renewable energy for Pentland Firth and Orkney waters (0–12 nm offshore)—an area long recognized for its ocean energy potential (Marine Scotland 2016).

Steps of marine spatial planning

Ideally, the development and implementation of marine spatial planning in-volves 10 essential steps (Ehler and Douvere 2009):

1 Identifying need and establishing authority;
2 Obtaining financial support;
3 Organising the process through pre-planning;
4 Organising stakeholder participation;
5 Defining and analysing existing conditions;
6 Defining and analysing future conditions;
7 Preparing and approving the spatial management plan;
8 Implementing and enforcing the spatial management plan;
9 Monitoring and evaluating performance; and
10 Adapting the marine spatial management process.

These 10 steps are not simply a linear process that moves sequentially from step to step. Many feedback loops should be built into the process. For example, goals and objectives identified early in the planning process are likely to be modified as costs and benefits of different management actions are identified later in the planning process. Analyses of existing and future conditions will change as new information is identified and incorporated in the planning process. Engaging stakeholders (Chapter 9) in marine spa-tial planning processes is an essential aspect of achieving successful planning outcomes (Pomeroy and Douvere 2008). The most important reason for this is that marine spatial planning aims to achieve multiple objectives (social, economic, and ecological) and should therefore reflect as many as possible ex-pectations, opportunities, or conflicts that are occurring in the marine spatial planning area. Bringing stakeholders together also facilitates the identifica-tion of collaborative opportunities between different groups (Chapter 3). The scope and extent of stakeholder engagement differs greatly from country to country and is often culturally influenced (Ehler and Douvere 2009). Stake-holder engagement will change the planning process as it develops over time. Planning is a dynamic process, and planners and stakeholders have to be open to accommodating changes as the process evolves over time.

Comprehensive marine spatial planning provides an integrated framework for management that guides, but does not replace, single-sector management. For example, marine spatial planning can provide important contextual in-formation for guiding protected-area or fisheries management, but does not replace it. As well as potentially having their own specific legislation associ-ated with them, marine plans have to include existing legal provision gov-erning the marine environment (Chapter 5). While marine spatial planning is often presented as a complicated process, fundamentally it answers four simple questions:

1   *Where are we today?* What are the baseline conditions? Where are we starting from as we begin planning?
2   *Where do we want to be?* What are alternative spatial scenarios of the future? What is the desired vision of the planning area 20 or more years from now?
3   *How do we get there?* What spatial management actions will move us toward the desired future?
4   *What have we accomplished?* Have the spatial management actions moved us in the direction of the desired vision? If not, how should they be adapted in the next round of planning?

Outputs of marine spatial planning

The principal output of marine spatial planning is a comprehensive spatial management plan for a marine area or ecosystem. The plan guides the whole system toward a "vision for the future". A marine spatial plan sets out priorities for the area and more importantly, defines what these priorities mean in time and space. Typically, a comprehensive spatial management plan has a 10- to 20-year horizon and reflects political and social priorities for the area. The comprehensive marine spatial plan is usually implemented through a zoning map, zoning regulations, and/or a permit system similar to a comprehensive regional plan on land (New York Department of State 2015). Individual permit decisions made within individual sectors, such as fisheries, oil and gas, or tourism sectors, should then be based on the zoning maps and regulations and be consistent with the comprehensive marine spatial management plan. Marine spatial planning aims to provide guidance for a range of decision makers responsible for particular sectors, activities, or concerns, so that they have the means to make decisions confidently in a more comprehensive, integrated, and complementary way than in traditional single-sector management.

Importance of performance monitoring and evaluation in adaptive marine spatial planning

Marine spatial planning is a continuing, adaptive process that should include performance monitoring and evaluation as essential elements of the overall management process. Rather than waiting until a spatial management plan has been developed to begin thinking about monitoring and evaluation, this step should be considered at the very beginning of the marine spatial planning process (Ehler 2014).

The need for an adaptive approach to marine spatial planning has been widely recognized in national and international policy documents. The USA's draft framework for coastal and marine spatial planning refers to the need for marine spatial planning to be "... adaptive and flexible to accommodate changing environmental conditions and impacts, including those associated with global climate change, sea-level rise, and ocean acidification, and new and emerging uses, advances in science and technology, and policy changes" (IOPTF 2009). One of the 10 principles for marine spatial planning, as defined in the European Commission's *Roadmap for Maritime Spatial Planning,* includes the "... incorporation of monitoring and evaluation in the planning process" and recognizes that "... planning needs to evolve with knowledge" (Commission of the European Communities 2008). Consistent with these marine spatial planning policy requirements, each of the marine spatial plans in Massachusetts (Commonwealth of Massachusetts 2009), Germany (German Maritime and Hydrographic Agency 2009), and Norway (Norwegian Ministry of Environment 2013)—often held up as models of good practice—includes references either to an adaptive approach or to monitoring and evaluation as essential elements of an adaptive approach.

Despite the importance of an adaptive approach to marine spatial planning, few have defined what such an approach actually entails (Douvere and Ehler 2011). An adaptive approach requires monitoring and evaluation of the performance of marine spatial plans, but little research has been done to determine how such performance monitoring and evaluation can lead to meaningful results and whether current marine spatial planning initiatives have the essential features, especially specific and measurable objectives, to allow it. However, the latter is crucial as more and more countries attempt to learn from existing marine spatial planning practices, and some countries begin their second- or third-generation marine spatial plans, e.g., the Netherlands (Dutch Ministry of Infrastructure and the Environment 2015).

Most marine spatial planning throughout the world claims to use an adaptive management approach—often defined as 'learning by doing' (Rist et al. 2013): what management actions work, which do not, and why? An adaptive approach to marine spatial planning and management is indispensable to deal with uncertainty about the future and to incorporate various types of change, including global change (e.g., climate change), as well as technological, economic, and political changes. For example, the 2010 Final Recommendations of the U.S. Interagency Ocean Policy Task Force (IOPTF) stated that "... marine spatial planning objectives and progress toward those objectives would be evaluated in a regular and systematic manner, with public input, and adapted to ensure that the desired environmental, economic, and societal outcomes are achieved" (White House Council on Environmental Quality, IOPTF 2010).

Climate change will influence the location of important biological and ecological areas and species (Molinos 2016), while technological change (and climate change) will considerably alter the exploitation of previously inaccessible marine areas such as the Arctic or the high seas (Aspen Institute 2011). Goals and objectives of marine spatial planning, and management plans and actions will inevitably have to be modified to respond to those changes, or plans quickly become ineffective, uneconomical, infeasible, and ultimately, irrelevant.

Success and failures of marine spatial planning—so far

While marine spatial planning remains a young process for marine governance, it certainly shows promise. It has become the principal approach to integrated and comprehensive marine management around the world (UNESCO-IOC 2017). Many examples now exist at all scales, and a steady stream of new applications appears regularly (Katona et al. 2017). However, it remains a work in progress, and it is essential to ask how do we know if marine spatial plans are "successful"? And what does "success" mean? How do we measure it?

"Success" in marine spatial planning can be defined in several ways. Completion of a marine spatial plan can be a success and even more of a success if it is approved and implemented. However, plans have been completed and not implemented due to a variety of reasons, most often a change in government, e.g., Canada's plans for its "Large Ocean Management Areas", including a plan for its portion of the Beaufort Sea that was approved, but not funded (Ricketts and Hildebrand 2011) and Australia's integrated plan for its southeast region that was completed in 2004, but not approved by government (Vince 2014). However, while only three marine spatial plans (Belgium, the Netherlands, and the German lander (state) of Mecklenburg-Vorpommern) were approved by 2006, 35 new plans were developed and approved between 2007 and 2016, most of which were in the latter half of that decade. By 2030, about a third of the surface area of the world's exclusive economic zones could be covered by approved marine spatial plans (UNESCO IOC 2017).

A more meaningful indicator of success is the delivery of actual outcomes specified in a marine spatial plan—a reduction or elimination of use conflicts, an increase of critical habitat protected, improvements in water quality, safe navigation, sustained fisheries, an increase in marine-related jobs and income, and so on. But delivery of these outcomes takes time—a decade or two in the case of responses of the marine environment and the economy to specific management actions. It will take another generation of marine spatial plans to determine if they have been successful in delivering desired outcomes.

Conclusion

Many marine spatial planning processes, particularly in Europe and the USA, have been driven by offshore energy, predominantly offshore wind. Further expansion of offshore energy, including other forms of ocean energy (wave, tidal), is expected over the next two decades. Since ocean energy projects can take up large areas of ocean space, they are likely to compete with other purposes for the same space. The possible negative impacts of offshore energy extraction and generation on other uses, such as marine transport, offshore aquaculture, fishing, and recreation, and the resulting conflict will depend on the spatial and temporal location of energy resources and infrastructure. Marine spatial planning offers a way of mitigating impacts and managing conflict. Certainly over the next 20 years, marine spatial planning will be up and running in the marine areas of most countries. Early and continuing engagement with these emergent marine spatial planning processes will certainly benefit the offshore energy sector.

Highlights

- Marine spatial planning identifies and resolves conflicts among competing uses of ocean space.
- Performance monitoring and evaluation is an essential part of adaptive management within a marine spatial plan.
- Offshore energy has been a principal driver for marine spatial planning processes in many countries.
- There is high potential for substantial further expansion of offshore energy, and in-depth consideration of it within marine spatial planning is essential.

References

Arkema, K.K., Abramson, S.C. and Dewsbury, B.M. (2006). Marine ecosystem-based management: From characterization to implementation. *Frontiers in Ecology and the Environment*, 4(10), pp. 525–532.

Aspen Institute Commission on Arctic Climate Change. (2011). *The Shared Future. Part Two: Marine Spatial Planning in the Arctic. The Aspen Institute: Washington, D.C.* https://assets.aspeninstitute.org/content/uploads/files/content/docs/pubs/Aspen_Climate_Change_Report_2011.pdf.

Commission of the European Communities. (2008). *Roadmap for Maritime Spatial Planning: Achieving Common Principles in the European Union. COM (2008) 791 Final. Brussels, 25.11.2008.* http://eur-lex.europa.eu/legal-content/EN/TXT/PDF/?uri=CELEX:52008DC0791&from=EN.

Commonwealth of Massachusetts, Office of Energy and Environmental Affairs. (2009). *Massachusetts Ocean Management Plan. The Commonwealth of Massachusetts: Boston.* [pdf] Available at http://public.dep.state.ma.us/EEA/eeawebsite/mop/final-v1/v1-complete.pdf.

Crowder, L. and Norse, E. (2008). Essential ecological insights for ecosystem-based management and marine spatial planning. *Marine Policy*, 32(5), pp. 772–778.

Crowder, L.B., Osherenko, G., Young, O.R., Airamé, S., Norse, E.A., Baron, N., Day, J.C., Douvere, F., Ehler, C.N., Halpern, B.S., Langdon, S.J., McLeod, K.L., Ogden, J.C., Peach, R.E., Rosenberg, A.A. and Wilson, J.A. (2006). Sustainability—resolving mismatches in US ocean governance. *Science*, 313(5787), pp. 617–618.

Douvere, F. and Ehler, C. (2011). The importance of monitoring and evaluation in adaptive maritime spatial planning. *Journal of Coastal Conservation*, 15(2), pp. 305–311.

Dutch Ministry of Infrastructure and the Environment. (2015). Policy Document on the North Sea, 2016–2021, including the Netherland's Maritime Spatial Plan, Appendix 2 to the National Water Plan. The Hague, the Netherlands.

Ehler, C. (2014). A Guide to Evaluating Marine Spatial Plans. *IOC Manuals and Guides*, 70(70 ICAM Dossier 8).

Ehler, C. and Douvere, F. (2007). *Visions for a Sea Change: Report of the First International Workshop on Marine Spatial Planning.* Intergovernmental Oceanographic Commission and Man and the Biosphere Programme. IOC Manual and Guides 46: ICAM Dossier 3. UNESCO: Paris.

Ehler, C. and Douvere, F. (2009). *Marine Spatial Planning: A Step-by-Step Approach toward Ecosystem-Based Management.* Intergovernmental Oceanographic Commission and the Man and the Biosphere Programme. IOC Manual and Guides 53. UNESCO: Paris.

European Parliament and the Council of the European Union (2014). *Directive 2014/89/EU of the European Parliament and the Council of 23 July 2014 Establishing a Framework for Maritime Spatial Planning.* http://eur-lex.europa.eu/legal-content/EN/TXT/PDF/?uri=CELEX:32014L0089&from=EN.

German Maritime and Hydrographic Agency (Bundesamt für Seeschifffahrt und Hydrographie). (2009). *Marine Spatial Plans for the North and Baltic Seas.* www.bsh.de/en/Marine_uses/Spatial_Planning_in_the_German_EEZ/index.jsp.

Halpern, B.S., Frazier, M., Potpenko, J., Casey, K.S., Koenig, K., Longo, C., Lowndes, J.S., Rockwood, R.C., Selig, E.R., Selkoe, K.A. and Walbridge, S. (2015). Spatial and temporal changes in cumulative human impacts on the world's ocean. *Nature Communications*, 6(1), p. 7615. doi:doi:10.1038/ncomms8615.

Katona, S.K., Polsenberg, J., Lowndes, J.S., Halpern, B.S., Pacheco, E., Mosher, L., Kilponen, A., Papacostas, K., Guzmán-Mora, A.G., Farmer, G. and Mori, L. (2017). Navigating the seascape of ocean management: Waypoints on the voyage toward sustainable use. Open Channels: Forum for Ocean Planning and Management. [online] p. 44. Available at www.openchannels.org/literaure/16817.

Lawrence, D., Kenchington, R. and Woodley, S. (2002). *The Great Barrier Reef: Finding the Right Balance.* Victoria, Australia: Melbourne University Press, p. 263.

Marine Scotland. (2016). *Pilot Pentland Firth and Orkney Waters Marine Spatial Plan.* www.gov.scot/Publications/2016/03/3696.

Mid-Atlantic Regional Planning Body. (2016). *Mid-Atlantic Ocean Action Plan.* www.boem.gov/Ocean-Action-Plan/.

Molinos, J.G., Halpern, B.S., Schoeman, D.S., Brown, C.J., Kiessling, W., Moore, P.J., Pandolfi, J.M., Poloczanska, E.S., Richardson, A.J. and Burrows, M.T., 2016. Climate velocity and the future global redistribution of marine biodiversity. *Nature Climate Change*, 6(1), pp. 83–88.

New York Department of State. (2015). *Zoning and the Comprehensive Plan.* p. 13. Division of Local Government Services, New York Department of State, New York. www.dos.ny.gov/lg/publications/Zoning_and_the_Comprehensive_Plan.pdf.

Northeast Regional Planning Body. (2016). *Northeast Ocean Plan.* http://neocean-planning.org/plan/.

Norwegian Ministry of Environment. (2013). *Integrated Management of the Marine Environment of the North Sea and Skagerrak (Management Plan). Report to the Storting.* www.regjeringen.no/contentassets/f9eb7ce889be4f47b5a2df5863b1be3d/en-gb/pdfs/stm201220130037000engpdfs.pdf.

Pomeroy, R. and Douvere, F. (2008). The engagement of stakeholders in the marine spatial planning process. *Marine Policy*, 32(5), pp. 816–822.

Ricketts, P.J. and. Hildebrand, L.P. (2011). Coastal and ocean management in Canada: progress or paralysis? *Coastal Management*, 39(1), pp. 4–19.

Rist, L., Selton, A., Samuelsson, L., Sandström, C. and Rosvall, O. (2013). A new paradigm for adaptive management. *Ecology and Society*, 18(4), p. 63. doi:10.5751/ES-06183–180463.

Smith, H.D. (2001). The industrialization of the world ocean. *Ocean and Coastal Management*, 44(9–10), pp. 563–566.

State of Oregon. (2013). *Oregon Territorial Sea Plan, Part 5: Uses of the Territorial Sea for the Development of Renewable Facilities or Other Related Structures, Equipment, or Facilities*. www.oregon.gov/LCD/docs/rulemaking/tspac/Part_5_FINAL_10082013.pdf.

State of Rhode Island. (2010). *Ocean Special Area Management Plan. 2 vols*. http://seagrant.gso.uri.edu/oceansamp/pdf/samp_crmc_revised/RI_Ocean_SAMP.pdf.

UNESCO-Intergovernmental Oceanographic Commission. (2017). *MSP around the Globe*. http://msp.ioc-unesco.org/world-applications/overview/.

United Kingdom Government, Marine Management Organisation. (2017). *Marine Planning in England*. www.gov.uk/government/collections/marine-planning-in-england.

Vince, J. (2014). Oceans governance and marine spatial planning in Australia. *Australian Journal of Maritime and Ocean Affairs*, 6(1), pp. 5–17.

White House Council on Environmental Quality, Interagency Ocean Policy Task Force (IOPTF). (2010). *Final Recommendations of the Interagency Ocean Policy Task Force, July 19*. www.whitehouse.gov/files/documents/OPTF_FinalRecs.pdf.

Chapter 2

Methods and utility of ecosystem service trade-off analysis for guiding marine planning of offshore energy

Joel Stevens, Sarah E. Lester and Crow White

Introduction

This era of unprecedented human population size and demand for electrical energy is contrasted by a growing societal concern for the environmental damage that accompanies natural resource extraction and conventional energy production. To meet the growing energy demand, but with reduced environmental cost, there are emphatic calls for extensive development of renewable-energy sources (Nunez, 2006). This challenge—and the opportunity to solve it—is perhaps most apparent in coastal regions where human population density tends to be highest worldwide (Østergaard, 2009). Congested coastlines tend to have limited land availability and affordability for the development of land-based renewable energy-production facilities, such as wind and solar (Wüstenhagen et al., 2007). However, there are promising opportunities offshore, where ocean wind- and wave-harnessing power facilities, benefiting from rapid technological and engineering advances, can support energy demand for nearby coastal populations (Jay, 2010).

The promise of commercial-scale ocean energy development is greatest along the thin slice of ocean environment known as the continental shelf that is relatively shallow and close to shore and thus cost-effective for development and power transmission to land (Hong and Möller, 2011). However, continental shelf areas are often already congested with existing ocean users, ('sectors'). Commercial and recreational fishing grounds, shipping lanes, biological and aesthetic conservation areas, military zones, oil and gas platforms, and other offshore uses can conflict with and be devalued by offshore ocean energy on the continental shelf (White et al., 2012). To limit such negative impacts, yet also facilitate development of offshore renewable energy facilities, management of continental shelf areas is shifting toward a more holistic, ecosystem-based approach known as 'marine planning' (Douvere, 2008; Jay, 2010). In concept, marine planning accounts for direct and indirect interactions among incumbent sectors, emerging sectors, and the environment in a rational, scientifically informed manner (Douvere and Ehler, 2009).

Marine planning in principle should produce balanced management plans able to both maintain healthy ecosystems and allow sectors to generate value. Although this goal is intuitive, identifying optimal marine plans often requires computationally intensive analysis of complex biological-economic models to elucidate interactions among sectors and with the ecosystem. Ecosystem service trade-off analysis is a promising approach for guiding this process; it explicitly accounts for sector interactions and interactions with the ecosystem to quantify trade-offs and to identify the optimal set of plans that balance the needs of both the ecosystem and its users (Guerry et al., 2012; Lester et al., 2013; White et al., 2012).

In this chapter, we characterise the history and framework for ecosystem service trade-off analysis, and its general methodological steps. We then outline various forms of ecosystem service trade-off analysis and their respective advancements in guiding marine planning. Finally, we outline several ways in which ecosystem service trade-off analysis can be improved in practice for marine planning.

Trade-off analysis framework and analytical steps

Ecosystem service trade-off analysis applies production- and portfolio-theory frameworks to develop holistic solutions to ecosystem service problems. Production theory was originally derived to optimise the production of goods by evaluating different inputs and quantifying subsequent changes to outputs. Thus, the goal is to maximise efficiency and sustainability by minimising the conflicts between overlapping assets within the production system, i.e., assessing the trade-offs (Johnson, 1974; Lester et al., 2013). Portfolio theory quantifies the amount of risk, or variance in expected value of alternative investment strategies, in relation to the expected return from their production or implementation (Halpern et al., 2011; Markowitz, 2014). Trade-off frameworks often apply concepts from decision theory to perform multi-objective analysis to maximise returns or production outputs. In the case of ecosystem service trade-off analysis, sectors and ecosystem services are optimised instead of production processes or financial investments (Lester et al., 2013).

Non-marketable ecosystem services often interact with (and sometimes serve as inputs for determining) the production of marketable services (i.e., fish habitat quality and fisheries profit). Consequently, the value of different objectives can be quantified in different units, such as health, monetary, or aesthetic value. Ecosystem service trade-off analysis is designed specifically to account for such multi-unit problems by allowing different sectoral values to be examined on their own axes in the trade-off analysis. Thus, ecosystem service trade-off analysis is not limited to strictly economic valuation of sectors, and thus can account for the full suite of issues in a marine planning process (Lester et al., 2010, 2013; White et al., 2012).

Analytical methods

For simplicity, we begin with a description of the framework for a two-sector system, let X_i^r and Y_i^r be raw outputs of the utility (value) of ecosystem services x and y to sectors X and Y from $i = 1, 2, 3, \ldots I$ management plans. The plans and outputs could be based on empirical measurements, or estimates developed from a bioeconomic simulation model. In either case, the utilities of the ecosystem services are specific to the different sectors, and they can conflict with each other such that an increase in one hinders production of the other. The objective thus is to identify or determine the management action that maximises the weighted sum of the utilities:

$$^{\alpha}Obj = \max_{M_i} \left\{ \alpha X_i^s + (1-\alpha) Y_i^s \right\}, \tag{2.1}$$

where α is a weighting parameter representing the overall, or societal-level, relative preference for the two sectors. The raw utilities can be in different units and of different magnitude; to account for these sectoral differences objectively in their valuation of the ecosystem, each sector's utility is scaled relative to the range of possible values for that sector:

$$X_i^s = \frac{X_i^r - \min\left(\mathbf{X}^r\right)}{\max\left(\mathbf{X}^r\right) - \min\left(\mathbf{X}^r\right)} \tag{2.2}$$

and

$$Y_i^s = \frac{Y_i^r - \min\left(\mathbf{Y}^r\right)}{\max\left(\mathbf{Y}^r\right) - \min\left(\mathbf{Y}^r\right)}, \tag{2.3}$$

such that they are bounded between 0 and 1:

$$\mathbf{X}^s \in [0,1] \text{ and } \mathbf{Y}^s \in [0,1]. \tag{2.4}$$

The objective function is executed in relation to a chosen range of weighting values, $\alpha = 0, \varepsilon, 2\varepsilon, 3\varepsilon, \ldots, 1$ where $0 < \varepsilon \ll 1$. The values of α represent the range of societal preferences for maximising one sector's utility versus the other's.

Evaluation of a given finite set of management options

When the analysis is limited to a given set of management options (e.g., a set of known historical scenarios, policy proposals, or modelled scenarios), the goal is to identify which of the set of given options is most effective at achieving the objective (Equation 2.1) for each weighting value α. In this situation, there are three important vectors: one representing the set of management options and the other two representing the associated raw utility values:

$$\mathbf{M} = \left[M_{i=1}, M_{i=2}, \dots M_{i=I} \right], \tag{2.5}$$

$$\mathbf{X}^r = \left[X_{i=1}^r, X_{i=2}^r, \dots X_{i=I}^r \right], \tag{2.6}$$

$$\mathbf{Y}^r = \left[Y_{i=1}^r, Y_{i=2}^r, \dots Y_{i=I}^r \right]. \tag{2.7}$$

Note that $\mathbf{M}$ is not necessarily the exhaustive list of all potential management options; it can be a subset of options that are under consideration or possible. In this case, the solutions to the objective function (in relation to each α) reveals the *perceived* efficiency frontier of the best management options relative to the finite set of given options considered. The efficiency frontier is considered perceived because there might be other management options not under consideration that better meet the objective function.

The trade-off analysis would proceed as follows. First, scale $\mathbf{X}^r$ and $\mathbf{Y}^r$ using Equations (2.2) and (2.3). Next, for each value of α, evaluate each pair of the scaled outputs using the objective function (Equation 2.1), and select the specific pair that best achieves the objective function (i.e., maximises it). The solution to these steps is the set of paired outcomes (sector utilities) and associated management options, among those given, that best maximise the weighted sum of the sectoral objectives to the extent possible:

$$^{EFg}\mathbf{M} = \left[{}^{\alpha=0}M_i, {}^{\alpha=\varepsilon}M_i, {}^{\alpha=2\varepsilon}M_i \dots {}^{\alpha=1}M_i \right], \tag{2.8}$$

$$^{EFg}\mathbf{X} = \left[{}^{\alpha=0}X_i, {}^{\alpha=\varepsilon}X_i, {}^{\alpha=2\varepsilon}X_i \dots {}^{\alpha=1}X_i \right], \tag{2.9}$$

$$^{EFg}\mathbf{Y} = \left[{}^{\alpha=0}Y_i, {}^{\alpha=\varepsilon}Y_i, {}^{\alpha=2\varepsilon}Y_i \dots {}^{\alpha=1}Y_i \right]. \tag{2.10}$$

EFg indicates that this solution represents the efficiency frontier of optimal management options among the given options made available for the analysis. If there are only two sectors under consideration, outputs for each for all evaluated management options can be visualised with a 2-D plot, and a 2-D efficiency frontier can be identified (Figure 2.1). Points that lie interior to the frontier are suboptimal, meaning that one sector's value can be increased without affecting another sector. Identifying the frontier is important for two reasons. First, solutions that have been identified to lie along the frontier can be directly used by planners to enhance ecosystem services and can be said to be optimal given the weighting parameters. Second, what is less intuitive is the importance of the shape of the frontier. Visually inspecting the shape produced by plotted solutions provides an understanding of the relationship and strength of the trade-off between sectors (see Lester et al., 2013 for further explanation of frontier shapes).

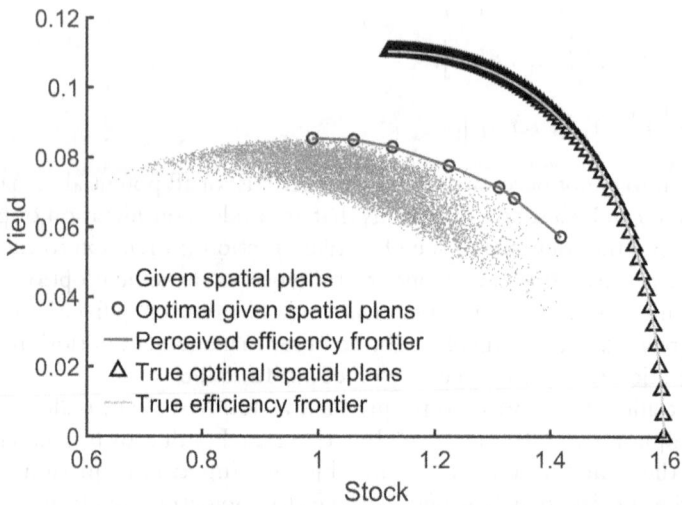

Figure 2.1 Illustrative case-study example of spatial plans, utility outputs, and associated efficiency frontiers. The case-study model follows the basic form of the spatial fisheries model presented by White and Costello (2014); in this case, there are 10 patches of fish populations connected by heterogeneous dispersal patterns. Sector utilities represent wild fish-stock biomass (e.g., valued by a conservation sector) and fisheries yield (e.g., valued by a fisheries sector). Utility values are mediated by fishery harvest effort in the 10 patches; a set of harvests define a particular spatial plan. 10,000 spatial plans were generated through random consideration of fishing effort, 0 < effort < 0.5 in each patch. Trade-off analysis of these given options reveals the perceived efficiency frontier. The optimal spatial plans on the true efficiency frontier were identified using an optimisation search algorithm of the patch-specific effort that maximised the objective function for each α. $\varepsilon = 0.01$ was used for generating both efficiency frontiers. MPAs = marine protected areas

Note that because only a limited set of management options is available to analyse, evaluating α using a small ε could reveal that the same management option is best at achieving the objective function for multiple, contiguous α values.

Searching across all potential management options

If in the planning scenario all possible options are under consideration, raw outputs of utility value are calculated for all the management options, and then they are evaluated in relation to the objective function as described by the steps above. The solution represents the true efficiency frontier of management options and associated utility values to each sector: $^{EF}\mathbf{M}$, $^{EF}\mathbf{X}$, and $^{EF}\mathbf{Y}$.

In many cases, it will not be computationally practical to evaluate all the possible management options. In this case, a heuristic optimisation algorithm is typically employed. For each value of α, the heuristic is used to find the management option that results in outputs that best achieve the objective function. Example heuristic search algorithms include greedy, genetic, and simulated annealing (Banos et al., 2011). Whatever the algorithm, the underlying bioeconomic model determining utilities X_i^r and Y_i^r is evaluated for each novel management option being considered by the algorithm. This can be time-intensive when the model is complex (e.g., a spatial dynamic model) and the search space vast. Consequently, it might be necessary to simplify the model to avoid the heuristic taking a prohibitively long time to finish evaluating the objective function. One such approach is to evaluate the system with a static model that ignores complex and computationally intensive spatial and/or temporal dynamics. Of course, this simplification comes with its own set of trade-offs (Brown et al., 2015).

After repeating these steps for each value of α, the solution represents an estimate of the true efficiency frontier of management options and associated utility values: $^{EFe}\mathbf{M}$, $^{EFe}\mathbf{X}$, and $^{EFe}\mathbf{Y}$, where EFe means that the true efficiency frontier is being estimated.

Maximising three or more outputs

To maximise three or more sector values, modify Equation (2.1) to have a weighting parameter for each utility. For example, in relation to three outputs, $^1X_i^r$, $^2X_i^r$, and $^3X_i^r$, the objective function is:

$$^{\alpha}Obj = \max_{M_i}\left\{\alpha_1\,{}^1X_i^s + \alpha_2\,{}^2X_i^s + \alpha_3\,{}^3X_i^s\right\}. \tag{2.11}$$

The ensuing steps are the same as described above for evaluating two sectors, except that now all different combinations of values of α are analysed in the objective function (instead of just $\alpha = 0, \varepsilon, 2\varepsilon, 3\varepsilon, \ldots, 1$), and the trade-offs and efficiency frontiers are n-dimensional (where n is the number of sectors).

Forms of ecosystem service trade-off analysis for guiding marine spatial planning

Conventional marine planning (i.e., without guidance from an ecosystem service trade-off analysis) is characterised by single-sector planning with a single-sector objective. This form of planning is still the norm (Albrechts, 2006) and does not explicitly account for trade-offs among sectors. However, it sometimes accounts for these trade-offs implicitly by imposing a constraint on the development of the single sector (e.g., so that it cannot dominate the entire seascape). In contrast, the most comprehensive form of marine planning that explicitly accounts for trade-offs among all interacting sectors is

Table 2.1 A categorisation of the four forms of strategic planning with case studies

	Sectors planned[a]		Sectors evaluated[b]			Case studies	Explicit trade-offs
	existing	emerging	existing	emerging			
conventional planning	no	one at a time	no	only the sector being planned	wind MPAs	Hong and Möller (2011) Eadie (1983)	no no
strategic planning of a single sector	no	one at a time	multiple	only the sector being planned	wave tidal wind	Galparsoro et al. (2012) Alexander et al. (2012) White et al. (2012)	no no yes
strategic planning of multiple sectors	no yes	multiple	multiple	multiple	aquaculture multiple	Stevens et al. (2015) Yates et al. (2015)	yes
clean slate strategic spatial planning	multiple	multiple	multiple	multiple	multiple sectors	Haruki Gulf marine spatial planning project (ongoing)	yes

[a]Sectors whose occupation within the study domain are optimised during the spatial planning process.
[b]For every proposed plan, impacts are quantified and minimised for each of these sectors.

characterised by simultaneous and strategic planning for all of the sectors—existing and emerging—across the seascape. This form of marine planning is rare—arguably non-existent to date. In practice, the process of marine planning is typically a compromise between the conventional and the comprehensive forms of marine planning. Below, we outline with empirical and theoretical examples this gradient in forms of spatial planning, their motivation and use in practice, and their estimated improvement in value over one another.

Conventional planning: Spatial planning of a single sector in relation to its own objective (i.e., its value) has been done for as long as people and governments have been able to establish, secure, and regulate development rights, e.g., property rights (Breheny, 1991). Recent large-scale examples include spatial planning of wind farms on land by energy industries to maximise their electricity production and industry profits (Banos et al., 2011; Hong and Möller, 2011; Østergaard, 2009), and by city planners to maximise water efficiency in new housing developments (McLean, 2004). There also are many examples (Table 2.1) of single-sector planning where the objective function is focused on non-economic utilities, such as ecological conservation in the design of public parks on land and marine protected areas in the sea (Eadie, 1983; Roberts et al., 2003). Although conventional planning is without regard to other sectors, the number of management options (i.e., potential spatial plans) is not limitless, but restricted by logistical constraints (e.g., seafloor depth), government regulations (Bryson and Roering, 1988; Douvere et al., 2007; Wüstenhagen et al., 2007; Yates et al., 2015), and the presence of existing users of the domain with non-revocable access rights (e.g., cargo ships in shipping lanes). While applying these constraints does consider inter-sectoral trade-offs because it essentially dictates some plans to be unacceptable due to their effect on the utilities of other sectors, the trade-offs are not necessarily quantified and are only accounted for implicitly (White et al., 2012).

Strategic single-sector planning: There has been a rise over the last decade in the development and use of trade-off analyses for guiding a single, emergent sector in relation to multiple, existing sectors in an ecosystem (Table 2.1) (Agardy et al., 2011; Lester et al., 2010; Polasky et al., 2008; Snyder and Kaiser, 2009; Stewart and Possingham, 2005; White et al., 2012; Yates et al., 2015). However, strategic single-sector planning is not considered 'simultaneous' because the existing sectors cannot be moved or re-planned (only the emerging sector can). By accounting for the value of an emerging sector and its effect on existing sectors, spatial planning can identify management options better able to maximise the values and minimise the conflicts among all sectors in the system. For example, White et al. (2012) did a strategic single-sector planning analysis to guide offshore wind-farm development in relation to existing fisheries, whale watching, and conservation sectors. Compared with conventional planning where trade-offs were not considered and wind farm designs were developed in relation to an objective function that considered only the value of the energy industry (in terms of industry profit), strategic single-sector planning was able to identify spatial plans that would avoid over a million dollars in lost revenue to existing fisheries, and also generated $10 million in extra revenue for the offshore wind energy sector than would be created using conventional planning (White et al., 2012). Although the analysis was only illustrative (no wind farms are being developed in the study area), it is a clear demonstration of the value of strategic spatial planning over conventional approaches that do not account for inter-sectoral trade-offs.

In other examples of strategic single-sector planning, ecosystem-service approaches have been used to develop informed plans for alternative energy development; however, minimal computational support has been done to evaluate trade-offs directly (Alexander et al., 2012; Galparsoro et al., 2012). Instead, collaboration between stakeholders and planners was used to develop spatial plans. In the case of Galparsoro et al. (2012), they developed an index that determined the suitability for wave-energy potential along the coast of Spain. The index itself is composed of individual geographic layers for energy potential constrained by additional layers based on technical, environmental, and socioeconomic impact estimates for each cell. Cells with a high suitability index have both a high energy potential and a low impact on existing sectors, and so on. Although the authors did not do a formal trade-off analysis, they measured the developmental impact on several existing sectors for their entire study domain. Alexander et al. (2012) used a similar approach to develop tidal energy arrays strategically along the coast of Scotland. In contrast to Galparsoro et al. (2012), Alexander et al. (2012) used stakeholders to identify interactively the areas of importance to them using the software *SeaSketch*. Such involvement is undoubtedly a vital part of the marine planning process (Douvere, 2008; Pomeroy and Douvere, 2008). However, without the aid of direct analytical evaluation, it is highly unlikely that collaboration with stakeholders alone will identify plans that lie along the efficiency frontier due to the high number of potential plans that are theoretically possible (Rassweiler et al., 2014; White et al., 2012). Instead, planners could have used available computing power to search through the millions of potential plans, and then have stakeholders review and respond to the optimal plans identified (Rassweiler et al., 2014).

Regardless of the execution of strategic single-sector planning, all have the major drawback of only accounting for the emergence of a single sector. In reality, there are a growing number of sectors vying for space in the oceans, e.g., offshore wave, wind, and tidal energy are all shifting offshore (Yates et al., 2015), and existing sectors could benefit from zoning or re-zoning. Marine planning frameworks must take this trend into account in addition to the impacts on existing sectors to allocate space optimally within future seascapes (Lester et al., 2013).

Strategic multi-sector planning: This form of marine planning is characterised by multi-sectoral planning with multiple objectives (Table 2.1). In this case, emerging sectors are planned to maximise their individual values, while existing sectors are also considered with the objective to minimise negative impacts (as in single-sector planning). However, the advance by multi-sector planning is that the process now is done for multiple sectors simultaneously. One example of this approach to planning is by Stevens et al. (2014, 2015) who aimed to guide the development of different types of offshore aquaculture farms in southern California in relation to a suite of existing sectors and the potential impacts of the aquaculture development (e.g., to wild capture fisheries, marine benthos, view-shed quality, and fish disease risk). They considered three types of aquaculture—finfish, mollusk, and algae—simultaneously to represent the major emerging marine aquaculture industries, and chose four existing sectors to represent those in the system predicted to be most impacted by the development of offshore aquaculture. They constructed interacting bioeconomic models for each of the seven sectors, and used the ecosystem service trade-off analysis framework to identify the set of aquaculture plans (location and farm type) best able to meet the multi-criteria objective function of weighted aquaculture and existing sectoral values. To our knowledge, this was the first spatial planning study to inform the planning of multiple interacting emerging sectors explicitly in relation to interacting existing sectors. Although preliminary, their study found that complete development of finfish or algal aquaculture would reduce an important wild fishery to ~ 93% of its current take within the region considered for aquaculture development, and full mollusc aquaculture development would reduce the fishery's value to 77% of its current value.

Fully comprehensive planning: Although multi-sector spatial planning is an advance toward a more comprehensive approach to spatial planning, it is still suboptimal in that it focuses entirely on planning of only a few sectors. It also assumes existing sectors' plans are fixed, except possibly in reaction to the development of an emerging use; e.g., fisheries displacement from emerging marine protected areas (Kittinger et al., 2014). Alternatively, fully comprehensive spatial planning not only considers trade-offs among existing and emerging sectors, but also the allocation of space for both types of sectors. This 'clean slate' approach represents fully comprehensive planning spatial planning and is expected to generate the greatest value over conventional planning.

Although fully comprehensive planning is theoretically optimal, it has yet to be done in practice because at least in part, its solutions would be economically and politically challenging to implement, especially for areas whose sectors are already established and/or have large financial investments in their current plans. Nonetheless, there are examples of well-established sectors being redesigned to reduce impacts on other sectors, e.g., altering shipping lanes entering a Boston (USA) harbour to reduce whale strikes (Kraus et al., 2005). On the other hand, areas with minimal human development should be more amenable to approximate fully comprehensive planning.

Development and implementation of fully comprehensive marine planning will likely require a thorough stakeholder engagement process in order to generate cooperation with the new plan. One such example of spatial planning that is attempting to approximate fully comprehensive planning is the Hauraki Gulf Marine Spatial Planning project in New Zealand (Jarvis et al., 2015). Starting in 2013 until late 2015, the project focused on encouraging all users of the Gulf to be actively involved in the development of a single plan. Similar to Alexander et al. (2012), stakeholders could identify areas important to their respective sector using the program SeaSketch. Following this stakeholder engagement phase (expected to conclude September 2015), a single marine plan within the Gulf will be developed which will completely redesignate space for every sector. The Hauraki Gulf project is an example of how a fully comprehensive planning process can be initiated without an explicit trade-off analysis. The Hauraki Gulf Marine Spatial Planning project is expected to generate an advance over conventional planning; however, the lack of a quantitative framework for calculating inter-sectoral trade-offs explicitly and identifying the efficiency frontier of optimal plans will likely result in the project choosing a suboptimal plan compared with what could have been accomplished through true fully comprehensive planning marine planning.

Lessons learned and moving forward

The benefits of ecosystem service trade-off analysis lie directly in its ability to quantify relationships between sectors to identify plans that maximise the overall value of the ecosystem and minimise inter-sectoral conflicts. But using an ecosystem service trade-off analysis approach for marine planning does not directly guarantee that optimal solutions will be identified. Here, we outline several criteria for marine planning processes to ensure that management goals are met and optimal spatial plans are identified.

It is certainly possible to understand sector interactions using qualitative methods alone. However, marine planning problems are often multidimensional and involve many possible solutions, which would most likely be underestimated using qualitative planning methods alone. Specifically, qualitative assessment of a system can generate spurious assumptions about its dynamics and lead to suboptimal marine planning (Rassweiler et al., 2014). Related to this, computational support of a trade-off analysis is essential for the efficient evaluation of the numerous alternative management plans under consideration in most marine planning problems. For example, White et al. (2012) considered wind farm development in coastal Massachusetts, USA, considering only eighty-four 2-km^2 cells. This modest search space contained >10^{25} alternative wind farm designs. Identifying optimal plans for each weighting scenario in the analysis would have been impossible without a heuristic search algorithm. In another example, Rassweiler et al. (2014) computationally evaluated bioeconomic models to guide spatial planning of marine protected areas among 135 coastal reef patches in southern California, USA. The analysis focused on trade-offs between fish biomass conservation and the economic value of fisheries. The authors used a heuristic search algorithm to identify optimal plans on the efficiency frontier that minimised the trade-off, which they compared with simulated 'stakeholder plans' developed based on rules of thumb guidelines for marine protected-area size and spacing. They found that plans generated using the guidelines were inferior at minimising sectoral trade-offs. Alternatively, allowing stakeholders to alter (by up to 25%) optimal plans on the efficiency frontier (i.e., those generated with the heuristic) did not substantially change the value of the plans and still resulted in them strongly outperforming the guidelines-generated plans.

These studies highlight two important messages for the future of spatial planning. First, computational evaluation of mathematical models is critical for explicitly and efficiently guiding complex planning processes. Second, although stakeholder involvement is necessary for successful marine planning (Alexander et al., 2012; Douvere, 2008; Foley et al., 2010; Pomeroy and Douvere, 2008), stakeholders alone are unlikely to generate optimal plans. Instead, what is needed is a collaborative exercise between stakeholders and scientists that *starts by considering the set of optimal plans on the efficiency frontier*. Achieving such a collaborative enterprise between scientists, who 'seed' the conversation with optimal plans determined with analytical models, and stakeholders, who improve those plans in relation to practical factors and intimate knowledge of the system not considered in the models, is expected to result in the most successful spatial plans.

The aforementioned studies by White et al. (2012) and Rassweiler et al. (2014) focused on planning for only one emerging sector at a time (wind turbines and marine protected areas). Yet, when comparing single-sector and multi-sector planning (Table 2.1), the value of marine planning is expected to increase with the number of sectors explicitly considered and planned in the analysis. This increase in value is expected in relation to both the number of emerging sectors, as well as existing sectors, included *and sited* in the analysis. Nonetheless, a clean-slate approach to planning, where existing sectors are re-planned along with emerging sectors, might not be feasible in heavily used areas due to high relocation costs, or existing government regulation. This could be a practical constraint that will always limit the potential value of marine planning. However, at the least, planning could simultaneously consider multiple emerging sectors (existing or expected). This might not obtain the greatest value of the system because existing stakeholders' respective values will not be optimised; however, it will ensure that future developments will be strategically located.

Strategic planning relies heavily on reliable bioeconomic models or extensive empirical data to assess ecosystem trade-offs accurately. In marine systems, dynamic models are particularly important for capturing the high spatial connectivity among ecological processes and sectoral responses (Mitarai et al., 2009). However, most dynamic models require considerable computational power to complete, which can be problematic if evaluating millions of possible management options. To speed up the planning process, static models can be substituted as long as the system is in an approximately steady state and the time horizon is short, i.e., a few years (Brown et al., 2015).

In summary, ecosystem service trade-off analysis holds great potential for developing optimal renewable energy marine spatial plans. To ensure that accurate and useful plans are developed, frameworks should be developed that integrate both collaboration with sector users and analytical rigour. Due to the shear complexity of spatial problems, identifying optimal plans must involve computational methods to distinguish the efficiency frontier. Scientists should also collaborate with stakeholders to sort through the suite of identified optimal plans, and choose those that reflect the preferences of the ecosystem user group. Frameworks should aim to incorporate as many emerging and existing sectors as possible to optimise the entire ecosystem, and finally, dynamic models should be used in trade-off frameworks when computationally feasible to simulate the spatial and temporal effects of plans accurately.

Highlights

- Ecosystem service trade-off analysis is an analytical tool to inform the identification of plans best able to maximise sector values and minimise inter-sectoral conflicts.
- The value of using ecosystem service trade-off analysis to guide marine planning is expected to increase with the number of sectors being considered and planned for in the system.
- Due to the large number of possible solutions in marine planning problems, computational search methods should be used to identify optimal plans.
- To promote a successful planning process with a high-value outcome, scientists and stakeholders should work collaboratively in the identification and fine-tuning, respectively, of optimal marine plans.
- The use of dynamic models is essential to predicting the costs and benefits of proposed plans in marine systems accurately.

References

Agardy, T., Di Sciara, G.N. and Christie, P. (2011). Mind the gap: Addressing the shortcomings of marine protected areas through large scale marine spatial planning. *Marine Policy*, 35(2), pp. 226–232.

Albrechts, L. (2006). Shifts in strategic spatial planning? Some evidence from Europe and Australia. *Environment and Planning A*, 38(6), pp. 1149–1170.

Alexander, K.A., Janssen, R., Arciniegas, G., O'Higgins, T.G., Eikelboon, T. and Wilding, T.A. (2012). Interactive marine spatial planning: Siting tidal energy arrays around the mull of Kintyre. *PloS One*, 7(1), p. 30031.

Banos, R., Manzano-Agugliaro, F., Montoya, F., Gil, C., Alcayde, A. and Gómez, J. (2011). Optimization methods applied to renewable and sustainable energy: A review. *Renewable and Sustainable Energy Reviews*, 15(4), pp. 1753–1766.

Breheny, M. (1991). The renaissance of strategic planning. *Environment and Planning B: Planning and Design*, 18(2), pp. 233–249.

Brown, C.J., White, C., Beger, M., Grantham, H.S., Halpern, B.S., Klein, C.J., Mumby, P.J., Tulloch, V., Ruckelshaus, M. and Possingham, H.P. (2015). Fisheries and biodiversity benefits of using static versus dynamic models for designing marine reserve networks. *Ecosphere*, 6(10), pp. 1–14.

Bryson, J.M. and Roering, W.D. (1988). Initiation of strategic planning by governments. *Public Administration Review*, 48(6), pp. 995–1004.

Douvere, F. (2008). The importance of marine spatial planning in advancing ecosystem-based sea use management. *Marine Policy*, 32(5), pp. 762–771.

Douvere, F. and Ehler, C.N. (2009). New perspectives on sea use management: Initial findings from European experience with marine spatial planning. *Journal of Environmental Management*, 90(1), pp. 77–88.

Douvere, F., Maes, F., Vanhulle, A. and Schrijvers, J. (2007). The role of marine spatial planning in sea use management: The Belgian case. *Marine Policy*, 31(2), pp. 182–191.

Eadie, D.C. (1983). Putting a powerful tool to practical use: The application of strategic planning in the public sector. *Public Administration Review*, 43(5), pp. 447–452.

Foley, M.M., Halpern, B.S., Micheli, F., Armsby, M.H., Caldwell, M.R., Crain, C.M., Prahler, E., Rohr, N., Sivas, D. and Beck, M.W. (2010). Guiding ecological principles for marine spatial planning. *Marine Policy*, 34(5), pp. 955–966.

Galparsoro, I., Liria, P., Legorburu, I., Bald, J., Chust, G., Ruis-Minguela, P., Gonzáles, M. and Borja, Á. (2012) A marine spatial planning approach to select suitable areas for installing wave energy converters (WECs), on the Basque Continental Shelf (Bay of Biscay). *Coastal Management*, 40(1), pp. 1–19.

Guerry, A.D., Ruckelshaus, M.H., Arkema, K.K., Bernhardt, J.R., Guannel, G., Kim, C.-K., Marsik, M., Papenfus, M., Toft, J.E. and Verutes, G. (2012). Modeling benefits from nature: Using ecosystem services to inform coastal and marine spatial planning. *International Journal of Biodiversity Science, Ecosystem Services & Management*, 8(1–2), pp. 107–121.

Halpern, B.S., White, C., Lester, S.E., Costello, C. and Gaines, S.D. (2011). Using portfolio theory to assess trade-offs between return from natural capital and social equity across space. *Biological Conservation*, 144(5), pp. 1499–1507.

Hong, L. and Möller, B. (2011). Offshore wind energy potential in China: Under technical, spatial and economic constraints. *Energy*, 36(7), pp. 4482–4491.

Jarvis, R.M., Breen, B.B., Krägeloh, C.U. and Billington, D.R. (2015). Citizen science and the power of public participation in marine spatial planning. *Marine Policy*, 57(2015), pp. 21–26.

Jay, S. (2010). Planners to the rescue: Spatial planning facilitating the development of offshore wind energy. *Marine Pollution Bulletin*, 60(4), pp. 493–499.

Johnson, R.M. (1974). Trade-off analysis of consumer values. *Journal of Marketing Research*, 11(2), pp. 121–127.

Kittinger, J.N., Koehn, J.Z., Le Cornu, E., Ban, N.C., Gopnik, M., Armsby, M., Brooks, C., Carr, M.H., Cinner, J.E. and Cravens, A. (2014). A practical approach for putting people in ecosystem-based ocean planning. *Frontiers in Ecology and the Environment*, 12(8), pp. 448–456.

Kraus, S.D., Brown, M.W., Caswell, H., Clark, C.W., Fujiwara, M., Hamilton, P.K., Kenney, R.D., Knowlton, A.R., Landry, S. and Mayo, C.A. (2005). North Atlantic right whales in crisis. *Science*, 309(5734), pp. 561–562.

Lester, S.E., Costello, C., Halpern, B.S., Gaines, S.D., White, C. and Barth, J.A. (2013). Evaluating trade-offs among ecosystem services to inform marine spatial planning. *Marine Policy*, 38(2013), pp. 80–89.

Lester, S.E., McLeod, K.L., Tallis, H., Ruckelshaus, M., Halpern, B.S., Levin, P.S., Chavez, F.P., Pomeroy, C., McCay, B.J. and Costello, C. (2010). Science in support of ecosystem-based management for the US West Coast and beyond. *Biological Conservation*, 143(3), pp. 576–587.

Markowitz, H. (2014). Mean–variance approximations to expected utility. *European Journal of Operational Research*, 234(2), pp. 346–355.

McLean, J. (2004). Aurora-delivering a sustainable urban water system for a new suburb. In: WSUD 2004: Cities as Catchments; International Conference on Water Sensitive Urban Design, Proceedings of. [online] Barton, A.C.T: Engineers Australia, p. 22. Available at: http://search.informit.com.au/documentSummary;dn=765264371313623;res=IELENG

Mitarai, S., Siegel, D., Watson, J., Dong, C. and McWilliams, J. (2009). Quantifying connectivity in the coastal ocean with application to the Southern California Bight. *Journal of Geophysical Research: Oceans (1978–2012)*, 114(C10), pp. 1–21.

Nunez, F. (2006). Assembly Bill 32: The California global warming solutions act of 2006. *California State Assembly*.

Østergaard, P.A. (2009). Reviewing optimisation criteria for energy systems analyses of renewable energy integration. *Energy*, 34(9), pp. 1236–1245.

Polasky, S., Nelson, E., Camm, J., Csuti, B., Fackler, P., Lonsdorf, E., Montgomery, C., White, D., Arthur, J. and Garber-Yonts, B. (2008). Where to put things? Spatial land management to sustain biodiversity and economic returns. *Biological Conservation*, 141(6), pp. 1505–1524.

Pomeroy, R. and Douvere, F. (2008). The engagement of stakeholders in the marine spatial planning process. *Marine Policy*, 32(5), pp. 816–822.

Rassweiler, A., Costello, C., Hilborn, R. and Siegel, D.A. (2014). Integrating scientific guidance into marine spatial planning. *Proceedings of the Royal Society B: Biological Sciences*, 281(1781), p. 20132252.

Roberts, C.M., Branch, G., Bustamante, R.H., Castilla, J.C., Dugan, J., Halpern, B.S., Lafferty, K.D., Leslie, H., Lubchenco, J. and McArdle, D. (2003). Application of ecological criteria in selecting marine reserves and developing reserve networks. *Ecological Applications*, 13(1), pp. 215–228.

Snyder, B. and Kaiser, M.J. (2009). Ecological and economic cost-benefit analysis of offshore wind energy. *Renewable Energy*, 34(6), pp. 1567–1578.

Stevens, J.M., Gentry, R.R., Maue, C.C., Bell, T.W, Kappel, C.V., Lester, S.E., Wendt, D.E. and White, C. (2014). Marine spatial planning makes room for offshore aquaculture in a crowded coastal zone Western Society of Naturalists. In: *AGU Fall Meeting*. Tacoma, Washington: American Geophysical Union.

Stevens, J.M., Gentry, R.R., Maue, C.C., Bell, T.W., Kappel, C.V., Lester, S.E., Wendt, D.E. and White, C. (2015). Marine spatial planning makes room for offshore aquaculture in crowded waters International Conference of Conservation Biology. In: *AGU Fall Meeting*. Montpellier, France: American Geophysical Union.

Stewart, R.R. and Possingham, H.P. (2005). Efficiency, costs and trade-offs in marine reserve system design. *Environmental Modeling & Assessment*, 10(3), pp. 203–213.

White, C., Halpern, B.S. and Kappel, C.V. (2012). Ecosystem service trade-off analysis reveals the value of marine spatial planning for multiple ocean uses. *Proceedings of the National Academy of Sciences*, 109(12), pp. 4696–4701.

Wüstenhagen, R., Wolsink, M. and Bürer, M.J. (2007). Social acceptance of renewable energy innovation: An introduction to the concept. *Energy Policy*, 35(5), 2683–2691.

Yates, K.L., Schoeman, D.S. and Klein, C.J. (2015). Ocean zoning for conservation, fisheries and marine renewable energy: Assessing trade-offs and co-location opportunities. *Journal of Environmental Management*, 152(2015), pp. 201–209.

Chapter 3

It starts with a conversation

Achieving conservation goals in
collaboration with the offshore
energy industry

Johanna Polsenberg and Anna Kilponen

Introduction

Industry collaboration can be emotive and divisive. Increasingly, marine conservationists are embracing—or perhaps simply accepting—that working directly with the offshore energy industry is an important way forward towards achieving some conservation goals (Kareiva and Marvier, 2012; Kark et al., 2015). However, many people still feel unsettled about collaborating directly with those they view as the cause of the very problems they seek to solve (Noss et al., 2013; Tuodolo, 2009).

The debate on industry collaboration has transferred from the pages of academic journals into mainstream media (Klein, 2013; Max, 2014). According to an article in *The Guardian* provocatively entitled "A pact with the devil? The challenges of partnering with the extractive industry", working with an extractive industry is viewed as one of the most controversial things a non-governmental organisation can do (Henley, 2012). "You have to have thick skin," says Patrick Laine, director of corporate partnerships at the World Wildlife Fund (United Kingdom), "... because it's absolutely inevitable that you will be attacked". Despite the negative attention that industry collaboration might bring, Laine states that collaboration is necessary: "... there is now the growing belief that whatever the natural antipathy between an NGO and an extractive, collaboration between the two needs to take place" (Henley, 2012). However, while many organisations strictly oppose working with an extractive industry, collaborative relationships between industry and many non-government organisations are becoming more common (Tuodolo, 2009).

The necessity of non-government organisations participating with industry is echoed by Dame Barbara Stocking, Oxfam's former chief executive. In discussing Oxfam's change of heart when it comes to the private sector, Dame Stocking said in 2013 that until the early 2000s, much of Oxfam's emphasis was on campaigning against the private sector, especially the extractive and pharmaceutical industries (Smedley, 2012):

We began to realise that we also had to work with the private sector. But also over the last few years the private sector has changed quite a lot too, with a better understanding of poverty and their engagement with it.

Unlike oil and gas, offshore renewable energy is not strictly 'extractive', and it generates a small fraction of the public antipathy targeted against the offshore extractive industry (Hooper, Hattam and Austen, 2017; Wever et al., 2015). Nevertheless, offshore renewable energy development does have environmental impacts (Chapter 8), and its greater environmental virtue in terms of CO_2 emissions certainly does not guarantee support (Firestone and Kempton, 2007; McNamara, 2015). When the Cape Wind project, a commercial offshore wind plant in the Nantucket Sound in Massachusetts, USA failed, 59 percent of respondents to a local poll stated they were happy that the project was unsuccessful (McNamara, 2015). This is not unusual; while there may be general public support, local stakeholder perceptions of offshore renewable energy are frequently negative (Gray et al., 2005; Haggett, 2011), and the success rate for renewable offshore energy project applications can be low (Toke, 2005).

Many developments that do succeed still face substantial opposition that causes long delays (Haggett, 2011), and there are calls for greater collaboration between developers, academia and the public sector to improve planning, management and governance processes (Bonar, Bryden, and Borthwick, 2015; Bruns and Gee, 2009). In this chapter we consider collaborative opportunities between the offshore energy industry and other stakeholders, particularly those interested in marine conservation.

Our underlying premise is that there are times when conservation can best be achieved in collaboration with industry. There are many on-the-ground, *in situ* opportunities wherein biodiversity, habitat, or overall ocean health conservation goals do not preclude and are not precluded by offshore energy extraction or production, and that in these many instances, collaborations with the offshore energy industry may be the quickest, best or even the only way to achieve certain conservation goals.

Industry has both access to more resources than non-government organisations and the public sector and, through its actions, the potential to have as great if not greater impact on biodiversity conservation and human well-being objectives (Glin, Mol and Oosterveer, 2013; Hole et al., 2005; Kleemann and Abdulai 2013). Alice Korngold, author of *A Better World, Inc.* (2014), forcefully posits that "Only global corporations have the vast resources, international scope, global workforces, and incentives to truly bring about the changes that are necessary in order to achieve global peace and prosperity". Korngold also highlights that, at least for the United States, which is the world's largest foreign-aid donor by volume (Organisation for Economic Co-operation and Development, 2014), private-sector investment as opposed to foreign aid or philanthropy is the major driver by far of global prosperity today.

Private capital also plays an important role in international efforts on global climate action, highlighted in the 21st Conference of the Parties to the United Nations Framework Convention on Climate Change (McInerney and Johannsdottir, 2016). Organisations like the World Economic Forum (2013), the World Bank, and the Global Environment Facility (a global fund supported by the United Nations and the World Bank) have also recognised the importance of mobilising private capital for both social and environmental good. Peter Seligmann, the Co-founder, Chairman and former CEO of Conservation International, says that because environmental issues are so massive and complex, non-governmental organisations cannot simply "throw stones from the sidelines" (Seligmann, 2011). Addressing environmental issues will require productive collaboration among all parties, governments, non-governmental organisations, and corporations alike (Bos, Pressey, and Stoeckl, 2015; Hamrick, 2016).

The private sector is being provided more evidence of a financial imperative to take sustainability into account in their business activities. In 2004, Former U.S. Vice President Al Gore and partner David Blood together founded an investment firm—Generation Investment Management LLP—based on their recognition, similar to that of Korngold's above, that companies and investors are in control of most of the capital needed to overcome the massive global challenges of climate change and environmental degradation, poverty and development, water and natural resource scarcity, pandemics and healthcare, and demographics, migration and urbanisation. After only seven years of investment activity and focused research, they could demonstrate that companies and investors that integrated sustainability considerations into their business practices had greater profitability (Gore and Blood, 2011).

Here we provide both a philosophy and a few operational frameworks around which professionals in the marine conservation community can build successful and productive collaborations with the offshore energy industry. Many instances exist where those working in marine conservation can and should identify common ground and shared goals with their industry counterparts. To do so, it will be necessary for both sides to understand and accept that motivations for a given collaboration will likely differ, but that once stated and kept fully transparent, those differences in motivation actually have little bearing on the achievement of positive conservation outcomes.

Incentivised to collaborate

In much the same way that the conservation community might be ethically reticent to collaborate directly with industries that impact the environment negatively, many in industry are just as uneasy about working closely with those they often view as directly opposed to their interests (Dahan et al., 2010).

Table 3.1 Potential benefits from collaboration

Conservation	Industry
New opportunities for funding and financing for projects	Improve business profile, visibility, legitimacy and reputation
Leverage skills, innovative solutions and technologies	Leverage skills and perspectives
Increase impact and scale	Increase stakeholder engagement
Expertise in capacity building, management and efficiency	Foresee long-term changes
Access to free marketing	Help set industry standards and environmental performance

Source: Alikhani et al. 2015; International Union for Conservation of Nature 2013; Global Environmental Management Initiative, 2008; Yaziji, 2006.

However, self-interest in industry has a financial bottom line and, more and more often, companies are recognising that their bottom line can be strengthened by working with conservation organisations (Table 3.1).

Three main variables appear to drive a company's willingness to collaborate with non-government organisations: (1) frequency of contacts, (2) perceived strategic fit, and (3) a company's commitment to corporate social responsibility (den Hond, de Bakker and Doh, 2015). Some companies also know they can leverage their relationship with non-government organisations to address social and political pressures, to strengthen legitimacy and reputation, and to anticipate and perhaps lessen negative actions by stakeholders (den Hond, de Bakker and Doh, 2015).

Environmental non-government organisations are often considered vital stakeholders, and collaborations with them are viewed as an essential form of stakeholder support (Henisz, Dorobantu and Nartey, 2014). In fact, increasing stakeholder support can enhance the financial value of a company (Henisz, Dorobantu and Nartey, 2014). And responsible industry views at least some forms of collaboration with non-government organisations as good business; they are incentivised to incorporate better environmental practices, techniques and standards into their business decisions, and believe that these actions can give rise to competitive edge and a higher profile (Martin-Mehers, 2016; Yaziji, 2006). As negative environmental impacts and concerns about sustainably become mainstream societal issues, the private sector can no longer afford to be reactive; instead, companies must embed these issues in their business strategies (Esty and Simmons, 2011). Companies that can differentiate themselves from competitors can obtain considerable financial benefits by adopting social and environmental practices, and lead them to new revenue streams, stronger brands and broader customer base (Esty and Simmons, 2011; Robinson, 2012; World Wildlife Fund & Carbon Disclosure Project, 2013). Indeed, the extractive industry is increasingly exploring new ways of integrating better environmental and social aspects into their business strategies (Ortega Lindsey, Janus and Murphy, 2014; Pedroni et al., 2012; Springer et al., 2012).

Companies have resources and access

Lori Anna Conzo, an environmental specialist in the Environment, Social and Governance Department of the International Finance Corporation, the private-sector lending-arm of the World Bank Group, recognises that industry involvement in conservation projects helps bring new resources to old problems (Table 3.1). These include working with construction engineers whose unique skills can offer innovative solutions for reducing impacts on fauna and flora from large-scale development projects (International Union for Conservation of Nature, 2013). Industry can also provide other expertise such as capacity building and management skills as well as infrastructure (Alikhani et al., 2015). The main characteristics driving the industries—the know-how, ambition, and excellent skills in problem-solving—have great potential to contribute to managing biodiversity more effectively (Conzo, cited in Alikhani et al., 2015).

Finding common ground to achieve shared goals when motivations differ

Oil spills are one of the most vivid and evocative images of the offshore energy industry gone wrong, but many less-extreme, cumulative impacts such as noise pollution, disruption of the migratory routes of marine mammals, loss of ecosystem services, and displacement of fisheries also need to be addressed (Chapters 6–8). Given that continued offshore energy development in the marine environment seems inevitable in the short term, the conservation community must engage proactively with industry to help reduce any negative impacts and take advantage of unrealised conservation opportunities (Kark et al., 2015).

Three steps can help structure the basis for collaboration (Figure 3.1) between parties—conservation and industry—that are normally on opposite sides of issues: (1) finding common ground, (2) being transparent about, but then largely setting aside, differing motivations, and (3) establishing shared goals. One of the clearest examples of common ground is risk. Whereas conservationists see environmental risk, industry sees operational risk. As much as most conservationists do not want sensitive habitats or species disrupted, the industry fears operation stoppages or delays due to the discovery of a previously unknown or unidentified environmental sensitivity. As much as a company fears bad publicity and the loss of the 'social license to operate' should a spill occur, damage to critical habitats concerns the conservation community.

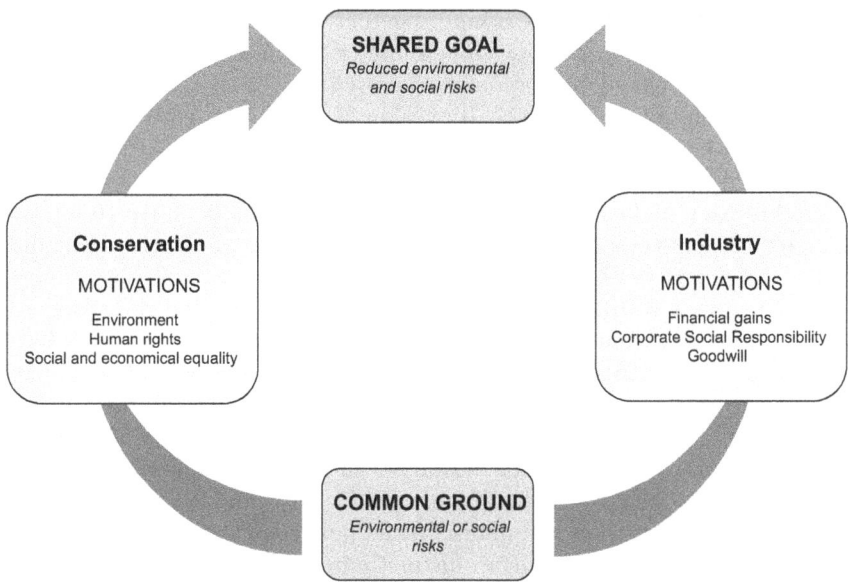

Figure 3.1 Framework for successful collaboration.

When structuring the collaboration, it is important to acknowledge the differing motivations and values of both parties, but then largely set them aside and focus on the aspects that parties have in common and shared interests and goals. The avoidance of a risk to the environment is generally the basis for the shared goals. For example, if there is an oil spill, there is a risk that seabirds could be killed (Antonio, Mendes and Thomaz, 2011), or behaviour of marine mammals could be altered due to offshore wind farms, particularly during the construction phase (Evans, 1998; Parsons et al., 2007). Reducing such risks becomes a clear goal shared by both the conservation community that wants to avoid environmental damage, and the industry that wants to avoid the financial losses due to operation stoppages, delays, or negative publicity caused by the damage or disruption to the environment (Crooks, 2016). Collaborations to avoid those risks are arguably stronger than acting independently (Seligmann, 2011).

For example, Sakhalin Energy Investment Company Ltd. and several non-government organisations with whom the company was traditionally at odds began working together due to the presence of the Critically Endangered western grey whale (*Eschrichtius robustus*) population in the seas around Sakhalin Island, Russia, where Sakhalin Energy was operating. Increased dialogue between the organisations, stakeholders and the company has resulted in positive outcomes for both conservation and business: the whale population has been increasing 3–4% per year since the collaboration started (Martin-Mehers, 2016), and Sakhalin Energy has been able to raise its profile and visibility. By working together with the non-governmental organisations, the company developed one of the most comprehensive industry monitoring and mitigation plans for seismic surveys, which has been used to inform policy processes all over the world (Martin-Mehers, 2016), and the Sakhalin Energy's *Western Grey Whale Protection Programme* was awarded the *Best Ecological Project of the Year* by the Russian Federation Minister of Natural Resources (Vladimirov et al., 2012) (see also Chapter 12, which describes an example of offshore wind development and right whale conservation).

Sometimes, it is difficult for people who have committed their lives to conservation to accept environmental risks as simply cost items on a balance sheet. While this is almost always true and money is undeniably an important incentive to change behaviour, due to the changing social and policy landscape, industry is increasingly being driven towards enhancing and improving its environmental and social practices and incorporating these into its strategies (Pedroni et al., 2012; Viana et al., 2016). We encourage the conservation community to acknowledge and accept that there can be shared goals that result in distinctly positive outcomes for conservation *and* for business, as the Sakhalin Island case (Martin-Mehers, 2016) demonstrates, even though the motivations for achieving those outcomes are completely different: conservation in one instance, and profit in the other.

Case study of collaboration

In 2012, the Wildlife Conservation Society and Tullow Oil developed a marine environmental research and oil industry-offset programme in Gabon, but also including The Republic of the Congo and Equatorial Guinea. The goal was to gather baseline data needed for the sustainable management of that marine ecosystem, collectively known as the 'Congo Basin Coastal Region'. The project was shaped around three shared objectives: to (1) broaden and deepen the baseline understanding of biodiversity and environmental sensitivities to minimise risk and to monitor changes, both ecological and industrial; (2) support marine spatial planning processes, in part to create effective and ecologically important marine protected areas; and (3) encourage both broader industry engagement with conservation and more rigorous national environmental standards. The mitigation hierarchy (described in more detail below) formed the core of the project. It consisted of a set of principles and standards for the achievement of no net loss of biodiversity during or following an activity that negatively impacts the environment, such as oil and gas development (Forest Trends, 2015).

An essential element of the partnership was that one of Wildlife Conservation Society's global marine mandates was to help governments meet the Aichi Convention on Biological Diversity's marine protected-area targets that calls for protecting 10% of coastal and marine areas globally (Convention on Biological Diversity, 2010). At the time, strong political will to establish a network of marine protected areas existed in Gabon (Moundounga Mouity and Ndjimba, 2012; National Geographic, 2014). Thus, in addition to meeting the company's voluntary decision to formulate best practices for its planned 2013 offshore operations and to minimise any negative environmental impacts, the collaboration directly supported both the Wildlife Conservation Society's core mandate and the host government's pre-existing commitment. In other words, there was no compromise on the part of the non-governmental organisation, which was funded by the company to meet the company's objectives; pre-existing and compatible goals were already present for all three parties. This collaboration played a major role in enabling the government of Gabon to make the decision to create a network of marine protected areas in November 2014, covering more than 46,000 km^2, or about 23% of Gabon's territorial waters and Exclusive Economic Zone, to help conserve whales, sea turtles, and other marine biodiversity and habitats (National Geographic, 2014) (Figure 3.2).

Aside from contributing to the more robust database used to plan a well-placed network of marine protected areas and sustainable fisheries zones created by the government, the collaboration provided a stronger baseline understanding of regional biodiversity, against which civil society and the government could better hold the industry accountable for negative impacts (Polsenberg, J., personal experience, 2013). While this might sound counterintuitive, it was actually an explicitly shared goal of the partnership (Polsenberg, J., personal experience, 2012). In the same way that the conservation community wanted to increase its capacity to hold poorly performing companies accountable, Tullow Oil—a company committed to a higher set of operating standards—did not want to be outcompeted by other companies who were in effect operating under a set of lower standards. Effectively, high environmental performance became a competitive advantage.

The collaboration also led to immediate changes to some of Tullow's and their partners' operating activities. For example, in 2013 Tullow Oil incorporated into their operational budget the costs of helping to determine whether their activities—both their current exploration as well as future development and production—might impact threatened olive ridley turtles (*Lepidochelys olivacea*). Further, on the sole advice of Wildlife Conservation Society scientists, they reversed the order in which they had planned to drill their exploratory wells. Although the original order would have been better for them from a geological, and thus potentially financial, point of view, reversing the order allowed a greater margin for response for protecting turtle nesting beaches should an oil spill have occurred during exploration (Polsenberg, J., personal experience, 2013).

Figure 3.2 Humpback whale surfacing in front of oil platform in Gabon, 2004. In 2010, the government of Gabon banned gas flaring from oil platforms. In 2014, Gabon further adopted sustainable development and environmental codes, which include both the ban on gas flaring as well as a requirement for setting aside funds for appropriate decommissioning of oil infrastructure. Credit: Tim Collins, WCS.

Starting with a conversation

Many guides, models and toolkits outline parameters and suggestions for how to structure non-governmental organisation-industry collaborations to share resources and expertise successfully across these traditionally divided groups, and there are many examples of successful and well-regarded partnerships between non-government organisations and corporations (Conservation International, 2006; Livesey, 1999; Martin-Mehers, 2016; Stafford et al., 1999). However, choosing to partner with an industry as high profile as the offshore energy industry, to which some of the world's worst environmental disasters are attributed (de Oliveira Neves et al., 2016; Hester et al., 2016), requires a more nuanced and careful approach compared to many other industry sectors. As such, the two most important elements to a successful collaboration will likely be the degree of and sensitivity around communication, both internally and externally, and the clarity and precision of the scope of work. In other words, successful communication throughout the life of the project will start with a conversation that takes place among a group of human beings. That is, the people from both sides involved in leading the collaboration have to approach it as objectively and with as few preconceptions as possible.

As with any healthy partnership, the partners need to get to know each other as people. Those on the conservation side migth be pleasantly surprised to find that many of their corporate counterparts are driven not only by their company's financial bottom line, but also by a range of social and environmental concerns (Darko et al., 2017; Gasbarro, Iraldo and Daddi, 2017), and a dedication to making the world a better place. Likewise, those on the corporate side might come to realise that they are dealing with professionals who, while they may have chosen to work outside a profit-driven field, understand the connection between the health of ecosystems and the business bottom line. As far as external communication, there will be considerations far beyond the remit of this chapter as to what gets shared publicly, by whom and when. In other words, lawyers will be involved. Rather than specifics, we discuss below some general approaches. At every step, the process will be facilitated or alternatively, impeded by the clarity and precision or lack thereof of the scope of work. Agreeing on the scope of work will probably be the most challenging step of the collaboration because it must happen at the outset, and that process will necessarily highlight some of the greatest differences between the partners. Third-party professional facilitators could play an invaluable role at this stage in making sure that the needs and concerns of all parties are addressed.

Once the parties have achieved the often large initial effort to establish a trusted and personal relationship, it is important also to ensure that that organisational relationship outlives the personal and does not come to depend only on the energies and commitment of just a few people. As with any long-term partnership, discovering, solidifying and maintaining a strong network of involved and interested people, on both sides, can help future-proof the collaboration. Evan Rapoport, Moonshots Product Manager at X, The Moonshot Factory (formerly Google X), spoke about sending an email to find out how many of his colleagues had an interest in oceans at the outset of his company's collaboration with Conservation International's Center for Oceans (personal communication, 2017). He was surprised and pleased to learn just how many colleagues had a deep interest in ocean conservation, even though their professional positions held no connection. Their input and enthusiasm helped solidify the collaboration internally.

Collaboration guides and models

While business-non-governmental partnerships can take many forms (Tennyson et al., 2008), collaborative projects can be categorised into three broad forms: (1) outcome-based, (2) issue-based and (3) project-based. Outcome-based collaborations can be generally thought of as public-private partnerships or alliances (Saul, Davenport and Ouellette, 2010). These sorts of collaborations have existed for a long time, especially between governments and the private sector in the development of public-sector infrastructure. However, they have recently garnered greater attention as mechanisms through which civil society can exert greater influence over and further their impact on development decisions (Robinson, 2012). The Global Environment Facility states that "... in order to bring transformational change to the global environment, the public and private sectors must identify new ways of working together" (Global Environment Facility, 2013). Specific to the marine environment, the Blue Ribbon Panel—a group of global experts convened by the World Bank to advise the Global Partnership for the Oceans—recommends well-structured public-private partnerships as an important approach to ocean resource development, management, maintenance and use that can help communities, economies and ecosystems thrive (Hoegh-Guldberg, et al., 2013).

Issue-based collaborations, commonly known as 'green alliances' (Stafford and Hartman, 2015), often form to market a more environmentally friendly product or process, while also drawing attention to a global problem. Ground-breaking collaborations in the early 1990s are those between Greenpeace and Foron Household Appliances (Stafford et al., 1999) to develop and market an ozone-safe refrigerant, and between the Environmental Defense Fund and McDonald's (Livesey, 1999) to eliminate McDonald's use of polystyrene, which was the single-largest user of polystyrene at the time.

On-the-ground, project-based collaborations have the greatest relevance to marine spatial planning. For most activities, the mitigation hierarchy provides a good, overarching starting point. The mitigation hierarchy is a set of principles and standards to achieve no net loss of biodiversity during or following an activity that negatively impacts the environment, such as oil and gas development (bbop.forest-trends.org/pages/mitigation_hierarchy; Business and Biodiversity Offset Programme, 2012). The hierarchy emphasises, in strict order of importance, doing all reasonable activities to avoid, minimise, and restore the impacts of a development project. Then, and only as a last resort, offsetting is proposed to compensate for any residual impacts associated with the development activity that cannot be avoided, minimised or restored. For example, biodiversity offsetting measures include activities that can demonstrate to deliver gains in biodiversity, such as habitat restoration or creation.

The mitigation hierarchy was designed primarily around forest projects, and informed by pioneering collaborations such as the Buffer Zone Project. This project united a logging company, an international conservation organisation, and the government of the Republic of Congo to protect the Nouabalé-Ndoki National Park from illegal poaching, to manage wildlife in logging concessions adjacent to the protected area, and to mitigate the negative effects of logging on biodiversity and the livelihoods of local residents (Poulsen and Clark, 2012). At a time when economic development in the marine realm is expanding, it is increasingly important to be able to address the impacts of such activities. While the mitigation hierarchy does provide a starting point to marine planning, the implementation in the marine environment has proven challenging (Jacob, Pioch and Thorin, 2016; United Nations Environmental Programme-World Conservation Monitoring Centre, 2015).

Creating successful collaborations

A framework for successful collaboration can be built around four core steps: (1) scoping what the benefits and motivations are, (2) structuring the partnership to the needs of both partners, (3) measuring the efficacy and the impact of the partnership, and (4) communicating the process and outcomes to both internal and external audiences according to the ground rules established (Figure 3.3).

SCOPE
- Identify common ground, i.e., the problem(s) that face both conservation community and industry
- Outline shared goals to address the problem(s) in common
- Discuss motivations for achieving the goals and determine if conflicting motivations require actions

STRUCTURE
- Establish strong and committed teams from each partner organisation
- Get the necessary buy-in from each organisation's leadership
- Decide how the partnership will be structured and if additional partners are needed
- Clearly identify and secure sources of funding

MEASURE
- Establish baseline conditions against which to measure the outcomes of the partnership
- Agree on the methods that will be used to monitor progress

COMMUNICATE
- Establish ground rules for public communication
- Share information clearly and transparently throughout the partnership
- Communicate across different levels within the organisations

Figure 3.3 Core steps for successful collaborations.

Scoping

In many cases, the very idea for a collaboration will have emerged *in situ* because conservation concerns and industrial activities overlapped in a given location, so common ground, i.e., the problem(s) facing both the conservation community and the industry, is identified at the outset of discussions. In situations where the question of why to partner remains, the *Guide to Successful Corporate-NGO Partnerships* (Global Environmental Management Initiative, 2008) outlines six possible, mutual benefits (Table 3.1): (1) Create business value and environmental benefits, wherein operational efficiencies also provide environmental benefits. (2) Raise the bar on environmental performance. By both raising their own environmental standards at the same time as helping non-government organisations establish government policies to enforce higher standards, good industrial players can create a well-earned competitive advantage. (3) Leverage skills and perspectives not available in the organisation. Non-government organisations can bring expertise, skills or resources as well as a valuable outside perspective to companies while taking advantage of resources, technologies and access not normally or feasibly available to them. (4) Build respect and credibility. Successful, transparent collaborations can improve the image and credibility of both organisations. (5) Provide independent validation. Non-government organisations can, and most likely will, closely scrutinise, and if needed, help improve an industrial partner's contribution to environmental and social benefits. (6) Help achieve a long-term vision. Most companies want to be around for the foreseeable future, and most non-government organisations want that future to be sustainable. Working together is often the best way for both to achieve their long-term goals.

Shared goals are therefore the potential outcomes needed to address the problems in common. In other words, what do both the conservation community and industry want to see happen, or not happen? For example, both parties in principle should want to avoid unnecessary habitat destruction, pollution, and direct and indirect impacts on human livelihoods—despite different motivations. From a purely business perspective, all of those actions are wasted time or products and create the potential for conflict later. Indeed, the step that causes perhaps the greatest discomfort for conservationists is accepting that motivations for achieving these goals can differ and might in fact not overlap at all. While an oil company might want to make sure that oil does not escape from their operations because they are interested in avoiding the loss of any of the product or gaining negative notoriety, conservationists work to avoid polluting water and coastal habitats perhaps out of a sense of moral duty and public service. Nevertheless, the goal—to keep oil out of the water and off the beaches—is the same. In essence, the motivations have no impact on the goal, and accepting that industrial self-interest can be used to benefit conservation goals is essential. Different corporate motivations for considering ecosystems in their business decision-making range from marketing and reputational motivations to operational and financing (Hanson et al., 2012).

There can be cases where motivations conflict with and create barriers to achieving a transparent and shared vision. In simpler cases, a modification or narrowing of the goals can be enough to ensure the integrity of the project. In others, collaboration might have to be abandoned. In all cases, due-diligence on the part of both parties is a necessary first step in helping to determine if the motivations, while probably different and often non-overlapping, might be great enough to preclude partnering. Devex Impact, a global initiative between Devex and the U.S. Agency for International Development, that connects and informs the global development community working at the intersection of business and global development to help all parties make an impact, stresses that most private and public organisations have a due-diligence process to "… evaluate potential partners and identify commonalities and risks—reputational and otherwise—that might result …" from a partnership (Robinson, 2012; Saldinger, 2015). Pippa Howard, the corporate partnerships programme director at Fauna and Flora International, which has many joint projects with major multinational extractive companies, concedes that there are "huge" dangers in managing the ongoing reputational risks of working with an extractive industry and that they face such risks everyday (Henley, 2012). However, despite the risks, many non-governmental organisations choose to partner with the industry because such partnerships have the potential to address negative impacts of industry operations, and find systemic and more sustainable solutions to social and ecological problems, particularly in poorly regulated countries (Bitzer and Glasbergen 2015; Dees and Anderson, 2003).

Structuring

Structuring the collaboration is likely to be no different to the management of most projects; however, open-minded individuals with a desire to learn and experience other professional cultures are required. Strong teams of individuals from each partner organisation who are committed and who have the necessary buy-in from their leadership are essential. It is also necessary to have individuals who recognise that work styles, speeds and expectations will likely vary widely across non-government organisations, academia and other scientific organisations, and the private sector. The partnership can be structured by (1) marrying the different styles and creating collective workgroups, (2) separating responsibilities, and (3) allowing each organisation to 'do what it does best' (or a combination of approaches). If gaps in skills or knowledge exist, additional partners might be needed. The Source Water Collaborative, a partnership union devoted to protecting America's drinking water, developed an extensive, comprehensive and broadly applicable

'How-to Collaborate' Toolkit that outlines how to initiate and enhance partnerships (Source Water Collaborative, 2015).

The primary question to be asked as far as funding the collaboration is
whether the non-profit partner will receive funds directly from the industry
partner. Peter Seligmann of Conservation International put forth that in such
partnerships, Conservation International's integrity, independence and effectiveness are maintained as long as corporate financial support is communicated clearly and transparently while ensuring the support is directed towards
key projects and the business partners are committed to best environmental
practices (Seligmann, 2011). Whereas, Patrick Laine, director of corporate
partnerships at World Wildlife Fund UK, says that they do not take money
from the oil and gas sector because they think that 80% of the known oil
and gas reserves need to stay under the ground (Henley, 2012). Laine admits,
however, that deciding not to collaborate with the extractive industry is easier for a big organisation, like World Wildlife Fund, and that the decision
would probably be much more difficult for a smaller charity (Henley, 2012).

Measuring

As with any project, it is good to establish baseline conditions against which
to measure the efficacy of the partnership and the impact of any modifications
in industrial behaviour or management actions. In a partnership, it is also
essential that the partners agree on the methods that will be used to monitor
progress or lack thereof. While private-sector projects usually have a simple
metric (profit) to measure success or failure, conservation projects often require more complex methods to assess whether desired outcomes have been
achieved, or undesired impacts avoided (Oliveira Júnior et al., 2016; United
Nations Environment Programme-World Conservation Monitoring Centre,
2015; Venegas-Li et al., 2016). Baseline conditions and achievements in the
marine environment measured on the scale of strategic environmental assessments as opposed to individual projects can be assessed using, for example,
the Ocean Health Index (Halpern et al., 2012; Ocean Health Index, 2015;
Selig et al., 2015), which can be specifically tailored to each relevant region.

Communicating

Establishing ground rules for public communication about the project is essential. In most cases, both organisations will likely insist that each be given
the opportunity to approve all public communication about the collaboration.
Laine, the director of corporate partnerships at World Wildlife Fund UK, be-
lieves that the most important facets in any partnerships are the right to criticise
and communicate the details of the financial relationship clearly and transparently (Henley, 2012). Indeed, many non-government organisations disclose
information about their corporate partners (including financial support) on

their websites or in separate reports (World Wildlife Fund, 2015). From the point of view of the industry partner, collaboration with a non-government organisation will likely serve to improve its reputation with stakeholders (C&E Advisory Services Limited, 2015), so they may also want to be able to communicate about it more than their conservation partner wishes. Strong internal negotiation based on the mutual trust that will hopefully build over the course of the partnership will be relied upon for these and many other decisions.

Both partners should seek to share their results widely with their colleagues and outside of their normal communication channels. Fauna and Flora International's corporate partnerships programme director emphasised that identifying a large business to work with can have far-reaching positive impacts not only within the industry, but also among host communities and government (Henley, 2012), which is the sort of information that should be disseminated among industry players and host countries.

Finally, the conservation community should recognise that the offshore energy industry makes investments in areas for decades and they have the funds, the technology, and the self-interest to help move conservation forward. By building a mutually beneficial, transparent, and trust-based relationship that is communicated clearly and effectively, it is possible to help them become long-term stewards of the regions most in need of protection.

Highlights

- Non-government organisation-industry collaboration is increasingly common, and it can produce positive outcomes for both conservation and business.
- Main incentives for collaborating include creating new business and environmental benefits, sharing skills and expertise, and building respect and credibility.
- Solid collaborations are built on common ground, shared goals and transparent motivations.
- Marine spatial planning exercises will only benefit from open, albeit challenging collaboration between the conservation community and the offshore energy industry.

References

Alikhani, I.A., Apfalter, S., Arizu, S., Hong, H., Kamezawa, T., Malca, V.D., Patel, U.D., Pinglo, M.E., Rajasingham, C.S., Scarpino, I., Siy, A.M. and De Silva, A.R. (2015). *World Bank Group Support to Public-Private Partnerships: Lessons from Experience in Client Countries*. FY2002–12. Washington, DC: World Bank Group. documents.worldbank. org/curated/en/405891468334813110/World-Bank-Group-support-to-public-private-partnerships-lessons-from-experience-in-client-countries-FY2002–12

Antonio, F., Mendes, R. and Thomaz, S. (2011). Identifying and modeling patterns of tetrapod vertebrate mortality rates in the Gulf of Mexico oil spill. *Aquatic Toxicology*, 105(1–2), pp. 177–179.

Bitzer, V. and Glasbergen, P. (2015). Business–NGO partnerships in global value chains: Part of the solution or part of the problem of sustainable change? *Current Opinion in Environmental Sustainability*, 12(1). pp. 35–40.

Bonar, P., Bryden, I. and Borthwick, A. (2015). Social and ecological impacts of marine energy development. *Renewable and Sustainable Energy Reviews*, 47(1). pp. 486–495.

Bos, M., Pressey, R.L. and Stoeckl, N. (2015). Marine conservation finance: The need for and scope of an emerging field. *Ocean and Coastal Management*, 114(1), pp. 116–128.

Bruns, A. and Gee, K. (2009.) From state-centered decision-making to participatory governance – planning for offshore wind farms and implementation of the water framework directive in Northern Germany. *GAIA - Ecological Perspectives for Science and Society*, 18(2), pp. 150–157.

Business and Biodiversity Offsets Programme. (2012). *Standard on Biodiversity Offsets. Business and Biodiversity Offsets Programme.* Washington, DC: Business and Biodiversity Offsets Programme. http://www.forest-trends.org/documents/files/doc_3078.pdf

C&E Advisory Services Limited. (2015). *Corporate-NGO Partnerships Barometer 2015.* p. 36. www.candeadvisory.com/sites/default/files/barometer_2015_final_3.pdf

Conservation International. (2006). *From Ship to Shore: Sustainable Stewardship in Cruise Destinations.* p. 54. brocku.ca/tren/courses/tren3p18/From%20Ship%20to%20Shore%20-%20Sustainable%20Stewardship%20in%20Cruise%20Destinations.pdf

Convention on Biological Diversity. (2010). *Aichi Biodiversity Targets 1–20.* www.cbd.int/sp/targets/default.shtml

Crooks, E. (2016). BP draws line under Gulf spill costs. *The Financial Times.* July 14, 2016. www.ft.com/content/ff2d8bcc-49e9-11e6-8d68-72e9211e86ab

Dahan, N., Doh, J., Oetzel, J. and Yaziji, M. (2010). Corporate-collaboration: Co-creating new business models for developing markets. *Long Range Planning*, 43(2–3), pp. 326–342.

Darko A., Chan, A., Owusu-Manu, D.-G. and Ameyaw, E. (2017). Drivers for implementing green building technologies: An international survey of experts. *Journal of Cleaner Production*, 145(1), pp. 386–394.

Dees, J. and Anderson, B. (2003). Sector-bending: Blurring lines between nonprofit and for-profit. *Society*, 40(4), pp. 16–27.

de Oliveira Neves, A., Nunes, F., de Carvalho, F. and Wilson Fernandes, G. (2016). Neglect of ecosystems services by mining, and the worst environmental disaster in Brazil. *Natureza & Conservação*, 14(1), pp. 24–27.

den Hond, F., de Bakker, F. and Doh, J. (2015). What prompts companies to collaboration with NGOs? Recent evidence from the Netherlands. *Business & Society*, 54(2), pp. 187–228.

Esty, D. and Simmons, P. (2011). *The Green to Gold Business Playbook: How to Implement Sustainability Practices for Bottom-Line Results in Every Business Function.* Hoboken, NJ: John Wiley & Sons, Inc., p. 456.

Evans, P. (1998). Biology of cetaceans of the North-east Atlantic (in relation to seismic energy). In: M.L. Tasker and C. Weir (eds.), *Proceedings of the Seismic and Marine*

Mammals Workshop, London, 23–25 June 1998. St. Andrews: Sea Mammal Research Unit, pp. 1–35.

Firestone, J. and Kempton, W. (2007). Public opinion about large offshore wind power: Underlying factors. *Energy Policy*, 35(3), pp. 1584–1598.

Forest Trends. (2015). *Mitigation Hierarchy. Business and Biodiversity Offsets Programme.* bbop.forestrends.org/pages/mitigation_hierarchy

Gasbarro, F., Iraldo, F. and Daddi, T. (2017). The drivers of multinational enterprises' climate change strategies: A quantitative study on climate-related risks and opportunities. *Journal of Cleaner Production* 160, pp. 8–26

Glin, L., Mol, A. and Oosterveer, P. (2013). Conventionalization of the organic sesame network from Burkina Faso: Shrinking into mainstream. *Agriculture Human Values*, 30(4), pp. 539–554.

Global Environmental Management Initiative. (2008). *Guide to Successful Corporate-NGO Partnerships.* www.gemi.org/resources/GEMI-EDF%20Guide.pdf

Global Environment Facility. (2013). *The GEF and the Private Sector, Global Environment Facility.* www.thegef.org/gef/PPP

Gore, A. and Blood, D. (2011). A manifesto for sustainable capitalism: How businesses can embrace environmental, social and governance metrics. *The Wall Street Journal*, December 14, 2011. https://www.wsj.com/articles/SB10001424052970203430404577092682864215896

Gray, T., Haggett, C. and Bell, D. (2005). Offshore wind farms and commercial fisheries in the UK: A study in stakeholder consultation. *Ethics, Place & Environment: A Journal of Philosophy & Geography*, 8(2), pp. 127–140.

Haggett, C. (2011). Understanding public responses to offshore wind power. *Energy Policy*, 39(2), pp. 503–510.

Halpern, B., Hardy, D., McLeod, K.L., Samhouri, J.F., Katona, S.K., Kleisner, K., Lester, S.E., O'Leary, J., Ranelletti, M., Rosenberg, A.A., Scarborough, C., Seig, E.R., Best, B.D., Brumbaugh, D.R., Chaplin, F.S., Crowder, L.B., Daly, K.L., Donney, S.C., Elfes, C., Fogarty, M.J., Gaines, S.D., Jacobsen, K.I., Karrer, L.B., Leslie, H.M., Neeley, E., Paulty, D., Polasky, S., Ris, B., St Martin, K., Stone, G.S., Sumaila, U.R. and Zeller, D. (2012). An index to assess the health and benefits of the global ocean. *Nature*, 488(7413), pp. 615–620.

Hamrick, K. (2016). *State of Private Investment in Conservation 2016. A Landscape Assessment of an Emerging Market. Ecosystem Marketplace - A Forest Trends Initiative.* www.jpmorganchase.com/corporate/Corporate-Responsibility/document/cr-es-investment-in-conservation-report-2016.pdf

Hanson, C., Ranganathan, J., Iceland, C. and Finisdore, J. (2012). *The Corporate Ecosystem Services Review: Guidelines for Identifying Business Risks and Opportunities Arising from Ecosystem Change. Version 2.0.* Washington, DC: World Resources Institute, p. 48. www.wri.org/sites/default/files/corporate_ecosystem_services_review_1.pdf

Henisz, W.J., Dorobantu, S. and Nartey, L. (2014). Spinning gold: The financial returns to stakeholder engagement. *Strategic Management Journal*, 35(12), pp. 1727–1748.

Henley, W. (2012) A pact with the devil? The challenges of partnering with the extractive industry. Guardian Sustainable Business. *The Guardian*. July 11, 2012. www.theguardian.com/sustainable-business/blog/pact-devil-challenges-partnering-extractive-industry

Hester, M., Willis, J., Rouhani, S., Steinhoff, M. and Baker, M. (2016). Impacts of the Deepwater Horizon oil spill on the salt marsh vegetation of Louisiana. *Environmental Pollution*, 216(1), pp. 361–370.

Hoegh-Guldberg, O., Aqorau, T., Arnason, R., Chansiri, T., Del Rio, N., Demone, H., Earle, S., Feeley, M.H., Gutierrez, D., Hilborn, R., Ishii, N., Lischewski, C., Lubchenco, J., Nguyen, K.A., Obura, D., Payet, H.E.R., Neroni Slade, T., Tanzer, J., Williams, J.H., Wright, D.J. and Xu, J. (2013). *Indispensable Ocean: Aligning Ocean Health and Human Well-Being - Guidance from the Blue Ribbon Panel to the Global Partnerships for Oceans*. Washington, DC: World Bank Group. www.globalpartnershipforoceans.org/sites/oceans/files/images/Indispensable_Ocean.pdf

Hole, D., Perkins, A., Wilson, J., Alexander, I., Grice, P. and Evans, A. (2005). Does organic farming benefit biodiversity? *Biological Conservation*, 122(1), pp. 113–130.

Hooper, T., Hattam, C. and Austen, M. (2017). Recreational use of offshore wind farms: Experiences and opinions of sea anglers in the UK. *Marine Policy*, 78(1), pp. 55–60.

International Union for Conservation of Nature. (2013). *Unlikely Alliances – Pioneering Collaboration between the Private Sector and Conservation*. Gland, Switzerland. www.iucn.org/what/experts/?13647/Unlikely-alliances--pioneering-collaboration-between-the-private-sector-and-conservation

Jacob, C., Pioch, S. and Thorin, S. (2016). The effectiveness of the mitigation hierarchy in environmental impact studies on marine ecosystems: A case study in France. *Environmental Impact Assessment Review*, 60(1), pp. 83–98.

Kareiva, P. and Marvier, M. (2012). What is conservation science? *BioScience*, 62(11), pp. 962–969.

Kark, S., Brokovich, E., Mazor, T. and Levin, N. (2015). Emerging conservation challenges and prospects in an era of offshore hydrocarbon exploration and exploitation. *Conservation Biology*, 29(6), pp. 1573–1585.

Kleemann, L. and Abdulai, A. (2013). Organic certification, agro-ecological practices and return on investment: Evidence from pineapple producers in Ghana. *Ecological Economics*, 93(1), pp. 330–341.

Klein, N. (2013). Time for big green to go fossil free. *The Nation*. www.thenation.com/article/time-big-green-go-fossil-free/

Korngold, A. (2014). *A Better World, Inc.* New York: Palgrave MacMillan.

Livesey, S. (1999). McDonald's and the Environmental Defence Fund: A case study of a green alliance. *Journal of Business Communication*, 36(1), pp. 5–39.

Martin-Mehers, G. (2016). *Western Gray Whales Advisory Panel: Stories of Influence*. International Union for Conservation of Nature, World Wildlife Fund, International Fund for Animal Welfare. p. 38. portals.iucn.org/library/sites/library/files/documents/2016–034.pdf

Max, D. (2014). Green is good. *The New Yorker*. May 12, 2014. www.newyorker.com/magazine/2014/05/12/green-is-good

McInerney, C. and Johannsdottir, L. (2016). Conference report. Lima Paris Action Agenda: Focus on private finance - note from COP21. *Journal of Cleaner Production*, 126(1), pp. 707–710.

McNamara, E. (2015). What really toppled cape wind's plans for Nantucket Sound. *The Boston Globe*. January 30, 2015. www.bostonglobe.com/magazine/2015/01/30/what-really-toppled-cape-wind-plans-for-nantucket-sound/mGJnw0PbCd-fzZHtITxq1aN/story.html

Moundounga Mouity, P. and Ndjimba K. (2012). *Le Gabon à l'épreuve de la politique de l'émergence*. Paris: Éditions Publibook, p. 306.

National Geographic. (2014). A massive new marine protected area network in Gabon. *Ocean Views*. November 12, 2014. voices.nationalgeographic.com/2014/11/12/a-massive-new-marine-protected-area-network-in-gabon/

Noss, R., Nash, R., Paquet, P. and Soulé, M. (2013). Humanity's domination of nature is part of the problem: A response to Kareiva and Marvier. *BioScience*, 63(4), pp. 241–242.

Ocean Health Index. (2015). *Ocean Health Index*. www.oceanhealthindex.org

Oliveira Júnior, J., Ladlea, R., Correia, R. and Batista, V. (2016). Measuring what matters – identifying indicators of success for Brazilian marine protected areas. *Marine Policy*, 74(1), pp. 91–98.

Organisation for Economic Co-operation and Development. (2014). *Aid to Developing Countries Rebounds in 2013 to Reach an All-Time High*. www.oecd.org/newsroom/aid-to-developing-countries-rebounds-in-2013-to-reach-an-all-time-high.htm

Ortega Lindsey, C., Janus, B. and Murphy, H. (2014). Growing Expectations: The Rising Tide of Sustainability Reporting Initiatives and Their Challenges for the Oil and Gas Industry. In: *Society of Petroleum Engineers International Conference on Health, Safety, and Environment, 17–19 March 2014*. California: Society of Petroleum Engineers.

Oxford Business Group. (2015). *Abu Dhabi Pumps Up Oil Investment*. www.oxford-businessgroup.com/news/abu-dhabi-pumps-oil-investment

Parsons, E., Clark, J., Ross, A. and Simmonds, M. (2007). The Conservation of British Cetaceans: A Review of the Threats and Protection Afforded to Whales, Dolphins and Porpoises in UK Waters. In: *Whale and Dolphin Conservation Society*. Chippenham: Whale and Dolphin Conservation Society, p. 123.

Pedroni, P.M., Springer, N., Spurgeon, J. and Romer, R. (2012). Integrating the Emerging Concept of Ecosystem Services into Oil and Gas Industry Environmental Management Practices. In: *Society of Petroleum Engineers/Australian Petroleum Production & Exploration Association Limited International Conference on Health, Safety and Environment in Oil and Gas Exploration and Production, 11–13 September 2012*. Perth: Society of Petroleum Engineers.

Poulsen, J. and Clark, C. (2012) Building Partnerships for Conservation. In: C.J. Clark and J.R. Poulsen (eds.), *Tropical Forest Conservation and Industry Partnership: An Experience from the Congo*. New York: Wildlife Conservation Society, pp. 21–62.

Saldinger, A. (2015). How international NGOs choose corporate partners. *Devex Impact*. www.devex.com/news/how-international-ngos-choose-corporate-partners-85878

Saul, J., Davenport, C. and Ouellette, A. (2010). (Re)Valuing Public - Private Alliances: An Outcomes-Based Solution. *United States Agency for International Development: Private Sector Alliances Division & Mission Measurement, LLC*. www.usaid.gov/sites/default/files/documents/1880/RevaluingPublicPrivateAlliances.pdf

Selig, E., Frazier, M., O'Leary, J.K., Jupiter, S.D., Halpern, B.S., Longo, C., Kleisner, K.L., Sivo, L. and Ranelletti, M. (2015). Measuring indicators of ocean health for an island nation: The ocean health index for Fiji. *Ecosystem Services*, 16(1), pp. 403–412.

Seligmann, P. (2011). Partnerships for the Planet: Why we must engage corporations. *The Huffington Post*. 19 May, 2011 (Updated July 19, 2011). www.huffingtonpost.com/peter-seligmann/conservation-international-lockheed-martin_b_863876.html?

Smedley, T. (2012). More NGOs finding fruitful collaborations with the private sector. *The Guardian*. August 7, 2012. www.theguardian.com/sustainable-business/ngos-collaboration-private-sector

Source Water Collaborative. (2015). Considering a collaborative effort, source water collaborative. *Source Water Collaborative*. www.sourcewatercollaborative.org/how-to-collaborate-toolkit/

Springer, N., Tait, R. and Parkerton, T. (2012). The Promise and Challenge of Ecosystem Services from an Industry Perspective. In: *Society of Petroleum Engineers. Australian Petroleum Production & Exploration Association Limited International Conference on Health, Safety and Environment in Oil and Gas Exploration and Production, 11–13 September 2012*. Perth: Society of Petroleum Engineers.

Stafford, E. and Hartman, C. (2015). The Paradoxes and Challenges of Creating Social Good Through Environmentalist-Marketer Collaboration. In: H.E. Spotts and H.L. Meadow (eds.), *Proceedings of the 2000 Academy of Marketing Science (AMS) Annual Conference SE -83*. Montreal: Springer International Publishing, p. 348. doi:10.1007/978-3-319–11885-7_83

Stafford, E., Polonsky, M. and Hartman, C. (1999). Environmental NGO-business collaboration and strategic bridging: A case analysis of the Greenpeace-Foron alliance. *American Marketing Association. Conference Proceedings*. p. 222. search.proquest.com/docview/199394100?accountid=14723

Tennyson, R., Harrison, T. and Wisheart, M. (2008). Emerging opportunities for NGO-business partnerships. *Feedback from the Cross Sector Partnership Project Involving Accenture Development Partnerships the Partnering Initiative World Vision*. p. 38 http://thepartneringinitiative.org/wp-content/uploads/2014/08/Emerging-opportunities-for-NGO-business-partnerships1.pdf

Toke, D. (2005). Explaining wind power planning outcomes: Some findings from a study in England and Wales. *Energy Policy*, *33*(12), pp. 1527–1539.

Tuodolo, F. (2009). Corporate social responsibility: Between civil society and the oil industry in the developing world, ACME. *An International E-Journal for Critical Geographies*, 8(3), pp. 530–541.

United Nations Environment Programme-World Conservation Monitoring Centre. (2015). *Marine No Net Loss: A Feasibility Assessment of Implementing No Net Loss in the Sea*. Cambridge: UNEP-WCMC.

Venegas-Li, R., Cros, A., White, A. and Mora, C. (2016). Measuring conservation success with missing marine protected area boundaries: A case study in the Coral Triangle. *Ecological Indicators*, 60(1), pp. 119–124.

Viana, L., Weikel, M., Blaha, J., Larsen, T., Ahumada, J. and Romer, R. (2016). Advancements in Biodiversity Baseline Assessments and Monitoring Approaches for Enhanced Oil and Gas Performance. In: *Society of Petroleum Engineers International Conference and Exhibition on Health, Safety, Environment and Social Responsibility, 11–13 April 2016*. Stavanger: Society of Petroleum Engineers.

Vladimirov, A., Ilyashenko, V., Oleyniko, E. and Chernyakhovsky I. (2012). *Sakhalin History/Gray Whales. The Sakhalin Story. 2012*. p. 108. www.sakhalinenergy.com/media/library/ru/publications/Gray_whales_book_web.pdf

Wever, L., Krause, G. and Buck, B. (2015). Lessons from stakeholder dialogues on marine aquaculture in offshore wind farms: Perceived potentials, constraints and research gaps. *Marine Policy*, 51(1), pp. 251–259.

World Economic Forum. (2013). *Foreign Direct Investment as a Key Driver for Trade, Growth and Prosperity: The Case for a Multilateral Agreement on Investment*. p. 26. www3. weforum.org/docs/GAC13/WEF_GAC_GlobalTradeFDI_FDIKeyDriver_Report_2013.pdf

World Wildlife Fund. (2015). *WWF Corporate Partnerships Report – 2015*. p. 20. http://d2ouvy59p0dg6k.cloudfront.net/downloads/final_fy2015_corporate_engagement.pdf [Accessed April 1, 2017]

World Wildlife Fund & Carbon Disclosure Project. (2013). *The 3% Solution: Driving Profits through Carbon Reductions*. p. 32. http://assets.worldwildlife.org/publications/575/files/original/The_3_Percent_Solution_-_June_10.pdf?1371151781

Yaziji, M. (2006). *Sleeping with the Enemy for Competitiveness Advantage: Corporation and NGO Partnerships*. Switzerland: IMD International, pp. 1–5. www.imd.org/research/challenges/upload/sleeping_with_the_enemy_for_competitive_advantage.pdf

Chapter 4

Challenges and opportunities for governance in marine spatial planning

Lucy Greenhill

Introduction

Marine spatial planning is a framework that aims to guide and support management of multiple and competing demands on marine resources to achieve economic, social and ecological objectives. Marine spatial planning enables a process of public negotiation regarding desired future conditions based on principles of integration, participation and adaptive management. The prevalence of marine spatial planning is increasing, with at least a third of the surface area of the world's exclusive economic zones expected to have government-approved marine spatial plans by 2030 (Ehler 2017). There is a generally accepted process of doing marine spatial planning (as set out in Ehler & Douvere 2009 and Chapter 1); however, approaches vary widely according to local conditions such as socio-political context, the emphasis placed on particular sectors and interests, the terminology used, whether it is legally binding, its relationship to other regulatory frameworks, the resources available for the planning process, and its prior prevalence in a region. Although there is a rapid expansion of marine spatial planning and its varied application, it can be unclear how it relates to, and ultimately influences, the governance of offshore energy and other marine activities.

Marine spatial planning is herein conceptualised as a framework for integrated governance, which must consider and relate to the broader framework of policy, planning and management of marine uses. Marine spatial planning includes the exercise of authority in terms of allocating space for marine activities guided by a broad range of policies relating to sector development (such as energy, climate change and food security), and informs the sector-specific licensing and management of marine activities within its boundaries. The functions of authorities responsible for marine spatial planning and decision-making processes intersect within it, and their respective roles are therefore of interest, including how they might change as marine spatial planning processes are established.

Both conceptual and operational problems exist in marine spatial planning in (a) understanding the respective roles of government, industry and civil society, (b) reaching consensus on the purpose of marine planning (including its definition and 'operationalising' concepts such as 'sustainability' and 'ecosystem-based management'), (c) understanding how marine planning relates to existing governance structures of decision making and management, (d) how to evaluate the effectiveness of marine spatial planning and enable ongoing improvements through adaptive management, and (e) how to ensure the knowledge and capacity to support the process.

Such issues are explored in this chapter, along with the challenges and opportunities for transitions to more effective governance of marine resources, to address ambitious maritime development policies such as 'Blue Growth' and the Integrated Maritime Policy of the European Union (EC 2009), or the 'Blue Economy', a concept more regularly used in the context of Small Island Developing States. While marine spatial planning does not replace existing decision-making processes, it provides an opportunity to reflect on the problems of the currently fragmented, sector-specific approaches and consider more co-ordinated, efficient and cost-effective marine governance.

Governance in the context of marine spatial planning

Understanding the influence of marine spatial planning on governance of marine activities is challenging, not least because the term 'governance' is difficult to define precisely, and it is growing in complexity, becoming "amorphous in recent years" (Turnpenny et al. 2005). Governance represents the structures and processes by which people in societies make decisions and share power (Lebel, cited in Folke et al. 2005). This includes the more obvious formal institutional arrangements through which legal authority is exercised (such as licensing of offshore wind-energy projects), but has expanded to address the increasing interaction between governments, market and civil society in new institutional arrangements (Arts & Tatenhove 2004), and the informal processes that also shape the manner in which problems are addressed (Juda & Hennesssey 2001). These wider elements are captured within the UNESCO definition of governance in relation to marine spatial planning to some extent, describing it as:

> … the process through which diverse elements of a society wield power and authority and thereby influence and enact policies and decisions concerning public life and economic and social development. Governments as well as the private sector and civil society carry out governance. However, governance is not the same as government.
>
> (Ehler 2014, p. 33)

More inclusive definitions of governance are appropriate to account for the increasing diversity and role of non-state actors and the progress away from traditional 'top down' governance in natural resource management. In this wider sense, marine spatial planning is an instrument for governance; however, interpretation of the term and the differences among actors in conceptions of governance present a problem in characterising and addressing the challenges in implementing marine spatial planning. Different views mean that there might not be a shared understanding regarding important features of governance, such as accountability, responsibility, power and ultimately, its purpose.

As academic 'meta–debates' on models of governance continue to develop (e.g., Hooghe, Liesbet, & Marks 2009; Ostrom 2010), and as marine spatial planning is developed as an applied instrument that guides decision making, it can be seen as an interface between these debates and developments in marine governance. Ongoing integration of new scientific knowledge, including new forms of knowledge generation, alongside practical experience, is therefore necessary. This applies similarly to the concurrent development of concepts of 'sustainability' (e.g., Robinson 2004) and 'ecosystem-based management' (e.g., Katsanevakis et al. 2011; Ruckelshaus et al. 2008), which are essential components of marine spatial planning.

Implementing marine spatial planning happens within a wider context of academic and practical learning regarding complex socio-ecological challenges and innovative governance. However, for marine spatial planning to be effective, there is a need to define the authoritative governance structure to ensure leadership and accountability for the delivery of a marine spatial planning process. But methods are also needed to reflect, adjust and transform governance as appropriate, integrating new knowledge to improve approaches—an important aspect of the 'adaptive capacity' of governance (Folke et al. 2005).

Communication and participation processes are essential to enable ongoing dialogue between actors in marine spatial planning, to achieve governance that is about "... political brokerage to negotiate potential futures, as much as it is about the exercise of authority" (Jentoft et al. 2007). The local context is important, and methods for analysing governance have been suggested as a precursor to planning, including the 'governance baseline' approach (Olsen & Olsen 2011) and 'governance profiling' (Juda & Hennessey 2001). These are useful frameworks for evaluating the current situation and setting the baseline for ongoing evaluation and adaptive management in marine spatial planning programmes.

Actors in governance in marine spatial planning

While some participants in marine spatial planning processes have a formal mandate (such as governments and planning authorities), others' roles can be informal and more difficult to set out, including civil society and the groups they form, including non-governmental organisations and other stakeholders. Roles also vary according to the socio-political context and the particular issues to be addressed. To establish legitimacy and credibility and to guide stakeholders in engaging with the process, clarity around the formal roles and authority of the principal actors is essential (Ehler & Douvere 2009), particularly for industry representatives whose activities and livelihoods are subject to the planning process. However, some institutional flexibility is also essential, with processes needed that enable deliberation and communication among participants to develop shared understanding of potential roles and responsibilities and how these might evolve through integrated planning.

The State

The state (or government) is a major actor in governance, and although its role will vary according to the legal framework of the particular marine planning process (Chapter 6), government authorities will approve produced plans. The 'power' held by state institutions in terms of the ability to shape the process, make decisions and influence outcomes, accords them some accountability in ensuring an effective process. Many national governments are required, through international commitments, to implement marine spatial planning, which might be set out in national legislation to deliver international commitments, such as in the Europe Union under the European Maritime Spatial Planning Directive (Directive 2014/80/EU). National governments can lead marine spatial planning efforts, or delegate to new dedicated planning authorities (such as the Marine Management Organisation in England), sub-national government bodies (e.g., state governments in the United States), or local planning bodies made up of local representatives (e.g., Regional Marine Planning Partnerships in Scotland). In most cases, government authorities will need to adopt the produced plans formally, and remain the primary 'decision makers' for centrally regulated activities via traditional, sector-specific licensing and management. Where the planning 'authority' is separate from the licensing authorities, the relationship between marine spatial planning processes in directing activities, and how consenting decisions are taken, might be unclear.

Where an overarching authority oversees sub-national/regional marine planning programmes, clarity regarding communication processes, information flow and balance of authority between the central body and the regional authority is essential to achieve national objectives, while accounting for regional 'uniqueness'. In such cases of multi-scalar ownership of marine planning, the state is required to maintain a focus on policy objectives that are relevant at multi-regional, national, or transnational scales (such as broad-scale conservation targets). The state also has a role in providing resources for marine planning, identifying areas in which consistency is required (e.g., for management of trans-boundary issues), support adaptive management through evaluation and comparison at national scaling, and developing strategic national research agenda to address knowledge gaps. Fundamental differences in institutional governance structures for marine spatial planning will lead to different approaches and perceptions of the process across nations, and will need to be considered particularly where integration between plans is required, to address cross-border challenges such as managing impacts on mobile marine species, or trans-boundary activities such as shipping.

The legal mandate afforded to the marine spatial planning process varies, with statutory and non-statutory marine planning being implemented. In Europe, the Maritime Spatial Planning Directive requires maritime Member States to develop marine spatial plans by 2021, whereas the development of marine spatial plans is 'voluntary' under the National Ocean Policy of the United States. A statutory process could be considered to lead to greater legitimacy, engagement and effectiveness in progressing towards objectives, since the lack of statutory basis assigned to the policy of Integrated Coastal Zone Management has been cited as a cause of its limited success in the 1990s (Potts et al. cited in Brennan et al. 2014).

Alternative governance models

The traditional role of governments is changing, with the decreasing role of a single, central decision-making body in relation to marine resource management (e.g., Hooghe, Liesbet, & Marks 2009). New forms of governance are emerging as appropriate for contemporary understanding of socio-ecological challenges (Arts & Tatenhove 2004), including pluralism, both in terms of institutional pluralism (involving multiple actors and governance bodies), and pluralism of participating groups (the participation of diverse social organisations) (Meadowcroft 2002). Growing diversity in models of governance includes development of bottom-up governance and co-management (e.g., Heylings & Bravo 2007), with closer interaction of managers and user groups (Young et al. 2007) and multiple centres of decision making (polycentricity) (e.g., Galaz et al. 2012; Ostrom 2010). These embody different concepts of resource ownership, authority, accountability and responsibility across different scales.

In addressing integrated resource governance within and across jurisdictional boundaries, marine spatial planning must interact with a range of resource management models, and its relationship to these needs to be investigated in each context. Marine spatial planning can provide a forum for discussing different management approaches and enable the consideration of other potential models, such as those of local ownership.

Civil society

Governance is influenced by the relationship between institutions and the citizens accountable for their existence, and so-called civil society are increasingly involved in institutional processes (Arts & Tatenhove 2004), extending the responsibility for governance to individuals. In addition to their political expression, civil society interacts directly with marine spatial planning as marine resource users, local communities, the beneficiaries, and consumers of ecosystem services. As such, civil society must be engaged in governance in marine spatial planning so that it is fair in the allocation and management of access to marine resources (which is "ultimately a matter of societal choice") (Ehler 2008, p. 841).

It has been suggested that considering the public as 'policy actors' rather than subjects of state-led governance is more appropriate for addressing sustainability in marine management. Engaging the concept of 'marine citizenship' could develop new relationships between the state and individuals, empowering the public in influencing governance at the policy level and increasing their role as policy implementation channels through altered behaviour (McKinley & Fletcher 2010). Indeed, the role of the public might justify greater attention, given that civic activity is increasing in policy areas where central government has reduced or retracted support, which includes planning in some areas such as the United Kingdom (Rees et al. 2012).

Public participation in the implementation of marine spatial planning is essential to plan multi-sector development scenarios that are most desirable in terms of the benefits they provide for society. This also ensures inclusion of spatial and non-spatial local knowledge into planning processes (e.g., Alexander et al. 2012) and enables the 'social learning' (Newig & Fritsch 2009) that is essential to guide governance in addressing policy objectives related to sustainable outcomes (Kenter et al. 2014; Robinson 2004). Dialogue and participatory/deliberative exercises are therefore a critical component of adaptive marine spatial planning, and civil society must be empowered to question and steer political processes.

Empowerment of civil society requires the broader education of individuals to be competent engagers, which is a two-way problem (Maynard 2002). Science and education is essential to improve knowledge and awareness of the marine planning process and how to engage with it (UNESCO 2014, p. 34). Such outreach empowers citizens by increasing their social and political capacity for participating and influencing decision making (Heylings & Bravo 2007).

Industry

As active participants with capital and vested interests in marine management, industry organisations and representatives are essential actors in marine governance, including formal and informal arrangements. Depending on the socio-political context and the importance of a sector to a particular national or local economy, industry representatives are likely to have considerable power in influencing marine spatial planning. In most cases, the purpose of marine spatial planning is to ensure that (sustainable) business ambitions of the private sector are fulfilled for the benefit of local or national economies, in support of strategies such as the Blue Growth agenda in Europe. Industry is therefore the main subject of marine spatial planning, which aims to control and enable industry activities to varying degrees and to manage conflict between different sectors. Thus, the success of marine spatial planning often depends on industry engagement and conduct.

Industry sectors operate at market-relevant scales (trans-national and global in relation to the offshore energy sector), and their involvement in developing marine spatial planning can provide valuable input with regard to wider economic processes (such as energy markets), as well as regional information about their current and future activities. Engagement at the local level in marine spatial planning is critical for understanding industry ambitions, to explore new forms of resource use (such as community-based, off-grid renewable energy initiatives and co-location of multiple activities), (e.g., Christie et al. 2014), and to engender greater sensitivity to local issues in project development such as social acceptance of wind farms (Haggett 2008). The role of industry in influencing marine governance needs to be acknowledged and clarified in marine spatial planning processes to ensure balanced representation of different interests, including other sectors, local communities and ecological aspects.

Other actors

New knowledge is needed to support marine spatial planning, as planners endeavour to operationalise complex concepts of sustainability, understand the socio-ecological system, and support effective governance. The integration of sectors and interests in marine spatial planning requires interaction with academic institutions and scientists of many disciplines. Strategic research programmes to support planning processes are essential, including research developed in collaboration with planning authorities, civil society, and industry representatives. Lastly, non-governmental organisations that campaign for social or environmental interests also have a role in steering governance because they can hold actors to account and drive improvements through "monitoring, questioning, alternative service delivery" and "vigilance" (De Santo 2011, p. 38; Robinson 2004, p. 378).

Partnerships and networks

Most problems to be addressed in coastal and marine situations cross boundaries and jurisdictions; thus, regional networks and multi-lateral partnerships are crucial to inform governance. Partnerships have a valuable role in research disciplines, to develop joint conceptualisation of issues and begin to address potential conflicts (Kidd & McGowan 2013). Multi-stakeholder partnerships have proliferated as new governance tools to address a range of social and environmental issues, and are advocated in global policy as essential to address the United Nations' Sustainable Development Goals. However, multi-stakeholder partnerships require certain 'conditions for success' including effective leadership, stringent goal-setting, sustained funding, monitoring and reporting, among others (Pattberg & Widerberg 2014). Some partnerships already exist and support cross-border marine spatial planning, for example at a trans-national scale (e.g., the Wadden Sea Forum and BaltSeaPlan (Gee et al. 2011)). Ongoing review of these partnerships is appropriate to document success and develop their viability, legitimacy and efficacy in marine spatial planning.

Recognising the fragmented nature of governance, the multitude of actors and complex relationships between them, it is useful to consider governance in marine spatial planning as a network (Kannen 2014). While clarity around the roles of essential 'nodes' of governance is necessary, the facilitation of collaboration among actors forms the basis of successful marine spatial planning, with partnerships between local authorities, state government, industry representatives, non-governmental organisations and a wide range of stakeholders, needed to develop and sustain progress towards policy objectives over the long term.

Governance challenges and opportunities in marine spatial planning

Guiding the governance process

Marine spatial planning is a tool that is evolving alongside our understanding of socio-ecological systems and the complex challenges to be addressed. The developing social context and conceptual confusion regarding 'governance', 'marine spatial planning' and 'sustainability', and the variety of approaches taken, mean that challenges are apparent at the fundamental level of agreeing on the purpose of marine spatial planning and what it can achieve. While some objectives are easily articulated (such as achieving a specific growth rate in the energy sector), other strategic objectives are more difficult to translate into meaningful targets to be considered in marine spatial planning. Many objectives relate to 'sustainability', a notably ambiguous term, and a lack of clarity in what is meant by sustainability in policy objectives can create a basis of mixed expectations of how marine spatial planning will balance ecological and economic objectives. As a holistic tool, marine spatial planning provides a platform for public negotiation of these priorities within the context of sustainability, giving the opportunity to explore and account for different worldviews, perspectives and outlooks (Robinson 2004). Governance can therefore enable and engage this negotiation in marine spatial planning through multi-actor dialogue, helping to develop clarity of purpose and shared consensus in long-term objective setting for sustainability in a regional context, while relating to wider goals, such as the United Nations' Sustainable Development Goals.

National policy objectives to be addressed through marine spatial planning, such as national renewable energy targets, are generally established through separate processes. Potential conflicting policy objectives developed out with marine planning and relating to specific sectors (e.g., renewable energy and aquaculture) arising from different political drivers (energy security, food security, conservation, etc.) can present challenges in marine spatial planning, as regional governance aims to balance and address competition for space within planning jurisdictional areas. To frame these discussions, tools such as visioning and scenario analysis can enable actors to look collectively to the future, anticipate issues and conflicts, find synergies and compromises, and explore roles and responsibilities (e.g., Reed et al. 2013). Scenario analysis in marine spatial planning can be used to integrate objectives in a regional context, providing feedback to policy makers on the feasibility of policy objectives and highlighting where there could be difficult conflicts between national priorities (such as energy security and food security driving the development of offshore renewable energy and aquaculture, respectively).

Reaching consensus on the purpose of marine spatial planning

Acknowledging that marine spatial planning will continue to develop and be applied differently in different contexts, consensus on its role and relevance in addressing marine governance in particular jurisdictions is needed to ensure clear and sustained engagement of actors and effective implementation. Perceptions of marine spatial planning range from more simplistic views that emphasise zoning (allocation of space) within its boundaries, to descriptions of it as an integrated and novel approach to address complex marine management challenges, through dialogue and information sharing. In planning for the Baltic Sea, it has been described as "... a way of taking the initiative and expanding our thinking beyond the actual circumstances. Marine spatial planning extends our planning horizon, allowing us to actively influence developments rather than wait for things to happen" (Gee et al. 2011, p. 8) and enables "... communication processes, which link diverse sets of information and span a dialogue between groups of society and across spatial scales" (Kannen 2014, p. 2139). Governance, in steering the process, can support the development of consensus, acknowledging the potential for divergent views among actors and addressing this explicitly through debate.

Integrating governance across boundaries and scales

Marine spatial planning is nested within a web of fragmented and evolving cross-scale governance, including existing institutional frameworks of administrative functions, property rights, ownership and access to resources, sector-specific management (including fisheries governance), the establishment of marine protected areas, land-use planning, market-based governance, and many other statutory and non-statutory mechanisms related to the use of coastal and marine resources. Marine spatial planning initiatives must relate to and interact with these multi-scalar and trans-boundary governance mechanisms to support countries in effectively and efficiently addressing overarching national and international policy objectives.

Establishing co-operation and multi-level integration between overlapping governance is therefore essential to develop integrated governance that addresses the complex task of being locally relevant and legitimate, while fulfilling higher-level objectives. Trans-national co-operation is often necessary. However, this amount of co-operation requires overcoming "political, social and economic heterogeneity" (Freire-Gibb et al. 2014, p. 176) and is challenging where there is no overarching authority to steer national governments. Even where there are clear drivers for co-operation, trans-national planning is difficult, particularly given increasingly fragmented and diverse politics between countries, such as the United Kingdom seeking to leave the European Union. For example, while there are obvious advantages to strategic planning of energy across shared sea basins such as the North Sea, i.e., to address energy security, intermittency issues and cost-effectiveness more efficiently, large challenges are presented in integrating electricity networks and markets between countries. Marine spatial planning in this case must relate to, and consider, governance operating at broader scales to overcome these difficulties at a plan scale. Cross-boundary partnerships and trans-national bodies can help integrate across jurisdictional limits and provide a mechanism to develop shared objectives of strategic benefit to multiple nations.

Simplifying planning, licensing and management

The influence of marine spatial planning on marine activities is important in clarifying its purpose, objectives and to justify its legitimacy to actors, particularly to industry. In this regard, marine spatial planning is not directly equivalent to 'management' of marine activities, i.e., the granting of permits, licenses, economic incentives, etc. that are done by specific sectors in accordance with separate regulatory requirements. The extent to which planning (as in the allocation of space and access to resources) affects these processes might not be clear, particularly where these processes are managed by separate authorities. Strategic planning of particular sectors can also be done separately (and at different scales), such as the sectoral planning of wind energy in Scotland (e.g., Scottish Government 2011). In some cases, different authorities will lead marine spatial planning, sectoral planning and implement regulatory functions, resulting in complexity regarding how the processes interact. Understanding 'institutional overlays' of roles and responsibilities in each region is necessary to relate marine spatial planning to other governance instruments (Young et al. 2007) and will influence the extent to which actors (industry and civil society) understand and engage with the process. Marine spatial planning provides an opportunity for proactively reviewing and integrating different governance bodies, as it can highlight overlap and redundancies, enabling simplification and rationalisation of the use of resources for cost-effective and fair planning, regulatory and licensing practices.

Evaluating priorities, synergies and conflicts

Marine spatial planning aims to address trade-offs between sectors, and implicit in this is the need to integrate existing sector-specific approaches, and compare and contrast different development options to evaluate and negotiate preferred scenarios. Understanding the 'cumulative' effects of multiple activities is required through existing regulatory mechanisms (associated with environmental impact assessments applied to projects and sectors), but these are often of poor quality (Chapter 9). Difficulties in assessing cumulative effects rigorously are wide-ranging and include: (1) ambiguities in regulatory frameworks, (2) inconsistencies in scientific approaches to the assessment of impacts, risk and importance (Maclean et al. 2014); unclear methods to account for uncertainty, including environmental variation, structural or process uncertainty (Allen et al. 2011), (3) different authorities and their approach to decision making, (4) difference in the political relevance, and (5) public opinion and awareness of particular sectors. Such inconsistencies make it difficult to compare when evaluating the relative benefits and disadvantages of multi-sector development options in marine spatial planning. Reviewing different requirements for predicting cumulative effects across sectors and regions through marine spatial planning using cross-sector scenario analysis could be a way to investigate these differences further. This could also make progress towards a common and consistent language and approach to decision making, and ensure fairness in governance of individual sectors.

Enabling adaptive governance

Ongoing monitoring and evaluation of governance processes through adaptive management is an important tool to monitor feedback from the system, enabling collective learning and improvement as understanding increases (Kaufman, cited in Ruckelshaus et al. 2008). Adaptive approaches are necessary to ensure that institutional arrangements are appropriately addressing the problems that led them to being established in the first place (Heylings & Bravo 2007). Different interpretations of adaptive management exist, differing primarily according to the structure applied to the process, i.e., from 'learn by doing' without the definition of specific and measurable responses, to explicit parameters identifying goals, hypotheses of causation and procedures for review, adaptation and alternatives (Allen 2011). Using objectives to derive 'indicators' to chart progress and to measure effective governance is recommended (Ehler 2014); however, while some indicators are measurable and specific, others relating to effectiveness of governance might require qualitative assessment.

Generally agreed principles of good governance exist (e.g., United Nations Development Programme 1997) and include empowerment, participation, credibility and accountability. The UNESCO *Guide to Evaluating Marine Spatial Plans* (Ehler 2014) suggests specific governance indicators for measuring performance of phases of the marine planning process, such as authority, financing, participation, the status of marine spatial management planning and implementation, stakeholder participation, compliance and enforcement, as well as the progress and quality of management actions and of the marine plan itself. Academic research addresses approaches to evaluating governance, using quantitative and qualitative indicators, and particularly in relation to marine protected areas (e.g., Jentof et al. 2007; Turner et al. 2014), providing a theoretical basis for approaching the evaluation of marine planning initiatives. Effort and investment is required to translate these into specific monitoring programmes at a regional level, and integrating between monitoring and assessment objectives at different scales through a marine spatial planning platform requires strong leadership and co-ordination, financing and interaction between different authorities, organisations and the scientific community.

Developing capacity for marine planning

A critical factor in the design of effective marine spatial planning is the long-term availability of resources: human, technical and financial, including the training and skills of planning practitioners, financial and computational resources, and the capacity of stakeholders and civil society to engage productively in the process. National economic pressures mean that many regions face increasingly constrained public resources, and assigning public funds to marine spatial planning could be difficult to justify, particularly as the economic benefit of marine spatial planning is difficult to articulate. Evaluating and quantifying the manner in which marine spatial planning benefits particular sectors, particularly in economic terms, would encourage stronger political support, and could include, for example, (1) cost savings from greater efficiency through more streamlined planning, licensing and monitoring across all sectors, (2) optimising marine resource use and identifying opportunities (including co-location of activities; Chapters 12–14), (3) more efficient and effective resolution of conflict, either between resource users, or planned activities and conservation objectives, (4) better use of stakeholder-engagement resources (e.g., through better integration of effort across different governance tools such as development of marine protected areas and marine spatial planning), and (5) more cost-effective data collection and monitoring through public and private collaboration. Accounting for this through cost-benefit analysis, cost-effectiveness analysis, and broader evaluation methods such as the Balance Sheet Approach (UK National Ecosystem Assessment 2014), using scenarios to assess benefits of marine planning, could demonstrate the added value of marine spatial planning and justify the investment of public and private resources.

Promoting interdisciplinarity

Effective marine spatial planning relies on data and information, and a fundamental challenge is ensuring development of appropriate knowledge to guide decision making and ongoing improvements. Holistic planning to achieve diverse objectives requires integration among disciplines, sectors and actors within one framework, and new forms of dialogue are necessary (Chapter 4). At an operation level, this includes the ability to integrate evaluation of potential socio-ecological effects of multiple sectors to make comparisons, and to address synergies and trade-offs (Chapter 3). Greater integration between scientific disciplines is needed to enable evaluation of the implications for the environment and society, and to support decision making in relation to social, economic and ecological objectives.

Traditional science approaches are of limited use in addressing the complex objectives of sustainability (Mayumi & Giampietro 2006; Ravetz 2006), and integration across academic disciplines, and with other forms of knowledge, is necessary to improve planning processes. Active participation of non-scientists (including formal and informal governance actors) is needed to develop science to deal with interdisciplinary questions related to socio-ecological systems. Although there are methodological and conceptual issues in developing new forms of knowledge, strategic research agendas are emerging that address interdisciplinary (and multi-disciplinary, trans-disciplinary) science, to consider new methods of knowledge co-production (e.g., Future Earth 2014). Recognising that action must be taken on the basis of incomplete scientific knowledge, continued knowledge development and transfer at the science-policy interface are important components of implementing and adapting governance through marine spatial planning.

Conclusion

Marine spatial planning is receiving increased attention as an emerging tool to address development of maritime economies, with appropriate consideration for social and ecological risks and benefits. At a global scale, it is seen as an important mechanism for achieving global policy relating to the marine area, including the United Nations' Sustainable Development Goals. As an integrated framework that is being increasingly applied, it presents opportunity for unprecedented multi-stakeholder and cross-sector engagement, which are essential to address the complex, interdisciplinary challenges of sustainable resource use.

Many of the challenges faced in the offshore energy sectors, as outlined elsewhere in this book, are proving testing challenges for planners, decision makers and industry alike, and many require the ability to consider the system in a more holistic, integrated way. Managing conflict, promoting co-location of activities, considering cumulative impacts and ecological interactions across wide scales all require a strategic approach. The emergence of marine spatial planning and its principles provide a useful area of focus for reframing these issues and development of new governance solutions.

While marine spatial planning offers opportunities to improve management of marine resources, multi-faceted and context-specific challenges remain. This chapter highlights some areas that could limit the effectiveness of marine planning, both conceptually (in terms of reaching consensus on what it is, and what it can achieve) and practically, particularly in determining overlap with existing governance of individual sectors. Marine spatial planning provides an opportunity for proactively reviewing current governance and can highlight overlap and redundancies, potentially enabling simplification and rationalisation of resource use for cost-effective and fair planning, regulatory and licensing practices. To enable ongoing improvements, resourcing and support of marine spatial planning are essential, recognising the needs for sustained provision of human and technical resources, and integrating new, multi-disciplinary forms of knowledge.

Addressing these interrelated challenges will benefit from collectively considering the traditional roles of the main actors in governance and marine spatial planning, to enable the adjustment and transformation necessary to develop innovative, efficient and cost-effective approaches. A balance is needed between defining authoritative, clear processes (necessary to support development of business and sector ambitions) and ensuring that traditional precedents do not hamper the ability to use marine spatial planning to develop more meaningful and effective approaches to governance of marine resource use. Framing the main challenges in the context of governance and relating these to operational marine management highlights the opportunities through practical and theoretical development of marine spatial planning, to enable more effective governance for marine resources.

Highlights

- Marine spatial planning presents an opportunity to reflect on marine governance and develop more co-ordinated, efficient and cost-effective planning and management.
- Clarity on roles and responsibilities is needed for delivery of marine spatial planning, but with flexibility to reflect, adjust and transform based on new knowledge and experience.
- Stakeholder participation in visioning, scenario analysis and setting objectives is critical to build trust and as a basis for managing conflict in marine planning.
- Dialogue to explore different perspectives related to 'sustainability', 'governance' and the role of marine spatial planning is a fundamental basis for marine spatial planning.
- Continued evaluation, monitoring and adaptation are essential to refine approaches based on experience, and integrating with the development of scientific knowledge.

References

Alexander, K.A., Janssen, R., Arciniegas, G., O'Higgins, T.G., Eikelboom, T. and Wilding, T.A. (2012). Interactive marine spatial planning: Siting tidal energy arrays around the Mull of Kintyre. *PLoS One*, 7(1), p. e30031.

Allen, C.R., Fontaine, J.J., Pope, K.L. and Garmestani, A.S. (2011). Adaptive management for a turbulent future. *Journal of Environmental Management*, 92(5), pp. 1339–1345.

Arts, B. and Tatenhove, J.V. (2004). Policy and power: A conceptual framework between the 'old' and 'new' policy idioms. *Policy Sciences*, 37(3), pp. 339–356.

Brennan, J., Fitzsimmons, C., Gray, T. and Raggatt, L. (2014). EU marine strategy framework directive (MSFD) and marine spatial planning (MSP): Which is the more dominant and practicable contributor to maritime policy in the UK? *Marine Policy*, 43(2014), pp. 359–366.

Christie, N., Smyth, K., Barnes, R. and Elliott, M. (2014). Co-location of activities and designations: A means of solving or creating problems in marine spatial planning? *Marine Policy*, 43(2014), pp. 254–261.

De Santo, E.M. (2011). Environmental justice implications of maritime spatial planning in the European Union. *Marine Policy*, 35(1), pp. 34–38.

Ehler, C. (2008). Conclusions: Benefits, lessons learned, and future challenges of marine spatial planning. *Marine Policy*, 32(5), pp. 840–843.

Ehler, C. (2014). A Guide to Evaluating Marine Spatial Plans. *IOC Manuals and Guides*, 70(70 ICAM Dossier 8).

Ehler, C. (2017). World-wide Status and Trends of Marine/Maritime Spatial Planning. In: *Keynote Address 2nd International Conference on Marine/Maritime Spatial*

Planning UNESCO, 15–17 March 2017. Paris: UNESCO. Available at: www.
msp2017.paris/presentations

Ehler, C. and Douvere, F. (2009). Marine spatial planning a step-by-step approach.
Intergovernmental Oceanographic Commission Manuals and Guides, 53(53), pp. 1–99.

European Commission. (2009). The Integrated Maritime Policy for the EU –priorities
for the next Commission. *Report from the Commission to the Council, the European Parliament, the European Economic and Social Committee and the Committee of the Regions –
Progress report on the EU's integrated maritime policy.* Available at: http://eurlex.europa.
eu/LexUriServ/LexUriServ

Folke, C., Hahn, T., Olsson, P. and Norberg, J. (2005). Adaptive governance of
social-ecological systems. *Annual Review of Environment and Resources,* 30(2005),
pp. 441–473.

Freire-Gibb, L.C., Koss, R., Margonski, P. and Papadopoulou, N. (2014). Governance strengths and weaknesses to implement the marine strategy framework directive in European waters. *Marine Policy,* 44(2014), pp. 172–178. Available at: http://
dx.doi.org/10.1016/j.marpol.2013.08.025

Future Earth. (2014). *Future Earth Strategic Research Agenda 2014.* Paris. [pdf] Available
at: http://futureearth.org/sites/default/files/strategic_research_agenda_2014.pdf

Galaz, V., Crona, B., Österblom, H., Olsson, P. and Folke, C. (2012). Polycentric
systems and interacting planetary boundaries—emerging governance of climate change–ocean acidification–marine biodiversity. *Ecological Economics,* 81,
pp. 21–32.

Gee, K., Kannen, A. and Heinrichs, B. (2011). *BaltSeaPlan Vision 2030: Towards the
Sustainable Planning of Baltic Sea Space.* The Baltic Sea Plan. Hamburg. http://www.
baltseaplan.eu/index.php?cmd=download&subcmd=downloads/2_BaltSeaPlan_
Vision2030.pdf

Haggett, C. (2008). Over the sea and far away? A consideration of the planning, politics and public perception of offshore wind farms. *Journal of Environmental Policy &
Planning,* 10(3), pp. 289–306.

Hennessey, L.J.T. (2001). Governance profiles and the management of the uses
of large marine ecosystems. *Ocean Development & International Law,* 32(1),
pp. 43–69.

Heylings, P. and Bravo, M. (2007). Evaluating governance: a process for understanding how co-management is functioning, and why, in the Galapagos Marine
Reserve. *Ocean & Coastal Management,* 50(3), pp. 174–208.

Jentoft, S., van Son, T.C. and Bjørkan, M. (2007). Marine protected areas: A governance system analysis. *Human Ecology,* 35(5), pp. 611–622.

Kannen, A. (2012). Challenges for marine spatial planning in the context of multiple
sea uses, policy arenas and actors based on experiences from the German North
Sea. *Regional Environmental Change,* 14(6), pp. 2139–2150.

Katsanevakis, S., Stelzenmüller, V., South, A., Sørensen, T.K., Jones, P.J., Kerr, S.,
Badalamenti, F., Anagnostou, C., Breen, P., Chust, G. and D'Anna, G. (2011).
Ecosystem-based marine spatial management: Review of concepts, policies, tools,
and critical issues. *Ocean & Coastal Management,* 54(11), pp. 807–820.

Kenter, J., Reed, M.S., Irvine, K., O'Brien, L., Brady, E., Bryce, R., Christie, M.,
Cooper, N., Davies, A., Hockley, N. and Fazey, I. (2014). *UK National Ecosystem
Assessment Follow-on. Work Package 6: Shared, Plural and Cultural Values of Ecosystems.*

[pdf]   Available   at: http://uknea.unep-wcmc.org/LinkClick.aspx?fileticket=
NPDIZw2mq6k%3D&tabid=82

Kidd, S. and McGowan, L. (2013). Constructing a ladder of transnational partnership working in support of marine spatial planning: Thoughts from the Irish Sea.
Journal of Environmental Management, 126(2013), pp. 63–71.

Liesbet, H. and Gary, M. (2003). Unraveling the central state, but how? Types of
multi-level governance. *American Political Science Review*, 97(02), pp. 233–243.

Maclean, I., Inger, R., Benson, D., Booth, C.G., Embling, C.B., Grecian, W.J.,
Heymans, J.J., Plummer, K.E., Shackshaft, M., Sparling, C.E. and Wilson, B.
(2014). Resolving issues with environmental impact assessment of marine renewable energy installations. *Frontiers in Marine Science*, 1(2014), p. 75.

Mayumi, K. and Giampietro, M. (2006). The epistemological challenge of self-
modifying systems: Governance and sustainability in the post-normal science era.
Ecological Economics, 57(3), pp. 382–399.

McKinley, E. and Fletcher, S. (2010). Individual responsibility for the oceans? An
evaluation of marine citizenship by UK marine practitioners. *Ocean & Coastal
Management*, 53(7), pp. 379–384.

Meadowcroft, J. (2002). Politics and scale: Some implications for environmental governance. *Landscape and Urban Planning*, 61(2), pp. 169–179.

Newig, J. and Fritsch, O. (2009). Environmental governance: Participatory,
multi-level–and effective? *Environmental Policy and Governance*, 19(3), pp. 197–214.

Ostrom, E. (2010). Beyond markets and states: Polycentric governance of complex
economic systems. *Transnational Corporations Review*, 2(2), pp. 1–12.

Pattberg, P. and Widerberg, O. (2014). Transnational multistakeholder partnerships
for sustainable development: Conditions for success. *Ambio*, 45(1), pp. 42–51.

Ravetz, J.R. (2006). Post-normal science and the complexity of transitions towards
sustainability. *Ecological Complexity*, 3(4), pp. 275–284.

Reed, M.S., Kenter, J., Bonn, A., Broad, K., Burt, T.P., Fazey, I.R., Fraser, E.D.G.,
Hubacek, K., Nainggolan, D., Quinn, C.H. and Stringer, L.C. (2013). Participatory scenario development for environmental management: A methodological
framework illustrated with experience from the UK uplands. *Journal of Environmental Management*, 128(2013), pp. 345–362.

Rees, S., Fletcher, S., Glegg, G., Marshall, C., Rodwell, L., Jefferson, R., Campbell, M.,
Langmead, O., Ashley, M., Bloomfield, H. and Brutto, D. (2012). Priority
questions to shape the marine and coastal policy research agenda in the United
Kingdom. *Marine Policy*, 38(2013), pp. 531–537.

Robinson, J. (2004). Squaring the circle? Some thoughts on the idea of sustainable
development. *Ecological Economics*, 48(4), pp. 369–384.

Ruckelshaus, M., Klinger, T., Knowlton, N. and DeMaster, D.P. (2008). Marine
ecosystem-based management in practice: Scientific and governance challenges.
BioScience, 58(1), pp. 53–63.

Scottish Government (2011). *Blue Seas–Green Energy. A Sectoral Marine Plan for Offshore Wind Energy in Scottish Territorial Waters, Part A, the Plan*. The Scottish Government, Edinburgh.

Turner, R.A., Fitzsimmons, C., Forster, J., Mahon, R., Peterson, A. and Stead, S.M.
(2014). Measuring good governance for complex ecosystems: Perceptions of coral
reef-dependent communities in the Caribbean. *Global Environmental Change*,
29(2014), pp. 105–117.

Turnpenny, J., Haxeltine, A., Lorenzoni, I., O'Riordan, T. and Jones, M. (2005). Mapping actors involved in climate change policy networks in the UK. *Tyndall Centre for Climate Change Research, Working Paper, 66.*

UK National Ecosystem Assessment (2014). UK National Ecosystem Assessment, Follow on Phase: Synthesis Report. Cambridge, UK.

Young, O.R., Osherenko, G., Ekstrom, J., Crowder, L.B., Ogden, J., Wilson, J.A., Day, J.C., Douvere, F., Ehler, C.N., McLeod, K.L. and Halpren, B.S. (2007). Solving the crisis in ocean governance: Place-based management of marine ecosystems. *Environment: Science and Policy for Sustainable Development*, 49(4), pp. 20–32.

Chapter 5

Legal aspects of marine spatial planning

Erik van Doorn and Sarah Fiona Gahlen

Introduction

Law is the tool with which planning strategies will be put into practice, because concerted action on a multilateral or global scale requires procedural rules that provide direction of how this should be done, along with associated enforcement mechanisms. The body of treaties, principles and customary law within international law that deals with activities at sea and the marine environment is the *law of the sea* (Churchill and Lowe, 1999). Core provisions of the law of the sea, such as the division in and status of different maritime zones, the freedom of the high seas—irrespective of certain limitations—or the general obligation to protect the marine environment, are globally considered as part of customary international law (Churchill and Lowe, 1999). Reciprocity, the idea that states should respect the same rights and obligations towards each other, is well-established in international law and helps to create stable expectations of behaviour between states even in the absence of binding or non-binding texts, even when mutual rights and obligations are not always consistently complied with in reality (Crawford, 2012). Hence, there is a global consensus on which rules could be based, both for the procedural aspects of assessing the compliance with certain criteria, and for governance mechanisms to make actual changes in the rights and obligations states owe to each other.

The lines along which this global consensus is established also create some important limits. First, international law is notoriously difficult to put into place because it is organised as a body of rules applicable between states or groups of states, and governing the behaviour of other actors, both on the sub-state (individuals, entities such as companies) and the supra-state levels (regions not defined by state borders or the international community as a whole) (Crawford, 2012). Second, states consider the largest part of the ocean—the high seas and the deep seabed—as the global commons that cannot be made subject to state sovereignty; in other words, subjects of international law cannot appropriate these parts of the ocean. Only a flag state (the state whose flag a ship flies) can traditionally exercise rights in these parts of the ocean, and only on the ships that fly its flag (Churchill and Lowe, 1999).

These arrangements generally require some form of international consensus on governance issues. This international unanimity is, in many matters, difficult or impossible to achieve (Churchill and Lowe, 1999).

The prescription of certain material standards or the regulation of given practices such as environmental standards for offshore drilling or other activities might be valuable to create legal certainty, but there are many cases in which it will be impossible to formulate global rules, or even global standards to come up with regional rules that would only differ according to geographical particularities. Hence, the importance of procedural measures should not be overlooked. Common mechanisms for decision-making within administrative bodies, representation rules, or mechanisms for dispute settlement will guarantee the cooperation between states and create a sound basis for more material cooperation.

Whereas marine spatial planning is a comparatively young form of ocean governance, the law of the sea dates back to the earliest times of navigation. In its beginnings, the law of the sea was a fragmented body of rules mostly derived from state practice, until the 1950s marked the negotiation of the Geneva Conventions on the Law of the Sea (Churchill and Lowe, 1999). Today, the 1982 United Nations Convention on the Law of the Sea (hereinafter, the Law of the Sea Convention) comprises rights and obligations for states at sea to a wide extent and deserves undoubtedly the place as the basic document for and starting point of almost all rules of the law of the sea (Churchill and Lowe, 1999). In other words, the Law of the Sea Convention is a framework convention on which new agreements concerning the law of the sea should be built. These subsequent agreements have so far covered topics such as shipping, resource extraction, and pollution on international and regional scales. More recent agreements tend to contain elements of marine spatial planning, e.g., with regard to an enhanced protection of natural resources in certain zones.

The Law of the Sea Convention is a worldwide success, currently having 168 parties (United Nations Treaty Collection, 2017), which effectively means that its provisions are generally accepted and are largely considered customary international law. However, the Law of the Sea Convention reflects markedly the interests and concerns of the time in which it was negotiated by emphasising the importance of traditional uses of the sea, such as navigation (Law of the Sea Convention, Parts II–IV, VII, X and XIII) and fisheries (Law of the Sea Convention, Arts. 61–72 and 116–120). New technologies or unprecedented human activities at sea are sometimes difficult to accommodate under the rules of the Law of the Sea Convention. Examples include: the difficulties with offshore carbon capture and storage technologies (Armeni, 2013; Much, 2013), and the specific legal regime for the exploration and exploitation of marine genetic resources in the high seas (Broggiato et al., 2014; Leary and Juniper, 2014). The negotiators of the Law of the Sea Convention had included the generation of wind and tidal energy, but the Law is not more specific than a mere mention of these uses (Law of the Sea Convention, Art. 56(1)(a)).

On the other hand, the duty to protect and preserve the marine environment is, as codified in the Law of the Sea Convention, one of the fundamental principles of the Convention (Maes, 2008) (together with states' sovereign right to exploit their natural resources) (Law of the Sea Convention, Parts V and VI). To achieve an integrated use of maritime zones, it will first be necessary to recognise what the Convention misses with regard to more recent technologies, and to explore if these can be filled by analogy, or if amendments or complements are needed. Because of the changes in the use of the ocean, marine spatial planning would be difficult if one did not look further than the Law of the Sea Convention. Still, the Convention and its marked focus on the environment is where everything would start.

Marine spatial planning and the law of the sea

In contrast to the procedures necessary for marine spatial planning, its goals—namely, integrated governance to achieve more sustainable use of marine space—are easily reconcilable with the Law of the Sea Convention, even though the latter does not explicitly mention it. The organisation of ocean governance according to geographic areas is frequently used in the Law of the Sea Convention, not least in the establishment of maritime zones itself, but also regarding the establishment of sea lanes and traffic-separation schemes by coastal states (Law of the Sea Convention, Art. 22), or the establishment of safety zones around artificial installations in the exclusive economic zone (Law of the Sea Convention, Art. 60(5)). One could even state that the zoning of the ocean through the Law of the Sea Convention is itself a form of a planning process (Young, 2015).

Sovereignty

Under the law of the sea, coastal states exercise territorial sovereignty over their internal waters and the territorial sea. The internal waters landward of the low-water baseline are entirely subject to the sovereignty of the coastal state. A right of free navigation only exists for the specific case of straits that were historically not regarded as part of the internal waters, but have been included by a state's later declaration of straight baselines (Law of the Sea Convention, Art. 8). However, there are few examples of such declarations having rendered straits part of the internal waters, the most prominent being Canada's declaration of straight baselines in the Arctic region. A coastal state can claim a territorial sea with a breadth up to 12 nautical miles seaward from the low-water baseline (Law of the Sea Convention, Art. 3). For the purposes of marine spatial planning, the sovereignty of archipelagic states might extend far beyond these 12 nautical miles from the coast because of their entitlement to draw archipelagic baselines, which create a special zone of archipelagic waters. Starting at the latter's outer limits, a territorial sea can be measured (Law of the Sea

Convention, Part IV). This territorial sea is subject to the powers of the coastal state in the same way as the land territory. In these waters, states are entirely free to adopt plans and to make use of this part of the sea. The customary right of innocent passage (Law of the Sea Convention, Arts. 2 and 17) is a crucial exception to coastal state sovereignty in the territorial sea. The Law of the Sea Convention defines passage as having to be continuous and expeditious, although it allows stopping and anchoring as long as it is incidental or caused by *force majeure* or distress of the passing ship or other vessels (Law of the Sea Convention, Art. 18). Regarding the planning of maritime traffic, the coastal state may adopt a whole range of regulations (Law of the Sea Convention, Art. 21; Maes, 2008). The coastal state may institute sea lanes and traffic-separation schemes for the purposes of safe navigation. Thus, when undertaking spatial planning exercises that include static structures such as offshore energy developments, states have to consider that the right of innocent passage even within 12 nautical miles from their coast and the existence of sea lanes might restrict the extent of potential offshore energy development sites.

Sovereign rights

Most marine spatial planning initiatives concern the area of the coastal states' exclusive economic zone, the waters up to 200 nautical miles from the low-water baseline over which the coastal state can claim 'functional' sovereign rights (Law of the Sea Convention, Art. 56). Typical stationary uses of the sea are commonly exercised here. For example, the generation of energy by wind turbines and the exploitation of oil and gas resources will need a certain allocation of space exclusively for that use. In the exclusive economic zone, the coastal state does not enjoy full sovereignty as it does in internal waters and the territorial sea, but can only exercise sovereign rights for certain purposes, namely the use of resources. For all purposes not regulated within the framework of the exclusive economic zone, especially navigation, the ocean beyond the territorial sea is to be regarded as 'high seas' (Law of the Sea Convention, Art. 56). This means that it is necessary to look at the rules for the high seas as well when a state plans an activity in the exclusive economic zone.

The 'territorialisation' of the exclusive economic zone is often lamented as decreasing the freedom of the high seas (Townsend-Gault, 2014), and it seems that a comprehensive marine spatial planning process could contribute to the impression that the freedoms of the exclusive economic zone are diminishing. On the other hand, planning initiatives enable states to accommodate all interests, including the uses by other states under the freedoms associated with these waters. Although some marine spatial plans contain areas for military exercises, if read literally, naval exercises fall within the freedoms applicable within the exclusive economic zone (Townsend-Gault, 2014).

The provisions of the Law of the Sea Convention that contain the rights regarding artificial islands, installations and structures in the exclusive economic zone provide a quintessential contribution to the process of marine spatial planning, not only for the exploitation of fossil fuels, but also that of energy generated by wind. Namely, the coastal state has the exclusive right to construct, authorise and regulate the construction, operation and use of these islands, installations and structures. Any abandoned or disused installations and structures should be removed (Law of the Sea Convention, Art. 60(3)).

The establishment of islands, installations and structures needs to be publicised, including the safety zones around them, which shall not exceed a distance of 500 metres. However, the islands, installations, structures themselves and their safety zones must respect the recognised sea lanes that are essential to international navigation (Law of the Sea Convention, Art. 60(3)–(7)). The same obligation applies if these installations are not in use for the exploitation of natural resources but for marine scientific research (Law of the Sea Convention, Arts. 260 and 261). In this regard, it is worth noting how Germany implemented this provision. The Federal Maritime and Hydrographic Agency of Germany, responsible for marine spatial planning in the exclusive economic zone, first analysed shipping movements before planning which activity could take place where. The areas used by shipping were hardly interfered with and received a prominent position in the marine spatial plan as a result (Nolte, 2010).

Diminishing freedom

The continuous fight between freedoms of the high seas and creeping coastal jurisdiction has always characterised the law of the sea. The absolute freedom of the high seas has been lost over the centuries, both functionally and spatially, due to the creation of exclusive economic zones and the obligation to exercise high-seas high-seas freedoms with "… due regard for the interests of other States" (Law of the Sea Convention, Art. 87(2)). Still, the ruling principle prescribes the obligatory lack of state sovereignty in areas beyond national jurisdiction (Law of the Sea Convention, Art. 89). This renders marine spatial planning on the high seas difficult, but could theoretically be done where decisions are taken without exercise of sovereignty of one single state (Ardron et al., 2008). For many kinds of offshore renewable energy, facilities are not yet able to operate in the high seas for technological reasons, wind turbines being one of them (Young, 2015). However, if they were one could ask the question to what extent the construction of facilities in the high seas would appear as an interference with the status as a commons (Lund, 2010).

The freedoms of navigation, overflight, laying of cables and pipelines, constructing artificial islands or other installations, fishing, and scientific research somehow presuppose a supranational agreement where space becomes scarce (Law of the Sea Convention, Art. 87(1)). That is probably why the latter four freedoms are subject to regulation in other parts of the Law of Sea Convention. Here again, examples emerge of legal regimes whose negotiators attempted to transcend the legal boundaries between coastal rights and high seas freedoms. It is, for instance, obvious that states need to manage fishes that spend part of their lives in exclusive economic zones and the other part in the high seas according to these maritime zones, but on the basis of compatibility of the different management systems (Law of the Sea Convention, Art. 116(b)). One proposal is to zone the high seas according to the impact of fisheries on a certain area (Ardron et al., 2008).

It can of course be regretted that the law of the sea does not contain more or clearer provisions on state cooperation, although the Law of the Sea Convention is consistent in its attribution of competences. However, it is still up to states to implement these powers. The examples of marine planning in state practice (Chapters 1, 12-15) show that states have, as a rule, taken the law of the sea underlying their governance into account.

Next to the freedom of navigation, the Law of the Sea Convention also contains important rules on the use of marine resources by the international community. In this respect, the Law of the Sea Convention presents a limit to the coastal state's power to do marine spatial planning in its waters. To avoid violating their obligations under the law of the sea, coastal states can only adopt marine spatial plans within their waters and in accordance with their competences under the Law of the Sea Convention and customary international law (Maes, 2008). The law of the sea sets out a careful balance between coastal states' interests in an independent governance of ocean waters adjacent to their coasts and other states' interests in free navigation and the use of marine resources by the international community. The unilateral implementation of a coastal state's domestic marine plan can upset this balance because of its potential interference with other states' rights and freedoms in the area where the marine plan will be applicable. Any changes in states' rights and obligations under the law of the sea must be dealt with as a matter of decision-making within the international community.

Even when marine spatial planning processes are done strictly within the powers of coastal states, success on a transboundary level still presupposes a certain cooperation between states in their planning activities and the exchange of scientific and other information. The law of the sea encourages cooperation of this kind in many respects. For example, the article in the Law of the Sea Convention on enclosed and semi-enclosed seas can be used as a basis for transboundary marine planning initiatives for these parts of the ocean, even if it does not contain a hard obligation (Maes, 2008). This provision encourages states to cooperate within a regional organisation and together with other international organisations regarding the management of marine living resources, the protection and preservation of the environment, and the coordination of scientific research policies, executing them jointly if possible (Law of the Sea Convention, Art. 123).

Transboundary marine spatial planning is still in its early stages in practice, perhaps reflecting the difficulty involved in cooperation in the concurrent governance of a wide range of activities across different administrations. In this context, the most successful initiatives came from regional agreements such as the Convention on the Protection of the Marine Environment in the North-East Atlantic (OSPAR Convention) or fisheries agreements such as the Convention on the Conservation of Antarctic Marine Living Resources (CCAMLR). It is naturally easier for regional agreements to adopt an ecosystem approach than for a globally applicable convention.

Marine spatial planning and the European Union

Many coastal states already use marine spatial planning to improve management and space allocation within the coastal zone (Calado et al., 2010; Day et al., 2008; Jay et al., 2013; Liu et al., 2011; Suárez de Vivero and Rodríguez Mateos, 2011; Turnipseed et al., 2009; Vince, 2006). Nevertheless, many other states still lack integrated, cross-sectoral approaches to marine resource management and conservation (Young, 2015). In some regions, for example within the European Union, regionally coordinated policies such as the Marine Strategy Framework Directive intend to provide harmonisation, guidance and direction to the local marine planning process (Directive 2008/56/EC). A framework directive contains binding objectives and clear legislative tasks for the member states. Directive 2014/89/EU of 23 July 2014, establishing a framework for marine spatial planning, enhances the importance of this form of ocean governance. The basis for marine spatial planning in the European Union is indebted to several legal drivers. Legislation concerning the environment, renewable energy and fisheries, and frameworks for cross-sectoral and integrated management influence each other and entail the necessity to allocate space for all of them (Qiu and Jones, 2013). Although the creation of a common marine space between all the member states of the European Union is the long-term goal of all the efforts, this vision is still far from being established (Costica, 2008).

There are also important impediments within European Union policies and competencies opposing the development of an integrated ocean governance tool. First, the interrelationship between different departments of the European Commission still characterises the approach that the European Union takes towards maritime affairs; the Marine Strategy Framework Directive is overseen by Directorate-General Environment, while the Integrated Marine Policy (Commission of the European Communities, 2007) and the Common Fisheries Policy are overseen by Directorate-General for Maritime Affairs and Fisheries.

Special protected areas, wind farms and the marine strategy

European Union law and policy is a major driver for spatial conservation; the Bird Directive (Directive 2009/147/EC) and the Habitats Directive

(Directive 92/43/EEC) are primarily responsible for the designation of a large number of protected areas in Europe over the last decades. Even though these protected areas were to form part of a network known as the Natura2000, designation was on an individual-site basis, with little or no thought of connectivity, and planning was fragmented (Opermanis et al., 2012; Yates, Payo-Payo and Schoeman, 2013). Nevertheless, these protected areas are still subject to European Union-wide mandatory measures and have hence been an important element in the marine spatial planning initiatives of member states (Qiu and Jones, 2013).

European Union law and policy also encourage the member states to progress on the generation of offshore wind energy, as a renewable energy and a more sustainable form of energy production than fossil fuel sources (Long, 2013; Qiu and Jones, 2013). Offshore wind farms can disturb the marine environment during their installation and normal operation (Chapters 8, 11 and 13; Long, 2013; Scovazzi and Tani, 2014). Still, this conflict around a technology that could mitigate the impact of anthropogenic climate change, but itself can damage the marine environment, is a good example for the difficult choices that will need to be made in the future. These are the kinds of choices that integrated governance through effective marine spatial planning could help to resolve.

Growing concerns about whether the sector-specific policies, such as those on special protected areas and the generation of offshore wind energy, could present adequate solutions to the pressure on the marine environment led to the adoption of specific overarching legislation. The European Union created its marine environmental policy by means of the Marine Strategy Framework Directive in 2008. Even before the entry into force of the Marine Planning Directive, it could be considered that the Marine Strategy Framework Directive was a sound legal basis for marine spatial planning initiatives in member states (Schaefer and Barale, 2011, p. 240). The overall aim of the Marine Strategy Framework Directive is to achieve a "good environmental status" by 2020 at the latest (Directive 2008/56/EC, Arts. 1(1) and 3(5)). However, criteria to define and assess 'good environmental status' are missing. The only explicit requirement named by the Directive is the establishment of "coherent and representative networks of MPAs" [marine protected areas], so that the Directive does recognise the importance of allocating space. Still, the exact interrelationship between marine spatial planning and the Marine Strategy Framework Directive, given its focus on environmental issues, has been heavily debated (Brennan et al., 2014, p. 363).

Marine spatial planning: the roadmap

A member state could arguably fulfil its obligations under the Marine Strategy Framework Directive and the Integrated Marine Policy (Commission of the European Communities, 2007) without proceeding on an actual planning process or without implementing the 'spatial' aspect of planning. Still, there could be pragmatic reasons that render spatial planning the most suitable instrument for a successful and sustainable governance of maritime activities. Stable and reliable legislation for licensing is certainly a prerequisite for offshore activities such as wind energy generation, which need space that the coastal state exclusively allocates to installations. With that in mind, the planning process could be part of the regulation of access to the market at sea (Long, 2013).

In 2008, the European Union officially emphasised the importance of marine spatial planning as a suitable instrument for an integrated ocean governance and adopted the Communication "Roadmap for Marine Spatial Planning: Achieving Common Principles in the EU" (Commission of the European Communities, 2008). The Roadmap contains 10 'key principles' that are meant to be the basis on which member states develop their marine plans (Schaefer and Barale, 2011). The Roadmap principles encompass *inter alia* the need to determine the area in which a state does marine planning according to geographical and biological criteria, the definition of clear objectives, transparency matters and stakeholder participation, and other principles that are at the core of marine planning as understood today (Chapters 1, 2 and 7). The principles coincide substantially with the ideas of the Marine Strategy Framework Directive and present a low common denominator regarding the guiding principles of planning. However, they still emphasise the transnational dimension of marine spatial planning in the European Union more than other European Union instruments on marine governance. The Roadmap is the first European Union communication not only mentioning 'marine spatial planning' as a means of ocean governance, but also giving detailed guidelines on how to adopt a plan and on the practical questions linked to the process (Costica, 2008).

The marine planning directive

In the feedback it received on the Roadmap, the European Commission identified some core deficits in marine governance within European Union waters (EC, 2013), and it seems that the member states still struggle with some of the original problems that were already identified before the adoption of the Integrated Marine Policy in 2007. The Commission therefore thought it advisable to adopt an instrument that would render the management of the ocean and the coastal zones through marine planning mandatory. It did so through the adoption of a framework directive, whereas the actual legislation and the adoption of plans are left to the member states. Considering that the European Union only has limited, defined competences and that legislative projects must be allocated under a common policy, it seems that the adoption of a more detailed directive would have been difficult. The Commission recalled that harmonisation should take place at

"... the lowest most appropriate level" and that the concrete planning should take into account local particularities.

Hence, Directive 2014/89/EU establishing a framework for marine planning only sets out minimum standards that the member states have to implement. Not surprisingly, the text mirrors the findings from the 2008 Roadmap closely, now codifying the largest part of the 10 principles into the Directive's articles. By calling for public participation (Directive 2014/89/EU, Art. 9), common data use and sharing (Directive 2014/89/EU, Art. 10), cooperation between member states and third countries, and the fulfilment of monitoring and reporting obligations, the Directive recalls most of the core criteria for successful marine planning that have been identified in the Roadmap, and even before that, in the respective domestic legislation and scientific studies on marine governance. The identification of a competent authority within each member state that will be responsible for the implementation of the Directive (Directive 2014/89/EU, Art. 13) will make it easier for the European Union to address the respective state and presents a solution to the problem of widespread competences in different maritime affairs in the member states. In itself, the Directive does not state anything that would be surprising after the extensive studies on marine spatial planning and the considerable experience within Europe. The Directive represents the final step by which the European Union now obliges the member states to proceed with a form of ocean governance that has been identified as useful and appropriate. Still, the Directive does not neglect the importance of governance at the national level and is nothing more than a framework setting out minimum requirements. Member states that already find themselves in the process of marine spatial planning will probably not need to change or adapt their plans. Still, the Directive is a way of compelling states to embark on this pathway if they have not yet done so, forcing them to work consciously on setting their national priorities at sea and reconciling different uses of maritime space.

Future opportunities

For the analysis of the international legal framework of marine planning, the Law of the Sea Convention appears to be the natural starting point. Depending on the location of offshore energy facilities, differentiated rules are applicable. Within the waters over which the coastal state can exercise sovereignty or sovereign rights, these facilities are fully under the jurisdiction of the coastal state, with the exception that no interference with navigational rights should occur (Leary and Esteban, 2009). To marine planners, the international rules of the law of the sea on, for instance, the environment and resource extraction might appear vague. These rules could provide the basis for a national marine plan, but would require adaptation if not amendment. Sooner or later, transboundary aspects might come into play. The European Union is positioned well to require cooperation. European marine spatial planning legislation started with conservation legislation aimed at developing protected areas (birds, habitats) and has evolved over time, based on need for increased integration, from sectoral to a Marine Spatial Planning Directive.

Marine spatial planning is only one of the tools for ocean governance (Costica, 2008); in other words, marine spatial planning is not the ultimate solution in every case. The advantage in many regards lies in the fact that one should view marine spatial planning as an ongoing process and subsequently has a higher potential to adapt to change. This aspect is not always reconcilable with law, which is much more static.

Highlights

- Marine spatial planning is a recent invention, but the international law of the sea has been influencing what nations can do in their own and international waters since 1982.
- A prime task for marine spatial planning is to find a balance between the freedoms of the high seas and exclusive rights of coastal states and to accommodate their particularities.
- Both material law, focusing on what a state could and should do where and when, and procedural law that provides directions of how this should be done, are of prime importance.
- The Law of the Sea Convention presents a limit to coastal states' marine spatial planning concerning the construction of installations for the generation of offshore energy.
- The example of the European Union's marine spatial planning efforts demonstrates the impact that bureaucratic difference can have on planning processes.

Acknowledgements

We thank the participants of the Future Ocean Spatial Planning workshop held in Kiel, Germany in March 2014.

References

Ardron, J., Gjerde, K., Pillen, S. and Tilot, V. (2008). Marine spatial planning in the high seas. *Marine Policy*, 32(5), pp. 832–839.

Armeni, C. (2013). Carbon dioxide storage in the sub-seabed and sustainable development: Please mind the gap. *Ocean Yearbook*, 27(1), pp. 1–27.

Brennan, J., Fitzsimmons, C., Gray, T. and Raggatt, L. (2014). EU marine strategy framework directive (MSFD) and marine spatial planning (MSP): Which is the more dominant and practicable contributor to marine policy in the UK? *Marine Policy*, 43(2014), pp. 359–366.

Broggiato, A., Arnaud-Haond, S., Chiarolla, C. and Greiber, T. (2014). Fair and equitable sharing of benefits from the utilization of marine genetic resources in areas beyond national jurisdiction: Bridging the gaps between science and policy. *Marine Policy*, 49(2014), pp. 176–185.

Calado, H., Ng, K., Johnson, D., Sousa, L., Phillips, M. and Alves, F. (2010) Marine spatial planning: Lessons learned from the Portuguese debate. *Marine Policy*, 34(6), pp. 1341–1349.

Churchill, R.R. and Lowe, A.V. (1999). *The Law of the Sea*, 3rd ed. Manchester: Manchester University Press.

Commission of the European Communities. (2007). An Integrated Maritime Policy for the European Union. COM 575 final.

Commission of the European Communities. (2008). Roadmap for Marine Spatial Planning: Achieving Common Principles in the EU. COM 791 final.

CCAMLR. (1980). Convention on the conservation of Antarctic marine living resources. *International Legal Materials*, 19(4), pp. 837–859.

Costica, F. (2008). La planification de l'espace maritime de l'Union Européenne. *Annuaire du droit de la mer*, 13, pp. 289–336.

Crawford, J. (2012). *Brownlie's Principles of Public International Law*, 8th ed. Oxford: Oxford University Press.

Day, V., Paxinos, R., Emmett, J., Wright, A. and Goecker, M. (2008). The marine planning framework for South Australia: A new ecosystem-based zoning policy for marine management. *Marine Policy*, 32(4), pp. 535–543.

EC. (2008). Impact Assessment Accompanying the document Proposal for a Directive of the European Parliament and of the Council establishing a framework for maritime spatial planning and integrated coastal management. SWD 65 final.

European Council. (1992). Council Directive 92/43/EEC on the conservation of natural habitats and of wild fauna and flora. *Official Journal of the European Union*, 35 (22 July 1992), pp. L 206/7–50.

European Parliament and European Council. (2008). Directive 2008/56/EC establishing a framework for community action in the field of marine environmental policy (Marine Strategy Framework Directive). *Official Journal of the European Union*, 51 (25 June 2008), pp. L 164/19–40.

European Parliament and European Council. (2009). Directive 2009/147/EC on the conservation of wild birds. *Official Journal of the European Union*, 53 (26 January 2010), pp. L 20/7–25.

European Parliament and European Council. (2014). Directive 2014/89/EU establishing a framework for maritime spatial planning. *Official Journal of the European Union*, 57 (28 August 2014), pp. L 257/135–145.

Jay, S., Flannery, W., Vince, J., Liu, W.H., Xue, J.G., Matczak, M., Zaucha, J., Janssen, H., van Tatenhove, J., Toonen, H. and Morf, A. (2013). International progress in marine spatial planning. *Ocean Yearbook*, 27, pp. 171–212.

Leary, D. and Esteban, M. (2009). Climate change and renewable energy from the ocean and tides: Calming the sea of regulatory uncertainty. *International Journal of Marine and Coastal Law*, 24(4), pp. 617–651.

Leary, D. and Juniper, S.K. (2014). Addressing the Marine Genetic Resources Issue: Is the Debate heading in the Wrong Direction? In: C. Schofield, S. Lee and M.-S. Kwon, eds., *The Limits of Maritime Jurisdiction*. Leiden: Martinus Nijhoff, pp. 769–785.

Liu, W.-H., Wu, C.C., Jhan, H.T. and Ho, C.H. (2011). The role of local Government in marine spatial planning and management in Taiwan. *Marine Policy*, 35(2), pp. 105–115.

Long, R. (2013). Offshore Wind Energy Development and Ecosystem-Based Marine Management in the EU: Are the Regulatory Answers Really Blowing in the Wind? In: M.H. Nordquist, J.N. Moore, A. Chircop and R. Long, eds., *The Regulation of Continental Shelf Development*. Leiden: Martinus Nijhoff, pp. 15–52.

Lund, N.J. (2010). Renewable energy as a catalyst for changes to the high seas regime. *Ocean and Coastal Law Journal*, 15(1), pp. 95–125.

Maes, F. (2008). The international legal framework for marine spatial planning. *Marine Policy*, 32(5), pp. 797–810.

Much, S. (2013). The Emerging International Regulation of Carbon Storage in Sub-Seabed Geological Formations. In: R. Caddell and D.R. Thomas, eds., *Shipping, Law and the Marine Environment in the 21st Century: Emerging Challenges for the Law of the Sea – Legal Implications and Liabilities*. London: Informa, pp. 255–275.

Nolte, N. (2010). Nutzungansprüche und Raumordnung auf dem Meer. *HANSA International Maritime Journal*, 147(9), pp. 79–83.

Opermanis, O., MacSharry, B., Aunins, A. and Sipkova, Z. (2012). Connectedness and connectivity of the Natura 2000 network of protected areas across country borders in the European Union. *Biological Conservation*, 153, pp. 227–238.

OSPAR (1992) Convention on the Protection of the Marine Environment in the North-East Atlantic. *International Legal Materials*, 32(4), pp. 1069–1100 [Abbreviated as: OSPAR Convention].

Qiu, W. and Jones, P.J.S. (2013). The emerging policy landscape for marine spatial planning in Europe. *Marine Policy*, 39, pp. 182–190.

Schaefer, N. and Barale, V. (2011). Marine spatial planning: Opportunities & challenges in the framework of the EU integrated maritime policy. *Journal of Coastal Conservation*, 15(2), pp. 237–245.

Scovazzi, T. and Tani, I. (2014). Off-shore Wind Energy Development in International Law. In: J. Ebbesson, M. Jacobsson, M. Klamberg, D. Langlet and P. Wrange, eds., *International Law and Changing Perceptions of Security. Liber Amicorum Said Mahmoudi*. Leiden: Brill Nijhoff, pp. 244–258.

Suárez De Vivero, J.L. and Rodríguez Mateos, J.C. (2011). The Spanish approach to marine spatial planning. Marine Strategy Framework Directive vs. EU Integrated Maritime Policy. *Marine Policy*, 38(1), pp. 18–27.

Townsend-Gault, I. (2014). The 'Territorialisation' of the Exclusive Economic Zone: A Requiem for the Remnants of the Freedom of the Seas? In: C. Schofield, S. Lee and M.-S, Kwon, eds., *The Limits of Maritime Jurisdiction*. Boston: Martinus Nijhoff, pp. 65–76.

Turnipseed, M. Crowder, L.B., Sagarin, R.D. and Roady, S.E. (2009). Legal bedrock for rebuilding America's ocean ecosystems. *Science*, 324(5924), pp. 183–184.

United Nations Convention on the Law of the Sea. (1982) *International Legal Materials*, 21(6), 1261–1354 [Abbreviated as: Law of the Sea Convention].

United Nations Treaty Collection. (2017). *Depositary. Status of Treaties Chapter XXI.6*. [online] Available from: http://treaties.un.org/Pages/showDetails.aspx?objid=0800000280043ad5 [Accessed: 12 January 2018].

Vince, J. (2006). The south east regional marine plan: Implementing Australia's oceans policy. *Marine Policy*, 30(4), pp. 420–430.

Yates, K.L., Payo-Payo, A. and Schoeman, D.S. (2013). International, regional and national commitments meet local implementation: A case study of marine conservation in Northern Ireland. *Marine Policy*, 38, pp. 140–150.

Young, M. (2015). Building the blue economy: The role of marine spatial planning in facilitating offshore renewable energy development. *International Journal of Marine and Coastal Law*, 30(1), pp. 148–174.

Chapter 6

Displacement of existing activities

*Andronikos Kafas, Penelope Donohue,
Ian Davies and Beth E. Scott*

Introduction

The ways in which sea space is used have changed over recent years, and the marine environment is becoming increasingly exploited (Baxter et al., 2011). As marine space becomes more congested, existing users (e.g., commercial fisheries) find themselves concerned about the loss of access and the associated management regimes applied to the shared resource (i.e., space) in response to competition with newly introduced uses (Day, 2002; Vink and van der Burg, 2006; Toonen and Mol, 2013; Jentoft and Knol, 2014; Tien and van der Hammen, 2015). Conflicts can arise from misunderstanding or lack of dialogue about local concerns and priorities, power imbalances between marine users, under-representation of local interests, and ineffective stakeholder engagement methods (ICES, 2017, Chapter 9). Conflicts can result in negative attitudes by existing marine users towards the new activity (e.g., the offshore energy sector), loss of trust in the planning authority, and stakeholder concerns that national priorities are taking precedence over local needs (FLOWW, 2014; Jentoft and Knol, 2014; ICES, 2017).

Understanding how emerging uses, such as the development of a new offshore energy site, can affect other existing activities such as fisheries through partial or full exclusion from an area (including the knock-on impacts of displacement) is essential for making sound planning and licensing decisions and developing effective mitigation strategies. As the commercial fishing industry is one of the most widespread and well-established marine users, it is the stakeholder group most often materially affected by displacement resulting from offshore energy developments (Jentoft, 2007). The introduction of offshore renewable energy could have negative implications for commercial fishing, particularly where changes occur or activity increases in sea space traditionally used as fishing grounds (Smith and Brennan, 2012; van Putten et al., 2012; Breen, Vanstaen and Clark, 2014; Tien and van der Hammen, 2015; Janßen et al., 2017). Fishing grounds (also referred to as 'prime fishing areas') are sea areas with fisheries employment that is functionally coherent in geographical, economic, and social terms (EC, 2015).

Fishing within offshore energy-development areas can be incompatible while they are being constructed and operating. As such, offshore energy areas act as restricted areas in which access for at least some types of fishing, such as trawling, is restricted and often effectively act as no-take zones (also called 'fishing closures') (Vize et al., 2008; FLOWW, 2014). Therefore, in this chapter we focus primarily on the impact of offshore energy developments on commercial fisheries. In general, many lessons learned from the perspective of the fisheries are transferable to other marine users experiencing competition in marine planning.

Displacement of fisheries

Fisheries around the world and in the EU

The commercial wild-capture, sea-fishing industry is of social, economic, and cultural importance (FAO, 2014). Fish flesh is highly nutritious, providing a large amount of animal protein in the diets of people around the world, with global per capita apparent fish consumption increasing from an average of 9.9 kg in the 1960s to 19.2 kg in 2012 (FAO, 2014). In 2012, the global production of marine capture fisheries equated to 79.7 million tonnes; capture fisheries offer employment opportunities to tens of millions and supports the livelihoods of hundreds of millions worldwide (FAO, 2014). For example, the fishing sector plays an important role in many European countries, supporting over 116,000 full-time-equivalent jobs directly employed in the fisheries sector, and over 115,500 full-time-equivalent jobs in the fish-products industry (STECF, 2013). Fisheries contribute more than 7.1 billion € of annual income in European Member States, with the total European Union fishing fleet exceeding 87,000 fishing vessels (STECF, 2013; EC, 2014). In coastal regions of some European countries, including Greece (e.g., Lefkada, Samos, and Kefallinia), United Kingdom (e.g., Shetland Islands, Western Isles, and Orkney Islands), Croatia (e.g., Zadarska županija and Dubrovačko-neretvanska županija), and Spain (e.g., El Hierro), more than half of the local jobs are in the fishing sector (EC, 2014).

The commercial fishing fleet comprises a diverse range of vessel types, sizes, and target species, including, for example, those targeting pelagic, demersal or whitefish, mixed demersal, and shellfish stocks (FAO, 2014). Fishing vessels deploy both mobile and static gear types. Mobile gears (e.g., trawling) involve pulling fishing nets through the water column or on the seabed behind one or more fishing vessels (FAO, 2005). Static gears (e.g., potting) are a passive fishing technique, deploying structures into which fish are guided or enticed through funnels that encourage entry but limit escape, which are mostly used to catch langoustines, crabs, and lobster (SeaFish, 2011). Combinations of a particular fishing technique category and a vessel length category are also referred to as 'fleet segments' (STECF, 2016). We illustrate below that different offshore renewable energy developments interact differently with different fishing fleets.

Fleet dynamics

The policy environment in which fisheries management operates is changing as competition for marine space increases (Pascoe, 2006; Janßen et al., 2017). It is often necessary for fishers to adapt their strategies when faced with change in the environment in which they operate (changes collectively referred to as 'fleet dynamics'; Hilborn and Kennedy, 1992; Fulton et al., 2011; van Putten et al., 2012). Fleet dynamics relate to changes both in fishing capacity of vessels (the physical dimension of fishing vessels measured in gross tonnage; horsepower is often used as index of fishing capacity; Sanders and Morgan, 1976), their fishing activity, including the effort intensity (a measure of the fishing activity of vessels based on fishing capacity and the time spent fishing; Sanders and Morgan, 1976), and allocation of effort in space and time (van Putten et al., 2012). These changes can result from choices made by individual fishers or wider groups of vessels (decision units) to react to management interventions. Nevertheless, fishers' behaviour shows a strong inertia, and typically some fishers continue to follow their long-established patterns of behaviour despite the new conditions (e.g., Marchal et al., 2002).

The extent of the impact of management measures on fisheries varies depending on local circumstances. Impact differs depending on the types of fishing gear used, preferred fishing locations, compatibility with the management measure (e.g., proposed new activity or colocation possibilities), as well as political, social, and economic contexts of the management intervention, involving regulatory authorities, local communities, ports, project developers, industry representation groups, and other affected parties. Potential direct spatial interactions between offshore energy developments and commercial fisheries include the loss of access to fishing grounds, obstruction of regular fishing vessel transit routes, and navigational safety concerns raised by fishing-gear entanglement and fouling on subsea infrastructure. The combination of all the above can force fishing to be displaced to alternative locations and thereby put increased fishing pressure on these alternative grounds. Yet, despite every situation being different and the need to take specifics into account, there are some consistent, overarching interactions that can help guide marine spatial planning (SeaPlan, 2015).

Fisheries displacement

We define fisheries displacement as changes in fishing patterns or fishing behaviour in space and/or time in response to new management measures. Changes in fishing patterns include the movement of fishing vessels (or effort) from traditional fishing grounds to alternative sea areas in response to spatial management measures (e.g., fishing closures). Adjustments in fishing behaviour include adopting new fishing methods and/or target species. Management measures can arise from policy and administrative interventions (e.g., landing obligation; European Union, 2014), marine conservation (e.g., marine protected areas; Lauck et al., 1998; Hilborn et al., 2004; Mascia et al., 2010), as well as new offshore, human-made developments such as offshore wind farms (e.g., BOWL, 2011), and other ocean energy developments (e.g., Aquatera, 2014).

The effects of fisheries displacement

Spatial restrictions on fishing activities resulting from offshore energy developments being located in traditional fishing grounds have the potential to cause displacement of the fishing fleet (e.g., van de Geer et al., 2013; Donohue and Murphy, 2015). Following environmental impact assessment definitions (European Council, 2014; IEMA, 2016), impacts are the changes resulting from an action, and effects are the consequences of impacts. We define direct effects as consequences of impacts observed within the restricted area, which are a direct result of displacement and realised immediately after restrictions come into force. Indirect effects are knock-on consequences observed adjacent to the restricted area, and take longer to be realised. An understanding of how fishers may respond to restrictions placed in previously fished areas is useful in determining the effects of the restrictions on the fishers, and the potential environmental implications for the remaining open areas. Fishing closures and resulting fisheries displacement can have a range of direct and indirect, positive and negative, economic, social, and environmental effects on individual fishers, the fishing industry, fishery-dependent coastal communities, and wider society. We provide a summary of the effects in Figure 6.1. Fishing displacement should not be a concern solely for the fishing industry and related industries, but should be shared by all stakeholders with an interest in the marine environment, such as the ocean energy industry.

Potential adverse economic effects

Fishing closures and resulting fisheries displacement can have negative direct and indirect effects on the economics of the commercial fishing industry and fishery-dependent coastal communities (Berman, 2007; FLOWW, 2014; SeaPlan, 2015; Tien and van der Hammen, 2015). The most direct effect of a fishing closure is the loss of fishing grounds realised immediately after the management measure comes into force. The current distribution of existing fishing activity is directly affected by a proposed fishing closure (e.g., a development during construction or installation), resulting in restricted or temporary loss of access to traditional fishing grounds (Gell and Roberts, 2003; Ledee et al., 2012; van de Geer et al., 2013). In the case of offshore wind farms, it is standard practice to apply a 500 m 'rolling' safety zone around working areas during construction, and during operation fishers still tend to avoid those areas (FLOWW, 2014; SeaPlan, 2015; Gray, Stromberg and Rodmell, 2016).

Impacted fishers are forced to move away from areas where fishing is restricted, leading to increased density of fishing vessels in adjacent areas where fishing is still allowed (also called 'overcrowding') (Holland and Sutinen, 2000; Murawski et al., 2005; Poos et al., 2009; Mangi, Rodwell and Hattam, 2011; van Putten et al., 2012; Alexander, Wilding and Heymans, 2013; Tien and van der Hammen, 2015; Bastardie et al., 2017). The movement of displaced vessels to other fishing grounds often generates conflicts among fishers (e.g., Hilborn, 2007; Rosenberg and Glass, 2007; Pita, Theodossiou and Pierce, 2013) and can lead to disputes such as the over-allocation of catch or gear entanglements (Sanchirico and Wilen, 2002; Gell and Roberts, 2003; Christie, 2004; Christie and White, 2007; Mangi, Rodwell and Hattam, 2011; van de Geer et al., 2013).

New fishing grounds, overcrowding, longer steaming, and lack of knowledge of new fishing grounds could reduce fishing efficiency in terms of the volume, quality or rate (catch per unit effort) of catch, and increase by-catch (Rijnsdorp, Piet and Poos, 2001; Murawski et al., 2005; Smith et al., 2010) in the new locations, which in turn result in the loss of income or profit. These effects are likely to be a result of the diversification of livelihood (Gell and Roberts, 2003; Diegues, 2008; Mascia, Claus and Naidoo, 2010). Fishers might have to travel longer distances to alternative grounds (Hilborn et al., 2004; Pomeroy and Douvere, 2008), incurring increased steaming costs due to greater fuel requirement (De Groot et al., 2014; Bastardie et al., 2015) and other capital costs potentially involved in alternative grounds. Capital costs can include, for example, new equipment to locate fish (Sanchirico and Wilen, 2002; van de Geer et al., 2013), or diversification of gears, with changes in rules governing a resource; this can require investment in new gear and the time to learn how to use it profitably (Mangi, Rodwell and Hattam, 2011; van de Geer et al., 2013).

Travelling farther to access alternative grounds, the likelihood of moving into less sheltered waters, and encountering storms or other hazards result in increased safety risks, and snagging potential with unknown seabed hazards or offshore energy installations (Sanchirico and Wilen, 2002; Hilborn et al., 2004; Pomeroy and Douvere, 2008; Mangi, Rodwell and Hattam, 2011; FLOWW, 2014). In theory, it can be argued that in the worst-case scenario of prolonged, direct economic losses, cumulative restrictions on fishing activity could result in the curtailment or cessation of fishing businesses (Jentoft and Knol, 2014; Bastardie et al., 2015; Janßen et al., 2017). However, there is no empirical evidence of this, and cessation of fishing activities during construction is temporary and is not generally considered by fishers to be overly restrictive (Gray, Stromberg and Rodmell, 2016).

Potential adverse social effects

In addition to economics, another notion of fisheries dependence is based on the social and cultural value of the fishing industry (Nuttall, 2000). For example, marginalised fishers might be forced to leave the fishery because of higher costs or lack of fishing grounds (Christie, 2004; Diegues, 2008; Sen, 2010; Shaw and Johnson, 2011), resulting in the loss of traditional fishing community. The disruption of formal and informal fishing-tenure systems can interfere with informal marine resource governance systems and result in the loss of local knowledge (Mascia, Claus and Naidoo, 2010; Wiber, Young and Wilson, 2012; van de Geer et al., 2013). The potential indirect social effects of fishing displacement can also impact cultural traditions or customary marine tenure, with restrictions on fishing changing the traditional use of areas and undermining traditional governance systems (Hilborn et al., 2004; Diegues, 2008; Hattam et al., 2014).

In some cases, fishing closures can have minimal impacts on the average incomes and financial profits of fishers and fish merchants, although to compensate for the loss of the fishing grounds (due to longer travel time and/or increased competition in available fishing grounds), fishers might have had to work longer hours to achieve this (Mangi, Rodwell and Hattam, 2011). This can impact fishers' welfare and their family life. If dependence on fish for food is high, reduced catches could result in changes to health of fishers and their families (Mascia, Claus and Naidoo, 2010). New governance arrangements can revoke the existing arrangements and erode decision-making authority, shifting the balance of power between different stakeholder groups and disempowering stakeholders (Christie, 2004; Pomeroy and Douvere, 2008; Mascia, Claus and Naidoo, 2010).

The potential, indirect social effects of fisheries displacement can have implications for other stakeholders who include the community associated with fishers, tourists (e.g., recreational fishers and divers), and other fishing-related businesses (e.g., handling and processing). A decrease in net profit for the fisher can result in a decrease in spending in the local community and/or the use of other fishing-related businesses (Farmery et al., 2014; Douet, 2016). Other potential indirect impacts also include the effects on shore-based industries that depend on commercial fisheries' landings and activities. When vessels have to change landing ports, indirect impacts might transfer through the supply chain and affect, for example, fish processors or gear manufacturers (Mangi, Rodwell and Hattam, 2011; Shaw and Johnson, 2011). Negative socio-economic changes might be particularly pernicious for smaller vessels, which often have less opportunity to fish elsewhere (Panayotou, 1988; Allison and Ellis, 2001). Furthermore, remote communities where the local economy depends (economically, socially, and culturally) on the fishing sector, can experience negative socio-economic changes more severely (Nuttall, 2000; Brookfield, Gray and Hatchard, 2005; EC, 2014).

Potential adverse environmental effects

Fisheries displacement can have direct and indirect environmental effects (Halpern and Warner, 2002; Dinmore et al., 2003; Gell and Roberts, 2003). In some cases, spatial restrictions can impact some gears (e.g., mobile gears), but might allow the proliferation of alternative gear methods such as static gears (Gray, Stromberg and Rodmell, 2016). Alternative gears can target different components of the stock (due to different gear selectivity) and therefore have different environmental footprints. In addition, fishers may attempt to counteract lost landings by moving fishing to adjacent areas (De Groot et al., 2014; Bastardie et al., 2015). Displaced fishing activity can therefore also present implications for the environment and/or fish stocks in these adjacent areas, if fishing pressure becomes disproportionately concentrated in the unrestricted areas (Dinmore et al., 2003). Harvesting the fish resource in alternative locations might run the risk of catching vulnerable elements of the stock, and change the abundance or distribution of targeted species (Sen, 2010). Kellner and Hastings (2009) have also suggested that fishing closures can displace fishing effort into smaller areas and change local community structure enough to facilitate invasion by exotic species. Displace fishing might also result in increased competition between fishers in the remaining area (Campbell et al., 2014; De Groot et al., 2014). Furthermore, displacement to areas that have previously been subjected to low fishing pressure can have negative environmental effects (Hilborn et al., 2004; Roberts, Hawkins and Gell, 2005). The increase in fishing activity, coupled with increased congestion in the non-restricted area, and displacement of vessels to previously unfished areas could have greater cumulative impacts on benthic communities (total benthic invertebrate productivity) and could lead to localised reductions in benthic biomass (Dinmore et al., 2003; Bastardie et al., 2015).

Potential positive effects

Nonetheless, the overlap of commercial fisheries with offshore energy developments can generate opportunities for coexistence and potentially bring socio-economic benefits to fishers. Restricting fishing can reduce collateral ecological impact (e.g., less damage to benthic habitats) (Kamenos, Moore and Hall-Spencer, 2004), engender less by-catch of protected and commercial species (Hilborn et al., 2004), and improve ecosystem structure and function (Hilborn et al., 2004). In addition, there is the potential for development structures to be used as direct habitat enhancement by adding substrata for settlement of juvenile invertebrates (Lacroix and Pioch, 2011). As a result, the abundance, recruitment, and size of individuals for target species in a restricted area can increase following development (Halpern and Warner, 2002; Hilborn et al., 2004; Chapter 8). The 'spill-over' of individuals across restricted-area borders can augment local catches in adjacent areas (Rowley, 1994; Russ and Alcala, 1996; Willis et al., 2003; Abesamis and Russ, 2005; Roberts, Hawkins and Gell, 2005). If catches associated with spill-over exceed those normally taken inside the restricted area, fishers can benefit from increased yield in the adjacent areas (Greenstreet, Fraser and Piet, 2009; Bastardie et al., 2015); however, the spill-over benefit depends on the size of the reserve (Allison, Lubchenco and Carr, 1998; McClanahan and Mangi, 2000). For non-restricted gear users (e.g., static gears), increased or better-quality catches within the restricted areas and decreased competition can subsequently increase income (Mangi, Rodwell and Hattam, 2011). Sustainable use of resources and the protection of commercially important habitats can help promote longevity for fisheries and businesses. Conventional fisheries management through catch and effort controls can fail because of uncertainty or errors in stock assessments, or inadequate institutional frameworks. Restricted areas can act as *de facto* marine protected areas and can buffer management uncertainty and the impact of such failures (Lauck et al., 1998; Hilborn et al., 2004; Diegues, 2008). Fishing closures can therefore enhance food security to consumers through environmental benefits, relocating fishing rights, thereby reducing local competition for fishing resources (e.g., Himes, 2003; Mangi et al., 2011).

Area restrictions might require some of the fishing industry to adapt and diversify into areas and species not previously fished. This can lead to the development of existing fisheries not currently at capacity, new areas and/or species not previously fished, targeting niche species (Shaw and Johnson, 2011), or developing new sectors such as aquaculture (Wiber, Young and Wilson, 2012) or coastal tourism (Hall, 2001); these actions can work to keep the local seas profitable, productive, and sustainable. The newly introduced activity (e.g., an offshore windfarm) can also offer employment opportunities to local fishing vessels through diversification into commissioned services, and/or covering associated transitional costs such as training, certification, or insurance.

Such opportunities can have wider socio-economic benefits to the local community and other stakeholders (Pascoe et al., 2014). For example, new investments in the area (e.g., new shore-side facilities; FLOWW, 2015) and opportunities for other businesses (e.g., recreational fishers and associated businesses; Westerberg et al., 2013) can arise. Increased benefits and higher catch rates can increase the popularity of the area, increasing tourism that can benefit the local community economically (Cass, Walker and Devine-Wright, 2010; Walker, Wiersma and Bailey, 2014). Alternative employment opportunities can assist in the development of cooperative working crews and passing knowledge to other sectors (Shaw and Johnson, 2011; Wiber, Young and Wilson, 2012).

	Direct	Indirect	
Negative	Loss of fishing grounds Overcrowding Increased conflicts between fishers Loss of income or profit, due to reduced fishing efficiency Increased steaming cost Capital costs such as those associated with diversification of gears Increased safety risks and snagging potential Curtailment or cessation of businesses	Locking up of productive biological resources Increased pressure on the environment and/or fish stocks in adjacent areas Loss of traditional fishing communities Loss of local knowledge Reduced Fisher welfare and health Disempowering of Fishers Undesirable impacts on cultural traditions or customary marine tenure	**Negative**
Positive	Increase in yield Increase in income Increase in food security	Reduce collateral ecological impact Buffer against management uncertainty Promote longevity for fisheries and businesses Development of fisheries into areas and for species that were previously not fished	**Positive**
	Direct	**Indirect**	

Figure 6.1 Summary of the potential direct and indirect effects of fishing closures and resulting fisheries displacement on fishers, based on international literature.

Managing fisheries displacement

Understanding fleet dynamics and the behaviour of fishers can contribute to the successful management of fisheries displacement (Wilen et al., 2002; Fulton et al., 2011; van Putten et al., 2012; Bastardie, Nielsen and Miethe, 2014). Yet, despite the potentially negative economic and social implications (Salas and Gaertner, 2004; Branch et al., 2006), little research has examined the change in fleet dynamics caused by policy or administrative intervention, or by the introduction of a new marine activity (van Putten et al., 2012). Fisheries displacement can be addressed either through proactive or reactive management. Proactive management involves preventative measures, such as avoiding spatial conflicts in the planning stage. The ultimate aim is to reduce the likelihood of displacement occurring. On the contrary, reactive management addresses conflicts after they have been expressed, and involves mitigation measures aiming to reduce the extent and impacts of displacement.

Preventive measures

An effective marine spatial planning process aims to accommodate space for emerging uses, such as offshore energy, and helps achieve overarching objectives like reducing society's dependence on fossil fuels while minimising disruption to and displacement of existing activities such as fishing. To do so, an understanding of the spatial requirements and potential overlaps between activities is essential. Avoiding prime fishing grounds when planning offshore energy development will help achieve stakeholder objectives while minimising the impact on commercial fishing activity (Helson et al., 2010). In the case of a proposed offshore energy site, analysis prior to site selection is usually done at a strategic spatial scale, usually by a competent marine spatial planning authority (e.g., Davies and Watret, 2011). Constraint mapping based on multi-annual activity data can assist with the high-level siting of an offshore energy development. Analytical tools such as fisheries-displacement models (e.g., Bastardie et al., 2014) can be used for refinement at a medium-level scale, and to check if the proposed zones are valid at the relevant temporal scales. The likely effects of displacement will be case-specific depending upon the local intensity, extent, and nature of fishing operations occurring there, the availability and ease of access to alternative productive fishing areas, a combination of personal characteristics and drivers affecting fisher behaviour, and the potential for incompatibility of fishing activities with the proposed development.

Central to addressing questions surrounding displacement of fishing effort is an understanding of traditional practices and activity of fishers based on robust quantitative and qualitative techniques (Ashley et al., 2012). These include direct stakeholder engagement (see Chapter 9) and assessment of the risk of displacement based on fine-scale, local-activity data, and modelling tools. However, effective engagement can be difficult to achieve due to lack of trust. Brown et al. (2001) proposed the use of trade-off analysis (Chapter 2) directly incorporated into the planning process, suggesting that this makes explicit the diverse perceptions and values of different groups, thereby creating opportunities for decision making and management based on consensus rather than conflict.

Negotiations are made easier by providing layers prepared by a trusted broker of the spatial footprint of fishing operations over time (e.g., Yates and Schoeman, 2013; Ojeda-Ruiz et al., 2015; Turner et al., 2015; Kafas et al., 2017); layers quantifying the overlap provide a better understanding of the interactions between activities (e.g., Vandendriessche et al., 2011; FLOWW, 2014; SeaPlan, 2015). Historically, it has been difficult to obtain activity data to quantify and assess the distribution of fishing effort; however, recent technological advances in new systems for vessel monitoring and enforcement, and improved stock assessments, mean that the availability, quality, and resolution of data have improved (Ashley et al., 2012). More specifically, larger

European Union vessels (≥12 m in length) are covered by the satellite-based vessel monitoring system and offer a minimum of bi-hourly locations (EC, 1997). Vessel monitoring system data are used primarily to enforce fisheries management, but they also attest the spatial distribution of fishing activity (Lee et al., 2010) and are valuable for assessing the distribution of fishing effort (Mills and Townsend, 2007; Witt and Godley, 2007; Fock, 2008; Stelzenmüller, 2008; Lee, South and Jennings, 2010; Jennings and Lee, 2011; Gerritsen, Minto and Lordan, 2013). Complementing vessel monitoring system data with information on catches further illuminates the spatial and temporal distribution of fishing activity and landings (Mullowney and Dawe, 2009; Bastardie et al., 2010; Gerritsen and Lordan, 2010; Vermard et al., 2010; Hintzen et al., 2012). More recently, vessel monitoring system data have been used to assess fishing displacement resulting from area closures (Dinmore et al., 2003; Murawski et al., 2005; Bastardie et al., 2017) and offshore wind farms (Vandendriessche et al., 2011; Bastardie et al., 2015).

Vessel monitoring system data come with limitations in terms of sharing and coverage, because sharing these data for reasons other than fisheries compliance and stock assessment is legally constrained due to commercial sensitivity (EC, 2008). It is a growing technology with full coverage of larger vessels in European waters, but it is not universally available around the world (Hinz et al., 2013). Moreover, there is no analogous system used to monitor smaller (<12 m) boat activity. Other remote-sensing applications include counting fishing vessels (primarily for compliance purposes) by airplane surveys (e.g., Satellite Applications Catapult, 2016). Recent developments include participatory mapping approaches to fill data gaps (Chapter 9); for example, an inshore fisheries mapping project in Scotland 'ScotMap' (Kafas et al., 2017) provided spatial information on the fishing activity of 72% of all registered commercial fishing vessels ≤15 m.

Using such activity data, the baseline for various proposed changes in the use of sea space (e.g., offshore renewables) can be characterised. An impact area of 12 nautical miles around a proposed development site is generally considered good practice in northern Europe for impact assessments in the case of mobile gears (demersal, and to a lesser extent, pelagic fisheries) as advised by the United Kingdom fishing industry (John Watt, personal communication, SFF, 2014; UKFEN, 2012; Gray, Stromberg and Rodmell, 2016). This impact area is large enough to identify any possible interaction between adjacent fisheries and the proposed change (and a subsequent restriction or displacement of fishing) over the course of a normal trawl of 4–5 hours, assuming a vessel was trawling at the average speed of 2.5–3.0 knots. The impact area does not need to be as large when considering the deployment of static gears where the range of interaction is likely to be <500 m (Gray, Stromberg and Rodmell, 2016). The size of the buffer around the proposed development footprint is best based on the particularities of local fisheries, as identified during consultation.

In addition to baseline characterisation, modelling fishing displacement has also been done mostly around Europe, North America, and Oceania (van Putten et al., 2012). Modelling involves developing and applying a scientifically defensible method to predict the likely cost to the fishing industry as a whole (van Putten et al., 2012), and less frequently to individual fishers (Bastardie, Nielsen and Miethe, 2014). However, prediction are still hampered by data limitations at appropriate resolution, as well as a limited understanding of the diversity of displacement responses (Tien and van der Hammen, 2015). A range of conceptual models have been developed to explain and predict the behaviour of fishing fleets and the impacts of displacement (see reviews in van Putten et al., 2012; Tien and van der Hammen, 2015). These models are primarily based on either micro-economic theory using profit maximisation (Robinson and Pascoe, 1997) or foraging behaviour based on maximising resource intake (Macarthur and Pianka, 2016). Profit and utility maximisation involves discrete-choice modelling and approaches based on random utility models, in which decisions are defined as discrete choice problems (e.g., Pradhan and Leung, 2004a, 2004b; Tidd et al., 2011) and both monetary and non-monetary inputs can be assessed. Foraging theory models are based on maximising resource intake at minimal energy output (Gillis and Peterman, 1998; Rijnsdorp, 2000; Abernethy et al., 2007), with applications focussing on the prediction of fishing-effort allocation in space (see van Putten et al., 2012 for an overview). The ideal free distribution approach is an example of models in which the relationship between fishers (predator) is based on the resource (prey) distribution (e.g., Rijnsdorp, 2000).

Following theoretical models, applications for empirical fleet dynamics have focused on identified decision variables of fishers, primarily using discrete-choice analyses (van Putten et al., 2012)—these explore the relationship between explanatory variables and observed patterns in the fleet (Hilborn and Kennedy, 1992; Fulton et al., 2011; van Putten et al., 2012; Bastardie et al., 2013). Such models have been used to predict both short-term (including location choice or effort allocation, and compliance; e.g., Pradhan and Leung, 2004a) and long-term decisions (e.g., exit or entry and strategic behaviour; van Putten et al., 2012). Most analyses using discrete-choice analyses to predict location choice are predominantly based on economic attributes expressed as expected profits from a fishing location (Branch et al., 2006). Many forms of discrete-choice models have been used, including generalised linear models (Cabrera and Defeo, 2001), nested logit (Eales and Wilen, 1986), mixed logit (Mistiaen and Strand, 2000), multinomial logit (Berman, 2007), conditional logit (Hutton et al., 2004), and probit models (Hatcher et al., 2000). Most of this research has focused on the location-choice behaviour of large-scale demersal and pelagic trawl fisheries in Europe and North America (van Putten et al., 2012). Examples include the impacts of fishing closures (marine protected areas in particular; e.g., Gell and Roberts, 2003; Mangi et al., 2011; Hattam et al., 2014) and the ecological impacts of trawl fisheries

(Rijnsdorp et al., 2001). The location-choice studies can refer to individual vessel as the basic decision unit, but also aggregations such as those formed by fleet segments, organisations, or collectives (van Putten et al., 2012). Assuming short-term location choices are primarily based on perceived economic gain (Pascoe, 2006; Allan & McGlade, 1987), profit maximisation is often used as a determinant of effort allocation (an unverified assumption in most cases), and provides a basis for the identification of a set of explanatory variables. Baseline characterisation coupled with modelling exercises can help to quantify the likelihood of displacement. The likelihood of displacement is expected to be greater in areas of popular and highly productive fishing grounds. If fishing is displaced to less productive fishing areas, where greater fishing effort is required to catch equivalent quantities of fish, this will lead to negative socio-economic effects, at least in the short term (e.g. Dinmore et al., 2003; Sen, 2010; Vandendriessche et al., 2011; Gray, Stromberg and Rodmell, 2016).

Mitigation measures

Where displacement is unavoidable and without the appropriate provisions to address the negative impacts of displaced effort, commercial fishers could exercise substantial opposition to proposed changes in access to sea space and might challenge any proposed changes via legal or other means. Management after a development site has been selected can include a variety of mitigation strategies for fisheries displacement. In the first instance, good practice in fisheries liaison and assessment of the spatial risk should aim to minimise the disruption and/or displacement (see *Preventive Measures*), but residual impacts on fishing activities can still persist. Residual impacts of fisheries displacement can be addressed by (1) assistance mechanisms (e.g., transitional business assistance), (2) additional fisheries management (e.g., reducing the effort in the relevant fisheries), or by (3) monetary settlement as a last resort.

Assistance mechanisms can include additional employment opportunities, technology improvements (e.g., technological developments for fishing gears), and/or covering the costs for gear diversification. Additional employment opportunities could support fishers' income through the provision of survey or guard vessel duty in support of the construction of offshore energy developments. Guard vessels might be required during the construction phase as well as during major maintenance and cable repair of an offshore energy development (FLOWW, 2014). However, these opportunities are not always directly accessible to fishers, particularly if additional personal qualifications or vessel certifications are required.

Mitigation through technology development for fishing gear could cover research and development costs to design gears that are less intensive in terms of environmental impact and/or to enable fishing in previously logistically unavailable locations (e.g. Catherall and Kaiser, 2014). Replacing gear to enable participation in an alternative fishery can help mitigate fisheries displacement arising through incompatibility of existing gears with the proposed development. However, fishers often disregard offers for gear diversification because there is usually a strong cultural attachment to fishing methods and strong views on the impact that the other has on the environment and stocks (Pollnac and Poggie, 2008; Urquhart and Acott, 2014). For example, changing from mobile trawling to other forms of static fishing, or vice versa, would not be contemplated by many fishers (FLOWW, 2014).

Monetary settlement such as that documented in the United Kingdom and United States, have taken the form of individual disruption settlements or fishing community funds (Swart et al., 2009; FLOWW, 2015; SeaPlan, 2015). Disruption settlements are upfront payments to fishers with recent catch history in areas they are no longer able to operate as a result of an offshore energy development. There often exists no legal basis for financial compensation associated with the loss of access to fishing grounds. Nonetheless, resolution between the interested parties (developer and fishers) is still pursued. Settlements are agreed on a mutual basis and aim to counterbalance or offset any residual fisheries-related impacts associated with offshore energy developments (Gell and Roberts, 2003; Swart et al., 2009; UKFEN, 2012; FLOWW, 2015; SeaPlan, 2015). The overall aim of any settlement is to achieve a position whereby fishing interests are neither advantaged nor disadvantaged by the development. Settlements should be based on loss of earnings and/or increased costs incurred, rather than loss of revenue (Swart et al., 2009; UKFEN, 2012; FLOWW, 2014; SeaPlan, 2015).

It is often debated whether disruption settlements should be linked with the retirement of the equivalent fishing effort to avoid the environmental issues of displacement (Greenstreet, Fraser and Piet, 2009). In theory, following such a principle could avoid the knock-on impacts of displaced activity on marine conservation, indirectly mitigate overcrowding, and help maintain the environmental sustainability of affected fisheries (Zeller et al., 2004). However, a strategy for the retirement of effort can result in a two-fold impact on the commercial fishing sector (i.e., loss of fishing grounds and a decrease in the total allowable catch) and cannot be considered as a mitigation measure from the perspective of fishers since it can further jeopardise the economic sustainability of affected fisheries (Gray, Stromberg and Rodmell, 2016).

Fishing community funds in the United Kingdom and the United States (FLOWW, 2014, 2015; Scottish Government, 2014; SeaPlan, 2015) have been implemented as a broader strategy to address residual impacts, or simply on the basis of a goodwill gesture in recognition of a new marine activity being accommodated within an area of existing fishing activity. Fishing community funds aim to promote longer-term community relations through fisheries' community-orientated projects, initiatives, or research activities (see examples in FLOWW, 2015). In some cases, this funding has been used to match other sources for larger-scale projects, or to return greater value to fishing communities.

Conclusion

Established activities such as commercial fishing are particularly likely to be subject to displacement by emerging uses of the sea such as offshore wind farms. Displacement can result in the use of alternative fishing grounds, alternative fishing gear, or targeting different species. The response of individual fishers or industry groups can be difficult to predict, although in most cases they can have negative economic consequences. Concentration of displaced effort can reduce commercial stocks and exacerbate environmental degradation. The scale and consequences of displacement can be difficult to predict in the absence of reliable and accurate information on fishing effort distribution, particularly for smaller (<12 m) vessels. It is important that marine spatial planning processes recognise and incorporate this heterogeneity in resilience between fishing fleet segments and the reliance of communities, and make appropriate management arrangements. Avoiding developments on prime fishing areas, coupled with constructive engagement at an early stage between renewables developers, is likely to deliver more towards minimising fisheries displacement, rather than reactive management at a later stage. Compensation is commonly sought as a consequence of displacement, and this can be applied at to individual vessels or at community scales.

Society is dynamic and so too are the demands that we place on our natural resources. Licences granted to offshore developments will last many decades, so the way in which we use and manage the marine environment must change as the needs and requirements of society evolve. A continued increase in collaboration and dialogue between all stakeholders interested in the marine environment and compliance with management measures are both required for the spatial planners to decide on informed basis, move forward, and produce a positive outcome for the proposed changes to sea use.

Highlights

- Offshore energy developments often impose spatial restrictions on existing marine users, such as commercial fisheries.
- Spatial restrictions result in changes in patterns or behaviour in space and/or time (displacement).
- Fisheries displacement can result in the use of alternative fishing grounds, alternative fishing gear, or targeting different species with a range of economic, social, and environmental effects.
- Avoiding prime fishing areas combined with early, continuous stakeholder engagement will deliver the most towards minimising fisheries displacement.
- Many lessons learned from the perspective of the fisheries are transferable to other marine users experiencing space competition in marine planning.

References

Abernethy, K.E., Allison, E.H., Molloy, P.P. and Côté, I.M. (2007). Why do fishers fish where they fish? Using the ideal free distribution to understand the behaviour of artisanal reef fishers. *Canadian Journal of Fisheries and Aquatic Sciences*, 64(11), pp. 1595–1604. doi:10.1139/f07-125.

Abesamis, R.A. and Russ, G.R. (2005). Density-dependent spillover from a marine reserve: Long-term evidence, ecological applications. *Ecological Society of America*, 15(5), pp. 1798–1812. doi:10.1890/05-0174.

Alexander, K.A., Wilding, T.A. and Heymans, J.J. (2013). Attitudes of Scottish fishers towards marine renewable energy. *Marine Policy*, 37, pp. 239–244. doi:10.1016/j.marpol.2012.05.005.

Allan, P.M. and McGlade, J.M. (1987). Modelling complex human systems: A fisheries example. *European Journal of Operational Research*, 30, pp. 147–167. doi:10.1016/0377-2217(87)90092-0.

Allison, E.H. and Ellis, F. (2001). The livelihoods approach and management of small-scale fisheries. *Marine Policy*, 25(5), pp. 377–388. doi:10.1016/S0308-597X(01)00023-9.

Allison, G.W., Lubchenco, J. and Carr, M.H. (1998). Marine reserves are necessary but not sufficient for marine conservation. *Ecological Applications*, 8(1 SUPPL.), pp. 79–92. doi:10.2307/2641365.

Aquatera. (2014). Scotrenewables Tidal Power Lashy Sound Tidal Array Scoping Report. Available at www.aquatera.co.uk/.

Ashley, M.C., Campbell, M., De Groot, J. and Rodwell, L.D. (2012). *Assessing fisheries displacement as a result of developing a UK network of MPAs and offshore energy development*. Report funded by NERC, MREKE, NFFO, Seafish.

Bastardie, F., Angelini, S., Bolognini, L., Fuga, F., Manfredi, C., Martinelli, M., Nielsen, J. R., Santojanni, A., Scarcella, G. and Grati, F. (2017). Spatial planning for fisheries in the Northern Adriatic: Working toward viable and sustainable fishing. *Ecosphere*, 8(2), p. e01696. doi:10.1002/ecs2.1696.

Bastardie, F., Nielsen, J.R., Andersen, B.S. and Eigaard, O.R. (2013). Integrating individual trip planning in energy efficiency – building decision tree models for Danish fisheries. *Fisheries Research*, 143, pp. 119–130. doi:10.1016/j.fishres.2013.01.018.

Bastardie, F., Nielsen, J.R. and Miethe, T. (2014). DISPLACE : A dynamic, individual-based model for spatial fishing planning and effort displacement—integrating underlying fish population models. *Canadian Journal of Fisheries and Aquatic Sciences*, 21(2013), pp. 1–21.

Bastardie, F., Nielsen, J.R., Ulrich, C., Egekvist, J. and Degel, H. (2010). Detailed mapping of fishing effort and landings by coupling fishing logbooks with satellite-recorded vessel geo-location. *Fisheries Research*, 106(1), pp. 41–53. doi:10.1016/j.fishres.2010.06.016.

Bastardie, F., Rasmus Nielsen, J., Eigaard, O.R., Fock, H.O., Jonsson, P. and Bartolino, V. (2015). Competition for marine space: Modelling the Baltic Sea fisheries and effort displacement under spatial restrictions. *ICES Journal of Marine Science*, 72(3), pp. 824–840. doi:10.1093/icesjms/fsu215.

Baxter, J.M., Boyd, I.L., Cox, M., Donald, A.E., Malcolm, S.J., Miles, H., Miller, B. and Moffat, C.F. (2011). *Scotland's Marine Atlas: Information for the national marine plan*. Marine Scotland and Scottish Natural Heritage. Edinburgh, Scotland.

Berman, M. (2007). Modeling spatial choice in ocean fisheries. *Marine Resource Economics*, 21, pp. 387–406.

BOWL. (2011). *Beatrice offshore wind farm environmental statement non-technical summary.* [pdf] Available at http://sse.com/media/113151/BOWL_OffshoreApplication-NonTechnicalSummaryApril2012.pdf.

Branch, T.A., Hilborn, R., Haynie, A.C., Fay, G., Flynn, L., Griffiths, J., Marshall, K.N., Randall, J.K., Scheuerell, J.M., Ward, E.J. and Young, M. (2006). Fleet dynamics and fishermen behavior: Lessons for fisheries managers. *Canadian Journal of Fisheries and Aquatic Sciences*, 63(7), pp. 1647–1668. doi:10.1139/f06-072.

Breen, P., Vanstaen, K. and Clark, R.W.E. (2014). Mapping inshore fishing activity using aerial, land, and vessel-based sighting information. *ICES Journal of Marine Science*, 72(2), pp. 467–479.

Brookfield, K., Gray, T. and Hatchard, J. (2005). The concept of fisheries-dependent communities: A comparative analysis of four UK case studies: Shetland, Peterhead, North Shields and Lowestoft. *Fisheries Research*, 72(1), pp. 55–69. doi:10.1016/j.fishres.2004.10.010.

Brown, K., Adger, W.N., Tompkins, E., Bacon, P., Shim, D. and Young, K. (2001). Trade-off analysis for marine protected area management. *Ecological Economics*, 37, pp. 417–434. doi:10.1016/S0921-8009(00)00293-7.

Cabrera, J.L. and Defeo, O. (2001). Daily bioeconomic analysis in a multispecific artisanal fishery in Yucatan, Mexico. *Aquatic Living Resources*, 14(1), pp. 19–28.

Campbell, M.S., Stehfest, K.M., Votier, S.C. and Hall-Spencer, J.M. (2014). Mapping fisheries for marine spatial planning: Gear-specific vessel monitoring system (VMS), marine conservation and offshore renewable energy. *Marine Policy*, 45, pp. 293–300. doi:10.1016/j.marpol.2013.09.015.

Cass, N., Walker, G. and Devine-Wright, P. (2010). Good neighbours, public relations and bribes: The politics and perceptions of community benefit provision in renewable energy development in the UK. *Journal of Environmental Policy & Planning*, 12(3), pp. 255–275. doi:10.1080/1523908X.2010.509558.

Catherall, C.L. and Kaiser, M.J. (2014). Review of king scallop dredge designs and impacts, legislation and potential conflicts with offshore wind farms Report to Moray Offshore Renewables Limited. *Fisheries & Conservation Report, Bangor University*, (39).

Christie, P. (2004). Marine protected areas as biological successes and social failures in Southeast Asia. *American Fisheries Society*, 42, pp. 155–164. doi:10.1016/S0002-9610(03)00290-3.

Christie, P. and White, A.T. (2007). Best practices for improved governance of coral reef marine protected areas. *Coral Reefs*, 26(4), pp. 1047–1056. doi:10.1007/s00338-007-0235-9.

Davies, I.M. and Watret, R. (2011). Scoping study for offshore wind farm development in Scottish waters. Edinburgh, Scotland.

Day, J.C. (2002). Zoning—lessons from the Great Barrier Reef Marine Park. *Ocean and Coastal Management*, 45, pp. 139–156.

De Groot, J., Campbell, M., Ashley, M. and Rodwell, L. (2014). Investigating the co-existence of fisheries and offshore renewable energy in the UK: Identification of a mitigation agenda for fishing effort displacement. *Ocean and Coastal Management*, 102, pp. 7–18. doi:10.1016/j.ocecoaman.2014.08.013.

Diegues, A. (2008). Marine protected areas and artisanal fisheries in Brazil. *SAMUDRA Monograph*, p. 54.

Dinmore, T.A., Duplisea, D.E., Rachham, B.D., Maxwell, D.L. and Jennings, S. (2003). Impact of a large-scale area closure on patterns of fishing disturbance and the consequences for benthic communities. *ICES Journal of Marine Science*, 3139(3), pp. 371–380. doi:10.1016/S1054-3139(03)00010-9.

Donohue, P. and Murphy, E. (2015). Displacement of commercial sea fishing as a result of policy and administrative activities in the marine environment. Scottish Marine and Freshwater Science Bulletin. Edinburgh.

Douet, M. (2016). Change drivers across supply chains: The case of fishery and aquaculture in France. *Transportation Research Procedia*, 14, pp. 2830–2839. doi:10.1016/j.trpro.2016.05.349.

Eales, J. and Wilen, J.E. (1986). An examination of fishing location choice in the pink shrimp fishery. *Marine Resource Economics*, 2(4), pp. 331–351.

EC. (1997). Commission Regulation (EC) No. 1489/97. Laying down detailed rules for the application of Council Regulation (EEC) no. 2847/93 as regards satellite-based vessel monitoring systems.

EC. (2008). No 199/2008 establishing a community framework for the collection, management and use of data in the fisheries sector and support for scientific advice regarding the common fisheries policy (2008/949/EC). *Official Journal of the European Union*, L346, pp. 37–88.

EC. (2014). Facts and figures on the Common Fisheries Policy – 2014 Edition. Brussels, Belgium. doi:10.2771/35745.

EC. (2015). European maritime and fisheries fund in the UK. Brussels, Belgium.

European Council. (2014). Directive 2014/52/EU of the European Parliament and of the Council of 16 April 2014 amending Directive 2011/92/EU on the assessment of the effects of certain public and private projects on the environment. Brussels: European Council.

FAO. (2005). Report of the Workshop on Standardization of Selectivity Methods Applied to Trawling in the Mediterranean Sea, Sète, France, 9–11 February 2005. Sète, France: Food and Agriculture Organization of the United Nations.

FAO. (2014). The state of world fisheries and aquaculture, Food and Agriculture Oraganization of the United Nations. doi:92-5-105177-1.

Farmery, A., Gardner, C., Green, B.S. and Jennings, S. (2014). Managing fisheries for environmental performance: The effects of marine resource decision-making on the footprint of seafood. *Journal of Cleaner Production*, 64, pp. 368–376. doi:10.1016/j.jclepro.2013.10.016.

FLOWW. (2014). FLOWW Best Practice Guidance for Offshore Renewables Developments : Recommendations for Fisheries Liaison January 2014. Edinburgh, Scotland, UK.

FLOWW. (2015). FLOWW Best Practice Guidance for Offshore Renewables Developments: Recommendations for Fisheries Disruption Settlements and Community Funds. London, UK.

Fock, H.O. (2008). Fisheries in the context of marine spatial planning: Defining principal areas for fisheries in the German EEZ. *Marine Policy*, 32(4), pp. 728–739. doi:10.1016/j.marpol.2007.12.010.

Fulton, E.A, Smith, A.D.M., Smith, D.C. and Van Putten, I.E. (2011). Human behaviour: The key source of uncertainty in fisheries management. *Fish and Fisheries*, 12(1), pp. 2–17. doi:10.1111/j.1467-2979.2010.00371.x.

Gell, F.R. and Roberts, C.M. (2003). The fishery effects of marine reserves and fishery closures. *Trends in Ecology and Evolution*, 18, p. 90. doi:10.1016/S0169-5347(03)00189-7.

Gerritsen, H.D. and Lordan, C. (2010). Integrating vessel monitoring systems (VMS) data with daily catch data from logbooks to explore the spatial distribution of catch and effort at high resolution. *ICES Journal of Marine Science*, 68(1), pp. 245–252. doi:10.1093/icesjms/fsq137.

Gerritsen, H.D., Minto, C. and Lordan, C. (2013). How much of the seabed is impacted by mobile fishing gear? Absolute estimates from Vessel Monitoring System (VMS) point data. *ICES Journal of Marine Science*, 70(3), pp. 523–531. doi:10.1093/icesjms/fst017.

Gillis, D.M. and Peterman, R.M. (1998). Implications of interference among fishing vessels and the ideal free distribution to the interpretation of CPUE. *Canadian Journal of Fisheries and Aquatic Sciences*, 55(1), pp. 37–46. doi:10.1139/f97-206.

Gray, M., Stromberg, P. and Rodmell, D. (2016). Changes to fishing practices around the UK as a result of the development of offshore windfarms – Phase 1. Crown Estate, UK.

Greenstreet, S.P.R., Fraser, H.M. and Piet, G.J. (2009). Using MPAs to address regional-scale ecological objectives in the North Sea : Modelling the effects of fishing effort displacement. *ICES Journal of Marine Science*, 66(1), pp. 90–100.

Hall, C.M. (2001). Trends in ocean and coastal tourism: The end of the last frontier? *Ocean & Coastal Management*, 44, pp. 601–618.

Halpern, B.S. and Warner, R.R. (2002). Marine reserves have rapid and lasting effects. *Ecology Letters*, 5(3), pp. 361–366. doi:10.1046/j.1461-0248.2002.00326.x.

Hatcher, A., Jaffry, S., Thébaud, O., Bennett, E., Economics, S.L., Aug, N., Hatcher, A., Jaffry, S., Thibaud, O. and Bennett, E. (2000). Board of regents of the University of Wisconsin system normative and social influences affecting compliance with fishery regulations normative and social influences affecting compliance with fishery regulations. *Land Economics*, 76(3), pp. 448–461.

Hattam, C.E., Mangi, S.C., Gall, S.C. and Rodwell, L.D. (2014). Social impacts of a temperate fisheries closure: Understanding stakeholders' views. *Marine Policy*, 45, pp. 269–278. doi:10.1016/j.marpol.2013.09.005.

Helson, J., Leslie, S., Clement, G., Wells, R. and Wood, R. (2010). Private rights, public benefits: Industry-driven seabed protection. *Marine Policy*, 34, pp. 557–566.

Hilborn, R. (2007). Managing fisheries is managing people: What has been learned? *Fish and Fisheries*, 8, pp. 285–296.

Hilborn, R. and Kennedy, R. (1992). Spatial pattern in catch rates: A test of economic theory. *Bulletin of Mathematical Biology*, 54(2–3), pp. 263–273. doi:10.1016/S0092-8240(05)80026-6.

Hilborn, R., Stokes, K., Maguire, J.-J., Smith, T., Botsford, L.W., Mangel, M., Orensanz, J., Parma, A., Rice, J., Bell, J., Cochrane, K.L., Garcia, S., Hall, S.J., Kirkwood, G., Sainsbury, K., Stefansson, G. and Walters, C. (2004). When can marine reserves improve fisheries management? *Ocean & Coastal Management*, 47(3–4), pp. 197–205. doi:10.1016/j.ocecoaman.2004.04.001.

Himes, A. (2003). Small-scale fisheries: Opinions of artisanal fishers and socio-cultural effects of MPAs. *Coastal Management*, 31, pp. 389–408.

Hintzen, N.T., Bastardie, F., Beare, D., Piet, G.J., Ulrich, C., Deporte, N., Egekvist, J. and Degel, H. (2012). VMStools: Open-source software for the processing, analysis and visualisation of fisheries logbook and VMS data. *Fisheries Research*. B.V., 115–116, pp. 31–43. doi:10.1016/j.fishres.2011.11.007.

Hinz, H., Murray, L.G., Lambert, G.I., Hiddink, J.G. and Kaiser, M.J. (2013). Confidentiality over fishing effort data threatens science and management progress. *Fish and Fisheries*, 14, pp. 110–117. doi:10.1111/j.1467-2979.2012.00475.x.

Holland, D. and Sutinen, J. (2000). Location choice in New England trawl fisheries: Old habits die hard. *Land Economics*, 76(1), pp. 133–149.

Hutton, T., Mardle, S., Pascoe, S. and Clark, R.A. (2004). Modelling fishing location choice within mixed fisheries: English North Sea beam trawlers in 2000 and 2001. *ICES Journal of Marine Science*, 61(8), pp. 1443–1452. doi:10.1016/j.icesjms.2004.08.016.

ICES. (2017). Workshop report on Conflicts and Coexistence in Marine Report (WKCCMSP). Geesthacht, Germany. doi:ICES CM 2015/SSGEPI:22.

IEMA. (2016). Environmental Impact Assessment Guide to Delivering Quality Development, Lincoln. UK. Available at www.iema.net.

Janßen, H., Bastardie, F., Eero, M., Hamon, K.G., Hinrichsen, H.-H., Marchal, P., Nielsen, J.R., Le Pape, O., Schulze, T., Simons, S., Teal, L.R. and Tidd, A. (2017). Integration of fisheries into marine spatial planning: Quo vadis? *Estuarine, Coastal and Shelf Science*, (2016), pp. 1–9. doi:10.1016/j.ecss.2017.01.003.

Jennings, S. and Lee, J. (2011). Defining fishing grounds with vessel monitoring system data. *ICES Journal of Marine Science*, 69(1), pp. 51–63. doi:10.1093/icesjms/fsr173.

Jentoft, S. (2007). Limits of governability: Institutional implications for fisheries and coastal governance. *Marine Policy*, 31(4), pp. 360–370. doi:10.1016/j.marpol.2006.11.003.

Jentoft, S. and Knol, M. (2014). Marine spatial planning : Risk or opportunity for fisheries in the North Sea ? *Maritime Studies*, 12(1), pp. 1–16.

Kafas, A., McLay, A., Chimienti, M., Scott, B.E., Davies, I. and Gubbins, M. (2017). ScotMap: Participatory mapping of inshore fishing activity to inform marine spatial planning in Scotland. *Marine Policy*, 79, pp. 8–18. doi:10.1016/j.marpol.2017.01.009.

Kamenos, N.A., Moore, P.G. and Hall-Spencer, J.M. (2004). Attachment of the juvenile queen scallop (*Aequipecten opercularis* (L.)) to maerl in mesocosm conditions; juvenile habitat selection. *Journal of Experimental Marine Biology and Ecology*, 306(2), pp. 139–155. doi:10.1016/j.jembe.2003.10.013.

Kellner, J.B. and Hastings, A. (2009). A reserve paradox: Introduced heterogeneity may increase regional invasibility. *Conservation Letters*, 2(3), pp. 115–122. doi:10.1111/j.1755-263X.2009.00056.x.

Lacroix, D. and Pioch, S. (2011). The multi-use in wind farm projects: More conflicts or a win-win opportunity? *Aquatic Living Resources*, 24(2), pp. 129–135. doi:10.1051/alr/2011135.

Lauck, T., Clark, C.W., Mangel, M. and Munro, G.R. (1998). Implementing the precautionary principle in fisheries management through marine reserves. *Ecological Applications*, 8(1), pp. S72–S78. doi:10.2307/2641364.

Ledee, E.J.I., Sutton, S.G., Tobin, R.C. and De Freitas, D.M. (2012). Responses and adaptation strategies of commercial and charter fishers to zoning changes in the Great Barrier Reef Marine Park. *Marine Policy*, 36(1), pp. 226–234. doi:10.1016/j.marpol.2011.05.009.

Lee, J., South, A.B. and Jennings, S. (2010). Developing reliable, repeatable, and accessible methods to provide high-resolution estimates of fishing-effort distributions from vessel monitoring system (VMS) data. *ICES Journal of Marine Science*, 67(6), pp. 1260–1271. doi:10.1093/icesjms/fsq010.

Macarthur, R.H. and Pianka, E.R. (2016). On optimal use of a patchy environment. *The American Naturalist*, 100(916), pp. 603–609. doi:10.2307/2458820.

Mangi, S.C., Rodwell, L.D. and Hattam, C. (2011). Assessing the impacts of establishing MPAs on fishermen and fish merchants: The case of Lyme Bay, UK. *Ambio*, 40(5), pp. 457–468. doi:10.1007/s13280-011-0154-4.

Marchal, P., Ulrich, C. and Pastoors, M. (2002). Area-based management and fishing efficiency. *Aquatic Living Resources*, 15, pp. 73–85.

Mascia, M.B., Claus, C.A. and Naidoo, R. (2010). Impacts of marine protected areas on fishing communities. *Conservation Biology*, 24(5), pp. 1424–1429.

McClanahan, T.R. and Mangi, S. (2000). Spill over of exploitable fishes from a marine park and its effect on the adjacent fishery. *Ecological Applications*, 10(6), pp. 1792–1805. doi:/10.1890/1051–0761(2000)010%5B1792:SOEFFA%5D2.0.CO;2.

Mills, C. and Townsend, S. (2007). Estimating high resolution trawl fishing effort from satellite-based vessel monitoring system data. *ICES Journal of Marine Science*, 64(2), pp. 248–255.

Mistiaen, J. and Strand, I. (2000). Location choice of commercial fishermen with heterogeneous risk preferences. *American Journal of Agricultural Economics*, 82(5), pp. 1184–1190.

Mullowney, D.R. and Dawe, E.G. (2009). Development of performance indices for the Newfoundland and Labrador snow crab (*Chionoecetes opilio*) fishery using data from a vessel monitoring system. *Fisheries Research*, 100, pp. 248–254. doi:10.1016/j.fishres.2009.08.006.

Munday, M., Bristow, G. and Cowell, R. (2011). Wind farms in rural areas: How far do community benefits from wind farms represent a local economic development opportunity? *Journal of Rural Studies*, 27, pp. 1–12. doi:10.1016/j.jrurstud.2010.08.003.

Murawski, S.A., Wigley, S.E., Fogarty, M.J., Rago, P.J. and Mountain, D.G. (2005). Effort distribution and catch patterns adjacent to temperate MPAs. *ICES Journal of Marine Science*, 62(6), pp. 1150–1167. doi:10.1016/j.icesjms.2005.04.005.

Nuttall, M. (2000). Crisis, risk and deskilment in North-east Scotland's fishing industry. *Fisheries Dependent Regions*, 2000, pp. 106–115.

Ojeda-Ruiz, M.Á., Ramírez-Rodríguez, M. and De La Cruz-Agüero, G. (2015). Mapping fishing grounds from fleet operation records and local knowledge: The Pacific calico scallop (*Argopecten ventricosus*) fishery in Bahia Magdalena, Mexican Pacific. *Ocean and Coastal Management*, 106, pp. 61–67. doi:10.1016/j.ocecoaman.2015.01.011.

Panayotou, T. (1988). *Management concepts for small-scale fisheries: Economic and social aspects*. FAO, Food and Agriculture Organization of the United Nations. Rome.

Pascoe, S. (2006). Economics, fisheries, and the marine environment. *ICES Journal of Marine Science*, 63(1), pp. 1–3. doi:10.1016/j.icesjms.2005.11.001.

Pascoe, S., Doshi, A., Dell, Q., Tonks, M. and Kenyon, R. (2014). Economic value of recreational fishing in Moreton Bay and the potential impact of the marine park rezoning. *Tourism Management*, 41, pp. 53–63. doi:10.1016/j.tourman.2013.08.015.

Pita, C., Theodossiou, I. and Pierce, G.J. (2013). The perceptions of Scottish inshore fishers about marine protected areas. *Marine Policy*, 37, pp. 254–263. doi:10.1016/j.marpol.2012.05.007.

Pollnac, R.B. and Poggie, J.J. (2008). Happiness, well-being and psychocultural adaptation to the stresses associated with marine fishing. *Human Ecology Review*, 15(2), pp. 194–200.

Pomeroy, R. and Douvere, F. (2008). The engagement of stakeholders in the marine spatial planning process. *Marine Policy*, 32(5), pp. 816–822. doi:10.1016/j.marpol.2008.03.017.

Poos, J.J., Bogaards, J.A., Quirijns, F.J., Gillis, D.M. and Rijnsdorp, A.D. (2009). Individual quotas, fishing effort allocation, and over-quota discarding in mixed fisheries. *ICES Journal of Marine Science*, 67(2), pp. 323–333. doi:10.1093/icesjms/fsp241.

Pradhan, N.C. and Leung, P. (2004a). Modeling entry, stay, and exit decisions of the longline fishers in Hawaii. *Marine Policy*, 28, pp. 311–324. doi:10.1016/j.marpol.2003.09.005.

Pradhan, N.C. and Leung, P. (2004b). Modeling trip choice behavior of the longline fishers in Hawaii. *Fisheries Research*, 68, pp. 209–224. doi:10.1016/j.fishres.2003.12.006.

Rijnsdorp, A. (2000). Effects of fishing power and competitive interactions among vessels on the effort allocation on the trip level of the Dutch beam trawl fleet. *ICES Journal of Marine Science*, 57(4), pp. 927–937. doi:10.1006/jmsc.2000.0580.

Rijnsdorp, A.D., Piet, G.J. and Poos, J.J. (2001). Effort allocation of the Dutch beam trawl fleet in response to a temporarily closed area in the North Sea. *Ices CM 2001/N:01*, pp. 1–17.

Roberts, C.M., Hawkins, J.P. and Gell, F.R. (2005). The role of marine reserves in achieving sustainable fisheries. *Philosophical Transactions of the Royal Society of London. Series B, Biological Sciences*, 360(1453), pp. 123–132. doi:10.1098/rstb.2004.1578.

Robinson, C. and Pascoe, S. (1997). Fisher behaviour: Exploring the validity of the profit maximising assumption. Research paper 110. Portsmouth.

Rosenberg, A.A. and Glass, C.W. (2007). Fishers' responses to management measures and their socio-economic effects. *ICES Journal of Marine Science*, 64(8), pp. 1612–1613. doi:10.1093/icesjms/fsm143.

Rowley, R.J. (1994). Marine reserves in fisheries management. *Aquatic Conservation: Marine and Freshwater Ecosystems*, 4(3), pp. 233–254. doi:10.1002/aqc.3270040305.

Russ, G. and Alcala, A. (1996). Do marine reserves export adult fish biomass? Evidence from Apo Island, central Philippines. *Marine Ecology Progress Series*, 132, pp. 1–9. doi:10.3354/meps132001.

Salas, S. and Gaertner, D. (2004). The behavioural dynamics of fishers: Management implications. *Fish and Fisheries*, 5(2), pp. 153–167. doi:10.1111/j.1467-2979.2004.00146.x.

Sanchirico, J.N. and Wilen, J. (2002). The impacts of marine reserves on limited-entry fisheries. *Natural Resource Modeling*, 15(3), pp. 291–310.

Sanders, M.J. and Morgan, A.J. (1976). Fishing power, fishing effort, density, fishing intensity and fishing mortality. *ICES Journal of Marine Science*, 37(1), pp. 36–40. doi:10.1093/icesjms/37.1.36.

Satellite Applications Catapult. (2016). Ascension Island uses 'Eyes on the Seas' to monitor new marine reserve, Satellite Applications Catapult News. Available at https://sa.catapult.org.uk/.

Scottish Government. (2014). Good Practice Principles for Community Benefits from Offshore Renewables. Glasgow, UK.

Seafish. (2011). *Static Gear technology note. Edinburgh.* [pdf] Available at www.seafish.org/media/Publications/SeafishGuidanceNote_StaticGear_201102.pdf

SeaPlan. (2015). Options for cooperation between commercial fishing and offshore wind energy industries. A review of relevant tools and best practices. Available at www.openchannels.org/seaplan

Sen, S. (2010). Developing a framework for displaced fishing effort programs in marine protected areas. *Marine Policy*, 34(6), pp. 1171–1177. doi:10.1016/j.marpol.2010.03.017.

Shaw, S. and Johnson, H. (2011). Identifying, communicating and integrating social considerations into future management concerns in inshore commercial fisheries in Coastal Queensland. Report to the Fisheries Research and Development Corporation Project No. 2008/073. Fisheries Research and Development Corporation, Canberra, and The Queensland Seafood Industry Association, Clayfield Qld.

Smith, G. and Brennan, R.E. (2012). Ocean & coastal management losing our way with mapping: Thinking critically about marine spatial planning in Scotland. *Ocean and Coastal Management*, 69, pp. 210–216. doi:10.1016/j.ocecoaman.2012.08.016.

Smith, M.D., Lynham, J., Sanchirico, J.N. and Wilson, J.A (2010). Political economy of marine reserves: Understanding the role of opportunity costs. *Proceedings of the National Academy of Sciences of the United States of America*, 107(43), pp. 1830–1835. doi:10.1073/pnas.0907365107.

STECF. (2013). The 2013 Annual Economic Report on the EU Fishing Fleet (STECF-13-15), Scientific, Technical and Economic Committee for Fisheries. Publications Office of the European Union. Luxembourg.

STECF. (2016). The 2016 Annual Economic Report on the EU fishing fleet. Scientific, Technical and Economic Committee for Fisheries. Publications Office of the European Union. Luxembourg.

Stelzenmüller, V. (2008). Spatio-temporal patterns of fishing pressure on UK marine landscapes, and their implications for spatial planning and management. *ICES Journal of Marine Science*, 65(6), pp. 1081–1091.

Swart, S., Friesen, B., Holman, A. and Aue, N. (2009). Annex IV 2016 State of the Science Report - environmental effects of marine renewable energy development around the world. *Ocean Energy Systems*, doi:10.1097/JNN.0b013e3182829024.

Tidd, A.N., Hutton, T., Kell, L.T. and Padda, G. (2011). Exit and entry of fishing vessels: An evaluation of factors affecting investment decisions in the North Sea English beam trawl fleet. *ICES Journal of Marine Science*, 68(5), pp. 961–971. doi:10.1093/icesjms/fsr015.

Tien, N.S.H. and Van Der Hammen, T. (2015). Fisheries displacement effects related to closed areas: A literature review of relevant aspects. *IMARES report*, C170/15.

Toonen, H.M. and Mol, A.P.J. (2013). Putting sustainable fisheries on the map? Establishing no-take zones for North Sea plaice fisheries through MSC certification. *Marine Policy*, 37, pp. 294–304. doi:10.1016/j.marpol.2012.05.012.

Turner, R.A., Polunin, N.V.C. and Stead, S.M. (2015). Mapping inshore fisheries: Comparing observed and perceived distributions of pot fishing activity in Northumberland. *Marine Policy*, 51, pp. 173–181. doi:10.1016/j.marpol.2014.08.005.

UKFEN. (2012). Best Practice Guidance for Fishing Industry Financial and Economic Impact Assessments. Edinburgh, Scotland, UK.

Urquhart, J. and Acott, T. (2014). A sense of place in cultural ecosystem services: The case of Cornish fishing communities. *Society & Natural Resources*, 27(1), pp. 3–19. doi:10.1080/08941920.2013.820811.

Van De Geer, C., Mills, M., Adams, V.M., Pressey, R.L. and McPhee, D. (2013). Impacts of the Moreton Bay Marine Park rezoning on commercial fishermen. *Marine Policy*, 39(1), pp. 248–256. doi:10.1016/j.marpol.2012.11.006.

Vandendriessche, S., Hostens, K., Courtens, W. and Stienen, E.W.M. (2011). Chapter 8. Monitoring the effects of offshore wind farms: Evaluating changes in fishing effort using Vessel Monitoring System data: Targeted monitoring results Photo RBINS/MUMM Trawling vessel in the Belgian part of the North Sea, in In Degraer, S., R. Brabant & B. Rumes (eds),*Offshore wind farms in the Belgian part of the North Sea: Selected findings from the baseline and targeted monitoring, RBINS-MUMM, Brussels*, pp. 83–92.

Van Putten, I.E., Kulmala, S., Thébaud, O., Dowling, N., Hamon, K.G., Hutton, T. and Pascoe, S. (2012). Theories and behavioural drivers underlying fleet dynamics models. *Fish and Fisheries*, 13(2), pp. 216–235. doi:10.1111/j.1467-2979.2011.00430.x.

Vermard, Y., Rivot, E., Mahévas, S., Marchal, P. and Gascuel, D. (2010). Identifying fishing trip behaviour and estimating fishing effort from VMS data using Bayesian Hidden Markov Models. *Ecological Modelling*, 221(15), pp. 1757–1769. doi:10.1016/j.ecolmodel.2010.04.005.

Vink, B. and Van Der Burg, A. (2006). New Dutch spatial planning policy creates space for development. *disP - The Planning Review*, 42(164), pp. 41–49. doi:10.1080/02513625.2006.10556946.

Vize, S., Adnitt, C., Staniland, R. and Everard, J. (2008). Review of cabling techniques and environmental effects applicable to the offshore wind farm industry. *Department for Business Enterprise & Regulatory Reform* (January).

Walker, B.J.A., Wiersma, B. and Bailey, E. (2014). Community benefits, framing and the social acceptance of offshore wind farms: An experimental study in England. *Energy Research and Social Science*, 3, pp. 46–54. doi:10.1016/j.erss.2014.07.003.

Westerberg, V., Jacobsen, J.B. and Lifran, R. (2013). The case for offshore wind farms, artificial reefs and sustainable tourism in the French Mediterranean. *Tourism Management*, 34, pp. 172–183. doi:10.1016/j.tourman.2012.04.008.

Wiber, M.G., Young, S. and Wilson, L. (2012). Impact of aquaculture on commercial fisheries: Fishermen's local ecological knowledge. *Human Ecology*, 40(1), pp. 29–40. doi:10.1007/s10745-011-9450-7.

Wilen, J.E., Smith, M.D., Lockwood, D. and Botsford, L.W. (2002). Avoiding surprises: Incorporating fisherman behavior into management models. *Bulletin of Marine Science*, 70(2), pp. 553–575.

Willis, T.J., Millar, R.B., Babcock, R.C. and Tolimieri, N. (2003). Burdens of evidence and the benefits of marine reserves: Putting Descartes before des horse? *Environmental Conservation*, 30(2), pp. 97–103. doi:10.1017/S0376892903000092.

Witt, M.J. and Godley, B.J. (2007). A step towards seascape scale conservation: Using vessel monitoring systems (VMS) to map fishing activity. *PLoS ONE*, 2(10). doi:10.1371/journal.pone.0001111.

Yates, K.L. and Schoeman, D.S. (2013). Spatial access priority mapping (SAPM) with fishers: A quantitative GIS method for participatory planning. *PLoS ONE*, 8(7), p. e68424. doi:10.1371/journal.pone.0068424.

Zeller, D., Russ, G. R., Steele, J. and Hoagland, P. (2004). Are fisheries 'sustainable'? A counterpoint to Steele and Hoagland. *Fisheries Research*. 67(2), pp. 241–245. doi:10.1016/j.fishres.2004.01.001.

Chapter 7

Tracing regime shifts in the provision of coastal-marine cultural ecosystem services

Kira Gee and Benjamin Burkhard

Introduction

Marine spatial planning has been defined by the European Union as a "... process by which the relevant Member State's authorities analyse and organise human activities in marine areas to achieve ecological, economic and social objectives" (EC, 2014). In most of the literature, it is understood as a strategic arbiter between competing demands and an approach based on a rational paradigm of space (e.g., Douvere and Ehler, 2009, Chapters 1 and 2). While the decision to locate a commercial activity in the sea is ultimately an economic one (Turner et al., 2014), the aim of marine spatial planning is to create a strategic framework that facilitates and steers such development, encouraging activities or co-use in some marine areas and prioritising other concerns such as environmental objectives in others. Sustainability is a guiding principle, expressed, for instance, in the EU Marine Spatial Planning Directive that calls upon Member States to consider economic, environmental *and* social aspects in their maritime spatial plans (EC, 2014, own emphasis).

An important part of achieving sustainable marine development is to evaluate the potential impacts of new developments, weighing any opportunities that might arise (such as employment) against expected vulnerabilities and risks (such as socio-economic or ecological risks; e.g., Nunneri et al., 2009). A wide range of ecological, spatial and increasingly, economic evidence is available to facilitate such assessment, including the sensitivity of ecosystems to particular pressures and impacts (e.g., Petersen and Malm, 2006; Punt et al., 2009; Busch et al., 2015) or the economic value or employment generated from offshore activities (e.g., Oil and Gas, 2014; Oxford Maritime, 2015).

One element, however, is still largely missing from the evidence base. This is the social – specifically the socio-cultural – dimension, in particular the cultural and non-material meanings and values associated with the sea. Cultural values and practices associated with the sea contribute to human well-being by enabling a wide range of immaterial benefits such as recreation, aesthetic appreciation, learning or spiritual experiences (Busch et al., 2011; Church et al., 2014; Kenter et al., 2014). Attachment to coastal places or practices associated with the sea can be strong in communities, potentially leading to conflict if these values are perceived to be under threat, for example, from industrial development (e.g., Martinez et al., 2013; Döring, 2005). Marine spatial planners are increasingly aware of the importance of cultural values associated with the sea, and the fact that these values can be threatened by changing marine activities (ICES, 2013). However, the repercussions or the impact of loss or degradation of marine cultural values are often missing from current risk and impact assessments.

This evidence gap has implications for broader systems-oriented perspectives on changing marine use. Given the multi-dimensionality of the marine environment and the multiplicity of land-sea interactions, risk and impact assessments might be limited in their value if they are restricted to just one dimension of inherently complex social-ecological, coastal-marine systems (Luisetti et al., 2014). A resilience perspective can add value here. Resilience can be understood as the ability of a system to absorb perturbations without experiencing instability and decline (Burkhard and Gee, 2012), which can be applied to various individual system components or 'domains' (e.g., a particular habitat; the economic system). However, the main benefit of the resilience perspective is its ability to consider the impacts of change across system domains at various spatial and temporal scales. Apparently small changes in one domain can trigger cascading effects across domains, causing the system to undergo non-linear shifts from one dynamic equilibrium to another (Holling and Gunderson, 2002; Redman and Kinzig, 2003; Walker et al., 2007). If this shift is large enough and the system is not able to bounce back to a comparable state after a disturbance, an altogether different system can emerge as the result of a regime shift in any of the system domains (Kinzig et al., 2006; Burkhard and Gee 2012). An extreme example from the Wadden Sea would be the severe storm surges that occurred in the medieval period, causing the loss of entire islands and impacting the coastal population in the process.

The resilience perspective, which can also be considered as a social-ecological systems perspective, suggests that the full implications of sea-use change, such as the introduction of offshore wind farms, can only be understood first by achieving a complete understanding of the system in question, including the socio-cultural dimension, and second by considering cascading effects across the different system domains. Due to the lack of tangible evidence, socio-cultural values and the benefits arising from them have often been neglected as a sub-domain in their own right. More importantly, lack of awareness of socio-cultural values also leads to neglecting the essential role these values can play in linking the ecological and socio-economic elements of the system. Here, we use the concept of 'cultural ecosystem services' set out in the Millennium Ecosystem Assessment (2005) as a way of expressing and classifying socio-cultural values in a marine setting. As we will show below, changes in such values and corresponding cultural ecosystem services can cause changes in the socio-cultural domain, acting as a trigger of potentially subtle, but potentially large changes in human well-being both for individuals and society.

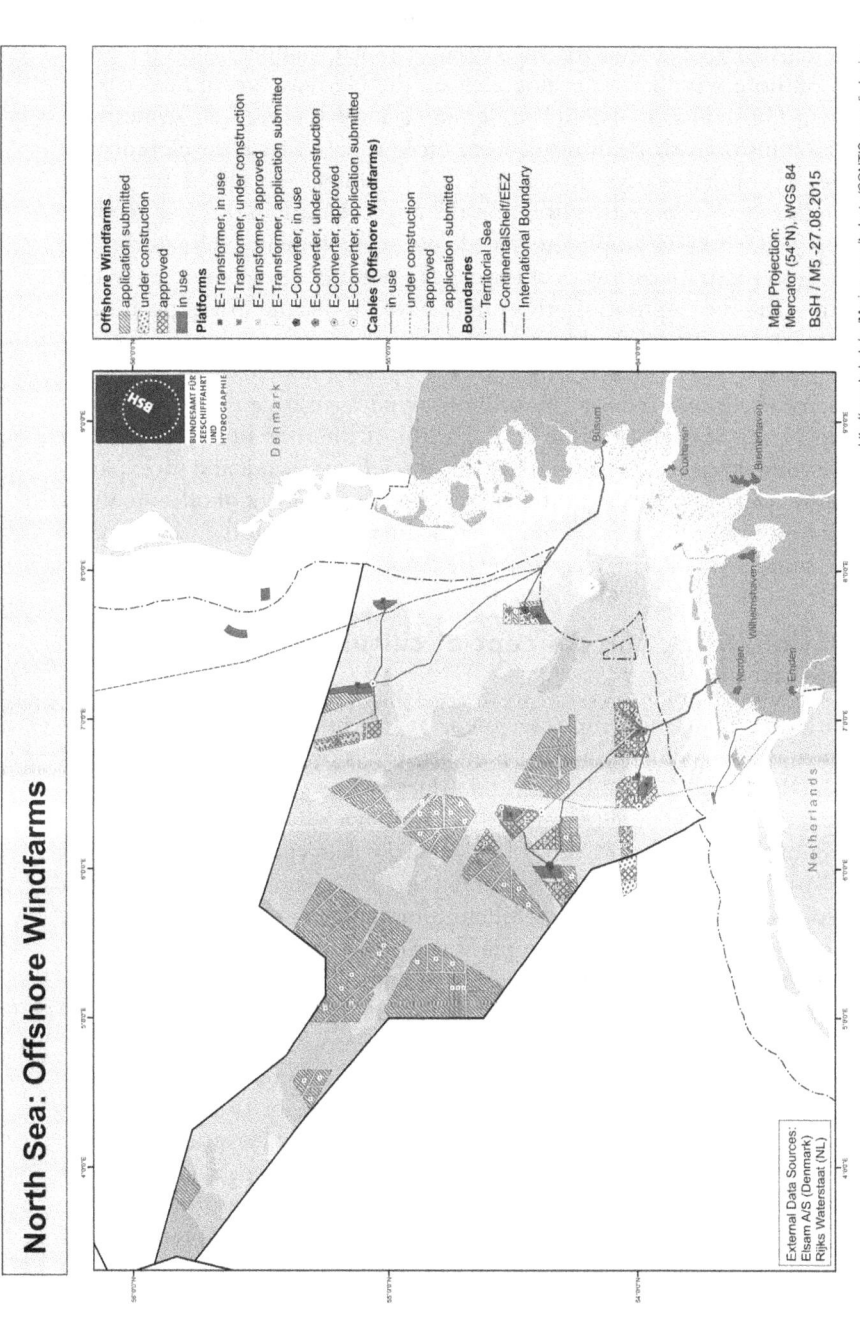

North Sea: Offshore Windfarms

Offshore Windfarms
- application submitted
- under construction
- approved
- in use

Platforms
- E-Transformer, in use
- E-Transformer, under construction
- E-Transformer, approved
- E-Transformer, application submitted
- E-Converter, in use
- E-Converter, under construction
- E-Converter, approved
- E-Converter, application submitted

Cables (Offshore Windfarms)
- in use
- under construction
- application submitted

Boundaries
- Territorial Sea
- ContinentalShelf/EEZ
- International Boundary

Map Projection:
Mercator (54°N), WGS 84
BSH / M5 -27.08.2015

External Data Sources:
Elsam A/S (Denmark)
Rijks Waterstaat (NL)

http://www.bsh.de/en/Marine_uses/Industry/CONTIS_maps/index.jsp

Figure 7.1 The case study area, encompassing the German Exclusive Economic Zone and the districts of North Frisia and Dithmarschen on the Schleswig-Holstein coast. The map also gives an overview of German offshore wind farms in the North Sea, showing those in use, under construction, approved and for which applications have been submitted (BSH, 2015).

Against this background, we make a case for greater consideration of socio-cultural values in assessing the impacts of new marine developments and the resilience of coastal-marine systems to such change. Using the prospect of offshore wind farm development on the German North Sea coast of Schleswig-Holstein as an example of sea-use change (Figure 7.1), we describe potential transitions that can result from offshore wind farm development at the level of cultural ecosystem services and the knock-on effects this could have on the socio-economic system on the coast. Although our case study is based on field work, we emphasise that elements of it remain exploratory. Our purpose is to illustrate pathways of impact as a result of interconnected system domains—and the implications this might have for marine spatial planning.

Our work is based on the results of the integrated research project *Zukunft Küste – Coastal Futures*, which aimed to develop a methodology for integrated assessment of changes induced by offshore wind farming on the Schleswig-Holstein North Sea coast (Lange et al., 2010). At the time of the project, no offshore wind farms had yet been built, so *Coastal Futures* worked with exploratory future scenarios assuming different scales of development of offshore wind farms in the case study area. A more detailed description of the systems perspective summarised in this chapter can be found in Burkhard and Gee (2012).

Operationalising the concept of cultural ecosystem services

Currently, including non-use (non-monetary or immaterial) values in environmental management and decision-making is debated in industrialised societies (Barrena et al., 2014). Cultural values are seen as important contributors to human well-being (Kumar, 2010) and they have gained prominence on account of the cultural ecosystem services concept, which is defined in the Millennium Ecosystem Assessment as the "... non-material benefits people obtain from ecosystems" (Millennium Ecosystem Assessment, 2005). Many other definitions have been put forward (Milcu et al., 2013), reflecting diverse understandings of the nature of cultural values and the ways in which they contribute to human well-being, but despite this variety, cultural ecosystem services are generally codified according to similar, broad categories comprising, for instance, sense of place, recreational, aesthetic, education/scientific, cultural/historical, inspirational, or spiritual services (Millennium Ecosystem Assessment, 2005). Standardised classifications have been proposed, but as cultural ecosystem services have specific meanings to different people (Barrena et al., 2014), these categories are increasingly understood as a broad framework that can and should be adapted to local contexts (ICES, 2013).

A characteristic of cultural ecosystem services is their intangibility, which is one explanation for their poor quantification and integration in management plans. Methodological gaps (Bieling and Plieninger, 2013) lead to difficulties in identifying cultural ecosystem services and a subsequent failure to evaluate them (Fletcher et al., 2014). Terminological issues compound this problem, because there is no clear-cut differentiation between cultural values, cultural ecosystem services, or the cultural benefits derived from services. Cultural ecosystem services are commonly understood as objects of value, such as the beauty of a place or cultural heritage, but more recently, they have also been defined as environmental settings – in other words, places or landscapes – that give rise to the cultural goods and benefits that people obtain from ecosystems (UK Ecosystem Assessment, 2010). Recreation, ecotourism and aesthetic values are the most frequently investigated subcategories of cultural ecosystem services (Milcu et al., 2013) because these are the "… most tangible of the intangibles" (Chan et al., 2012). The case study to which we refer in this chapter is also based on the Millennium Ecosystem Assessment definition of cultural ecosystem services as the non-material benefits people obtain from ecosystems.

In the marine context, challenges of quantifying and valuing cultural ecosystem services are magnified due to the intangible nature of the medium itself, and the observation that human uses tend to be fleeting or delocalised. As the sea is no classic dwelling place, human relationships to the sea can be more remote, but are still potentially intense in the case of certain user or interest groups (e.g., fishermen, recreational users). A problem is that intangible benefits might not always be related to distinct marine places, but refer to a more general experience or 'sense' of the sea, much like other remote places of perceived wilderness that are not personally visited, but still highly valued.

Many studies have elicited cultural ecosystem services or worked with the concept in a marine setting. For example, Liquete et al. (2013) discussed coastal protection as an ecosystem service, while Jobstvogt et al. (2014) looked at cultural ecosystem services values in the context of marine protected areas, and Ruiz-Frau et al. (2013) assessed the economic importance and spatial distribution of cultural ecosystem services related to biodiversity and recreation (such as seabird watching). Nevertheless, the problem persists of how to assess the relative importance of cultural ecosystem services in the context of other values, and how to incorporate such evaluation subsequently in decision-making (ICES, 2013). In a systems context, no clear method has yet been applied to trace changes in the provision of marine cultural ecosystem services and to evaluate the impacts of these changes across system domains. It is the latter issue we address in this chapter.

The Schleswig-Holstein case study example

Offshore wind farming in the German North Sea

Offshore wind farming is an example of a recent marine development with the potential to impact the social-ecological systems concerned (Gill, 2005; Punt et al., 2009). In Germany, offshore wind farming is part of a wider drive to expand renewables. Initially, this aimed to reduce CO_2 emissions, reduce dependency on energy imports, and boost job creation; it was also thought that economically weak coastal regions could benefit from an expansion of wind energy (BMU, 2007). The official policy of "Energiewende" (energy transition), introduced in 2010 to move towards an energy system based on 60% renewable energy by 2050 (BMWi, 2010), gave an added boost to renewables, for example, by amending the feed-in tariff for renewable electricity and increasing investment in research. Renewed attention was also given to the offshore wind sector. Funding was made available to construct the first 10 offshore wind turbines to understand the technical risks and facilitate financing (BMWi, 2010); at the same time, the country's first marine plan for the Exclusive Economic Zone came into force in 2009 with large-scale priority areas set aside for offshore wind.[1] The government's current energy concept aims to achieve an installed capacity of 25,000 MW in the North Sea and Baltic Sea by 2030 (BSH, online at www.bsh.de).

Germany currently[2] has a total of 947 grid-connected, offshore turbines that collectively deliver 4,107 MW (Sun and Wind Energy, 2017). Offshore expansion is set to continue with about 1,400 MW in 2017 followed by a steady average of around 1,000 MW per year until 2019 (Sun and Wind Energy, 2017). Figure 7.1 shows the spatial extent of offshore wind farming in the German part of the North Sea. Note the large areas for which planning applications have been submitted. If these plans are realised, offshore wind farms could cover a large part of Germany's Exclusive Economic Zone, introducing new dynamics to the social-ecological system. Below, we use the coastal districts of Dithmarschen and North Frisia on Schleswig-Holstein's West Coast to illustrate potential system responses. North Frisia is located just below the Danish border; Dithmarschen borders North Frisia to the south around the town of Büsum (Figure 7.1).

Coastal districts of Dithmarschen and North Frisia

To understand local responses to the potential introduction of offshore wind farming, some background information on the two case study districts is necessary. The history of Dithmarschen and North Frisia and the lives of their inhabitants are inextricably linked with the sea. The greater part of today's Dithmarschen and North Frisia is a human-made landscape, arising from centuries of land reclamation, intermittent losses and renewed protection by a succession of sea dykes. Local relationships to the coast and sea can be summarised as a mixture of respect and deeply felt attachment, which is only partly comparable to notions of the 'romantic' or 'picturesque' sea. The latter began to emerge in the 17th Century (Corbin, 1994) and retain a strong presence in the contemporary imagery of the North Sea coast as a holiday destination. The Wadden Sea, protected as a National Park since 1985, became an important element in the region's self-perception only in recent years, adding a widespread view of the sea as a largely natural environment.

Socio-economically, both districts are structurally vulnerable. Ranked as some of the remotest regions of Schleswig-Holstein, unemployment rates are high and the average household income is low compared to the rest of Germany (Statistisches Amt für Hamburg und Schleswig-Holstein, 2007). Demographic factors (e.g., a high proportion of older residents) also add structural vulnerability, as does the embeddedness of the region in a hierarchical, institutional framework and formal modes of governance (Bruns and Gee, 2009). Economically, the districts rely strongly on tourism in terms of employment and income generated (Business Development Corporation Nordfriesland, www.wfg-nf.de), although agriculture and onshore wind farming also play a role. The Wadden Sea National Park and seascape is a factor in marketing (Hasse, 2007); as a result, some local people and tourism operators are suspicious of any potential threats to the seascape (Gegenwind Sylt, www.gegenwind-sylt.de).

Onshore wind farming became an economic player in the 1990s due to the confluence of various enabling factors, including the presence of local innovators and locally based wind-farm enterprises (Suesser et al., 2017). The density of onshore wind turbines in Dithmarschen is one of the highest in Schleswig-Holstein (Grüner Kreisverband Dithmarschen, 2009), and the region is gradually developing towards an energy-based economy (Suesser et al., 2017). Together with a high proportion of citizen-owned wind farms (Business Development Corporation Nordfriesland, www.wfg-nf.de), this goes some way towards explaining the generally high acceptance of onshore wind farming in the region.

At the same time, residents perceive life in a remote rural region and the beauty of the landscape as strong contributors to the quality of life and their well-being (Gee 2010; Ratter and Gee, 2012). There is therefore some tension between the willingness to embrace new, home-grown economic opportunities, as occurred in the case of onshore wind, and the desire for the region to retain its traditional values and assets, not least those bound up in the seascape. The Wadden Sea National Park and surrounding seascape partly serve as a counterpoint to the highly developed mainland that some residents consider over-developed.

Conceptualising system change

So, how might the social-ecological system 'West Coast' respond to the introduction of offshore wind farming? Figure 7.2 is a conceptual diagram of the argument we present here. It shows the possible points at which the social-ecological system could either demonstrate a resilient response, or where a regime shift might occur—points that are also important to marine spatial planning in that it should take into account the different system trajectories that might ensue. The first potential regime shift could take place in the marine ecosystem as a result of introducing hard wind turbine structures into the sea (Chapter 8). This could lead to changes in the provision of ecosystem

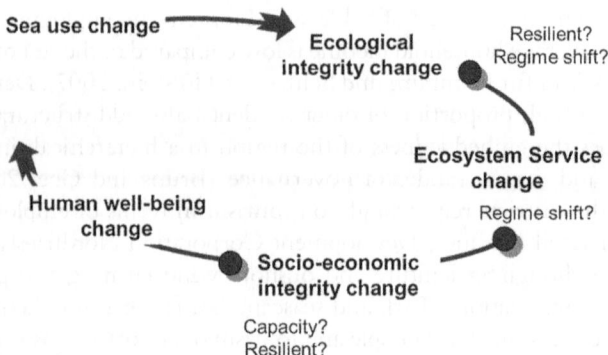

Figure 7.2 Conceptual model of the social-ecological system's response to the introduction of offshore wind farming.

services, including cultural ecosystem services. If substantial enough, these shifts could in turn impact the socio-economic integrity of the system, although this would depend on the capacity of the socio-economic system to absorb such change. Impacts on human well-being would be the final stage.

Establishing the resilience of the marine ecosystem

To assess the ecological impacts of introducing hard structures into the sea, the Coastal Futures project applied the concept of ecological integrity, which uses ecological indicators related to (1) the energy budget of the system (energy capture, entropy production), (2) the matter balance of the system (nutrient cycling, nutrient loss, storage capacity), and (3) structural components (biotic diversity, abiotic heterogeneity) as measures of ecosystem state and change (Lenhart et al., 2010; Burkhard et al., 2011). Ecological models showed that some of the selected indicators are sensitive to disturbances during the construction of offshore wind farms, but that many ecosystem processes and structures soon return to a near-original state after construction; this indicates resilient system behaviour in the short term. There were some indications that an artificial reef system might develop in the long term, but model simulations did not cover periods long enough to illustrate such long-term dynamics sufficiently. One exception was seabird diversity (see also Chapter 8), where permanent habitat loss due to avoidance was likely for selected species (Lenhart et al., 2010). If artificial reefs did indeed develop in the long term, this could benefit commercial fish species (Wilhelmsson et al., 2006) and through this, the fishery sector's welfare (Lenhart et al., 2010). Thus, one of the tasks of marine spatial planning might be to assess trade-offs between different ecosystem services and values, for instance, weighing seabird habitat losses against potential increases in commercial fish diversity (Lange et al., 2010; see also Chapter 2, 10 and 11).

Establishing the resilience of coastal-marine cultural ecosystem services

Eliciting cultural values associated with the sea in the case study area

To assess cultural ecosystem services in the case study area and the impacts of offshore wind farming, a survey of local residents in selected coastal communities was done as part of the Coastal Futures project ($n = 387$) (see Gee, 2010, for methodological details). To establish the perceived impacts of offshore wind farming, the first step was to establish a baseline understanding of the values commonly associated with the sea and the benefits local residents are drawing from it—presently still without offshore wind farms. The Coastal Futures survey approached this through closed (Likert scale) and open questions, asking local residents in the selected communities to indicate their views and relationships with the sea and what local sea experiences meant to them in everyday life. Responses to open questions were coded into thematic categories and also analysed quantitatively.

Results indicate that physical landscape elements (e.g., the tide, sand, dunes), subjective impressions (e.g., the beauty of the open horizon), past and present experiences, and practices associated with the sea (e.g., informal recreation, past activities related to land reclamation) all contribute to the perception of 'sea'. Some values ascribed to the sea are effectively landscape values (such as the aesthetic qualities of the marshlands), but others, such as the naturalness of the sea, the elementary qualities of the marine environment, or the unpredictability of the sea, are distinctly marine. One aspect particularly appreciated by local residents was the visual aesthetic quality of the seascape. Asked to describe what was treasured most about life on the West Coast, the wide-open horizon came second in the list, topped only by "fresh air and marine climate". Asked then to rank a number of items in terms of importance, the wide, expansive horizon was considered important to life on the West Coast by 58% of respondents (compared to only 13% who considered attracting profitable companies and industry as important). Other important qualities are the recreational and health benefits provided by the sea, the perceived 'naturalness' and 'last wilderness' qualities of the marine environment, and symbolic associations with the sea (such as freedom). The belief that the sea cannot truly be tamed and always represents a latent danger is a value in its own right, allowing identity to be constituted around the communal practice of sea defence, in particular dyking.

Overall, the seascape[3] emerges as an important value, but also as an environmental setting for a range of socio-cultural values and practices. The current expression of the seascape – a wide expanse and largely natural space only rarely interrupted by visible human influence – seems to make an important contribution to the personal well-being of local residents, linked to feelings such as "here I can breathe", "experience of peace and calm" and "let go of everyday stress".[4]

Establishing the potential impacts of offshore wind farming on the value base in the case study area

To establish the potential impacts of offshore wind farming to the values identified above, the second part of the Coastal Futures survey asked respondents to describe the risks and opportunities they associate with offshore wind farming, again based on a mixture of open and closed questions (Gee, 2010). At the time of surveying, no visible offshore wind farms had been built in the case study area, so the survey measured expected rather than encountered risks, and does therefore not reflect real experience.[5] The purpose of the survey was to establish what values residents mostly feel threatened by offshore wind farming and to identify competing values that might lead to mental trade-offs and with these different degrees of acceptance of offshore wind farming.

Overall, attitudes to offshore wind farming revealed a conflict between renewable energy values on the one hand and seascape values on the other. For both supporters and opponents of offshore wind farming, the seascape emerged as an important reference point. Supporters (40.5% of total respondents) did appreciate the seascape in its current 'undeveloped' form, but were willing to accept potential changes to the seascape in return for renewable energy. They considered offshore wind farming important in combating climate change and therefore a contributor to preserving the local area for future generations. Opponents (53.5%) also supported renewables in principle, but they did not consider offshore wind farming a suitable option. Many specifically opposed the 'industrialisation' of the sea they thought it would inevitably bring. Interestingly, this opposition not only referred to potential visual impacts of offshore wind farms, but also to more subtle threats to the perceived 'naturalness' and ecological value of the seascape, suggesting that visibility of offshore wind farms is not the only determining factor. Existence value (the benefit received from knowing that a particular environmental resource exists) plays an important role here, and some opponents expressed a strongly felt duty to "look after the sea" and protect it from undue human influence. Some described the expected impact of offshore wind farming on existence values (and by implication, on their personal well-being) in no uncertain terms, calling the prospect of offshore wind farming "cruel" or even "rape of the sea".

Establishing the response of cultural ecosystem services as a result of introducing offshore wind farming in the case study area

So, what are the impacts of introducing offshore wind farming on cultural ecosystem services in the case study area? To illustrate the range of impacts that could result, the values drawn from the survey were first used to adapt the cultural ecosystem services categories from the Millennium Ecosystem Assessment (2005) to produce a localised classification (Table 7.1). Additional expert assessment by scientists familiar with the case study area was then brought in to supplement survey results and describe potential impacts. Again, these are necessarily subjective assessments of potential developments based on the Coastal Futures results, rather than actual observations. Table 7.1 indicates that potential losses could occur with respect to intrinsic sea values, i.e., a loss of habitat and species values by losing seabird diversity, for example. Losses could also occur in the visual aesthetic category, although the latter would depend on the visibility of offshore wind farms from the coast. Only two of the respondents described offshore wind turbines as beautiful in their own right, indicating that a different idea of seascape beauty linked to a shift in attitude to offshore wind farms might take time to emerge. In all other categories, both losses and gains could result. For example, offshore wind farming could bring new recreational opportunities (such as tourist trips to offshore wind farms), give the region a stronger sense of place as an energy region, or add to existing knowledge systems. Losses in these categories would occur if offshore wind farming displaces existing values without offering a viable alternative. Acceptance of the resulting value shifts thus ultimately depends on whether the gains (e.g., renewable energy or biodiversity) are considered more valuable than the losses (e.g., the traditional seascape).

Knock-on effects on the socio-economic system

Offshore wind farming would lead to a regime shift in the seascape in that it would be modified from its current status as 'natural' towards a more visibly 'cultural' seascape comprising both industrialised and natural elements. A return of the seascape to its previous state would not be possible unless offshore wind farms were dismantled again. At the individual level, the interpretation of this regime shift as more or less dramatic is likely to vary among people, although careful siting of offshore wind farms at some distance from the coast could help preserve some of the former attributes of the seascape, allowing it to still be interpreted as a largely natural seascape. As described in Table 7.1, a semi-industrial seascape could conceivably offer a greater range of cultural ecosystem benefits as both traditional and novel functions of the seascape could be realised. If, on the other hand, the shift goes too far towards a purely industrial seascape, the natural seascape and its associated values (such

Table 7.1 Cultural ecosystem services (adapted to the local case study area from Millennium Ecosystem Assessment (2005) based on survey responses and expert assessment), and their response to the introduction of offshore wind farming

Cultural ecosystem service	Rationales behind expected system responses
Visual aesthetics	If visible, offshore wind farms would add a new element to the seascape, affecting the visual qualities of the sea (wide open horizon).
Seascape character	Offshore wind farms shift the character of the seascape from a largely natural to a more industrial landscape.
Sense of place	Offshore wind farming is incompatible with the desire to keep the sea free of industrial structures and challenges the traditional view of the seascape/area. Aspects of control and decision-making processes could be important. Indirect impacts, e.g. helicopter flights, could detract from what are now considered essential elements of the area, e.g. peace and quiet, remoteness.
Cultural heritage	In the long term, offshore wind farms could become an accepted element of the cultural landscape. Short-term impacts could include the destruction of archaeologically important sites (bad siting of offshore wind farms).
Habitat and species value	Offshore wind farms could fundamentally change natural habitats and impact on bird and mammal species. offshore wind farms is perceived as a potential threat to intrinsic natural values.
Regional image	Offshore wind farm can make a positive contribution to the region's image in helping it to modernise. Offshore wind farming can also be negative if it is seen to detract from essential traditional qualities. The nature of the impact depends on the choices the region makes for its future.
Inspiration	In adding a new element to the environment, offshore wind farms can act as a source of inspiration or detract from previous sources of inspiration. The effect is likely to be stronger the more visible offshore wind farms would be.
Informal education	Offshore wind farming represents a new topic that can be added to existing informal education issues.
Knowledge systems	In the mid- to long term, offshore wind farming could bring new local and scientific knowledge to the region, i.e. technology transfer, accumulation of more and different types of knowledge.
Recreation	Impacts could include new recreational activities such as trips to wind farms. offshore wind farms may impact on personal feelings of well-being. If visible, leisure may be affected, e.g. enjoying the open horizon.

as existence value) would be irretrievably lost. Survey results show that some local residents would equate this with a perceived loss of quality of life, primarily on account of a rather conservative value base and a deeply felt conflict between offshore wind farming and intrinsic sea values ("the sea should not be industrialised"). It is fair to assume that such losses could be acute initially, but it is also likely that habituation would occur and that the sense of loss would lessen as people gradually adapt. The adaptive capacity of individuals – or the willingness to accept such change – will partly depend on the emergence of alternative cultural ecosystem benefits that might compensate for the perceived losses, or the ability and willingness to embrace a new regional identity based on renewable energies (Burkhard and Gee, 2012). Wider social change could also play a role that begins to regard offshore wind farming not only as a necessary imposition, but also an intrinsic and desirable part of home.

Impacts could also manifest themselves at the socio-economic level. Adopting onshore wind farming early has contributed to diversifying the local economy in the case study area, decreasing its structural vulnerability by offering employment and opportunities for young people to remain in the area. These same opportunities have not been seized for offshore wind farming. Currently, the economic benefits of offshore wind farming (such as port facilities, servicing and maintenance of offshore wind turbines) are mainly generated outside the case study region, for example, in other coastal regions along the Lower Saxony coast. As a result, the region has missed out on an opportunity to stabilise its economic system further. Whether tourism would take a downturn on account of offshore wind farming is difficult to predict; it might ultimately depend on the scale of offshore wind farm development and how offshore wind farming is used in communication and regional marketing. Given the high overall support of renewable energies in Germany, it is conceivable that the image of a 'natural seascape' could coexist with that of a 'renewable energy seascape' or 'renewable energy region'. Tourist numbers did not decline on account of onshore wind farming in the case study region (Vogel, 2005); tourists could therefore be expected to make this transition readily. They could even generate new opportunity for the socio-economic system, for example, by taking boat tours specifically to visit offshore wind farms.

Linking results to marine spatial planning

What are the lessons this case study holds for marine spatial planning? A first lesson is that cultural values can, and indeed should, be included in marine spatial planning. In our case study, a local residents' survey described the personal values people associate with the sea, which were then categorised as cultural ecosystem services in a locally tailored classification. In our case, the seascape emerged as a key to understanding cultural ecosystem services, although we did not fully unravel this conceptually or suggest whether the seascape should be understood as a cultural ecosystem service in itself or is best described as a cultural ecosystem services-providing unit. More work to improve the conceptual understanding of seascape would be helpful, for example, by drawing on recent landscape research and related concepts such as place. Approaches such as seascape character assessments (LUC, 2015) or 'culturally significant' area assessments (ICES, 2013) can be valuable first steps towards a more comprehensive inclusion of seascape-related, place-based, cultural values in marine spatial planning. We note, however, that not all cultural values associated with the sea are distinctly place-based, and that other value ascriptions such as moral values or notions of justice could be equally important.

Apart from the diversity of values, our case study underlines the importance immaterial values can have to individuals and the role of value conflicts in determining acceptance of sea-use change. Irrespective of whether these values are classed as cultural ecosystem services, the sea is a complex mental construct composed of many layers of perception, experience and objects of value; various benefits arise from the sea and the seascape that clearly impact individual quality of life. Which values are rated more highly than others partly depends on an individual's internal held values and 'moral compass', and partly on pragmatic situational considerations – such as whether they can profit financially from offshore wind farming (Gee, 2010). Marine planning could benefit from improved understanding of, (1) the values people ascribe to different elements and experiences of the sea, (2) the risks these values might be exposed to on account of new developments, and (3) potential value conflicts and trade-offs between values necessary for understanding people's willingness and ability to accept a change in cultural ecosystem services and individual benefits to be drawn from the sea. As we showed, value conflicts can arise within individuals, such as when values related to the 'natural' seascape conflict with the belief that renewable energy is important and necessary, but also among individuals and groups for many of the same reasons. If a particular value is important enough (perhaps on account of its uniqueness or long-standing traditional significance), and the threat to that value is strong enough (e.g., its irreplaceable loss), action can ensue, such as organised protest or other forms of resistance, although our case study did not go as far as investigating the likelihood of such action in the case of offshore wind.

Our case study also shows that cultural ecosystem services, and the benefits they yield, represent an important link between the ecological and socioeconomic domains of social-ecological systems. As such, they are important for understanding the full dynamics that arise from the introduction of new sea uses. A social-ecological systems perspective can lead to better understanding of the consequences of change, highlighting how impacts in one domain can lead to a resilient response or regime shifts in other system domains. Our case study did not provide a detailed investigation of the factors contributing to resilience in each domain, as our intention was merely to show pathways of impact; this could be a subject for further research.

The case study example has also shown that risk and vulnerability analyses need to be more than economic cost-benefit analyses. They also need to account for immaterial values and structural vulnerabilities in the socio-economic sphere because both can play important roles in shaping a local community's response to change. Although not shown in our specific case study, this became apparent when a management plan for the Wadden Sea National Park was implemented in the 1990s without sufficient local consultation; local fishermen in particular perceived this as a threat to their livelihoods and heritage in times when they were already economically vulnerable.

Overall, our chapter highlights both the opportunities and limits of marine spatial planning. Opportunities exist to give greater recognition to cultural values, allowing marine spatial planning to improve its role as a balancing force and steward of sustainable marine development. Local communities and communities of interest can have deep emotional connections to the sea; presently, these are easily overlooked in marine spatial planning because they do not have an organised voice or the bargaining power of other sectors. As an arbiter of all marine stakeholders (see Chapters 1, 4 and 9), marine spatial planning could do more to account for the sea as part of everyday life worlds by proactively eliciting marine cultural ecosystem services and the benefits arising from these to local communities and wider society. It could do so in an organised way, using cultural value baseline surveys and sensitivity analyses to build up solid evidence for planning decisions, much like it uses ecological or other baseline knowledge on ecosystem sensitivity in planning decisions. Bringing to the table such essential local knowledge might help to counter yet another problem, which is that competencies for spatial and ecological decision-making in the sea usually lie with the regional and national, and not the local, level. There is a danger of focusing on the 'bigger picture' of *blue growth* (defined by the European Union as the long-term strategy to support sustainable growth in the marine and maritime sectors), at the expense of other values in coastal regions. It might be difficult for the affected regions to compensate for these losses at least in the short term. If it were to address cultural ecosystem services rigorously, marine spatial planning could be in a good position to redress the disparity between national and local issues, as well as between a more rational and more emotional understanding of the sea.

At the same time, the systems perspective also shows the limits of marine spatial planning in that a marine plan only provides a framework for developments in coastal regions. Regions could do more to understand that they too can act as drivers of change, and that they might even benefit from regime shifts in the sea.

Highlights

- A resilience perspective can help to understand the full implications of sea use change in inherently complex social-ecological, coastal-marine systems.
- More prominently accounting for socio-cultural values can strengthen risk and impact assessments related to sea-use change.
- Systematic assessment of local cultural values can strengthen the evidence for marine spatial planning and bridge the gap between local and national interests.

Notes

1 This was followed in 2012 by an offshore grid plan (BSH, online).
2 As of December 2016.
3 We use the term 'seascape' as analogous to 'landscape', denoting a marine area as perceived by people, whose character is the result of the action and interaction of natural and/or human factors (adapted from the European Landscape Convention (2000), Art. 1).
4 Mention was also made of the role of the seascape in tourism, e.g., in connection with the Wadden Sea National Park, indicating that material benefits are also derived from the seascape.
5 The survey can be assumed to reflect people's experience with onshore wind farming.

References

Barrena, J., Nahuelhual, L., Báez, A., Schiappacasse, I., and Cerda, C. (2014). Valuing cultural ecosystem services: Agricultural heritage in Chiloé Island, southern Chile. *Ecosystem Services*, 7, pp. 66–75. doi:http://dx.doi.org/10.1016/j.ecoser.2013.12.005.

Bieling, C., and Plieninger, T. (2013). Recording manifestations of cultural ecosystem services in the landscape. *Landscape Research*, 38(5), pp. 649–667. doi:10.1080/01426397.2012.691469.

BMU (Federal Ministry for the Environment). (2007). Offshore wind power deployment in Germany. Federal Ministry for the Environment, Nature Conservation and Nuclear Safety (BMU) and Offshore Wind Energy Foundation. January 2007.

BMWi (Federal Ministry of Economics and Technology). (2010). *Energy concept for an environmentally sound, reliable and affordable energy supply*. 28. Available at www.osce.org/eea/101047?download=true.

Bruns, A. and Gee, K. (2009). From state-centered decision-making to participatory governance: planning for offshore wind farms and implementation of the Water Framework Directive in Northern Germany. *GAIA*, 18(2), pp. 150–157.

Burkhard, B., Opitz, S., Lenhart, H., Ahrendt, K., Garthe, S., Mendel, B., and Windhorst, W. (2011). Ecosystem based modeling and indication of ecological integrity in the German North Sea – case study offshore wind parks. *Ecological Indicators*, 11, pp. 168–174. http://dx.doi.org/10.1016/j.ecolind.2009.07.004.

Burkhard, B., and Gee, K. (2012). Establishing the resilience of a coastal-marine social-ecological system to the installation of offshore wind farms. *Ecology and Society*, 17(4), pp. 32. Available at www.ecologyandsociety.org/vol17/iss4/art32/.

Busch, M., Buisson, R., Barrett, Z., Davies, S., and Rehfisch, M. (2015). Developing a habitat loss method for assessing displacement impacts from offshore wind farms. JNCC Report 551, Peterborough. (ISSN 0963 8091). Available at http://jncc.defra.gov.uk/page-6987.

Busch, M., Gee, K., Burkhard, B., Lange, M., and Stelljes, N. (2011). Conceptualizing the link between marine ecosystem services and human well-being: The case of offshore wind farming. *International Journal of Biodiversity Science, Ecosystem Services & Management*, 7(3), pp. 190–203. doi:10.1080/21513732.2011.618465.

BSH online. *Spatial offshore grid plan*. Available at www.bsh.de/en/Marine_uses/BFO/index.jsp

BSH. (2009). Ordinance on spatial planning in the German Exclusive Economic Zone in the North Sea (AWZ Nordsee-ROV) of September 21st 2009 (unofficial translation). http://www.bsh.de/en/Marine_uses/Spatial_Planning_in_the_German_EEZ/documents2/ordinance_north_sea.pdf

BWE, online. (2015). *Offshore Wind Energy Half-Year Figures 2015 in Germany - Expansion on schedule: 1, 765 megawatts more on line*. Available at www.wind-energie.de/en/press/press-releases/2015/offshore-wind-energy-half-year-figures-2015-germany-expansion-schedule

Chan, K.M.A., Satterfield, T., and Goldstein, J. (2012). Rethinking ecosystem services to better address and navigate cultural values. *Ecological Economics*, 74, pp. 8–18. doi:http://dx.doi.org/10.1016/j.ecolecon.2011.11.011.

Church, A., Fish, R., Haines-Young, R., Mourato, S., Tratalos, J., Stapleton, L., Willis, C., Coates, P., Gibbons, S., Leyshon, C., Potschin, M., Ravenscrof T, N., Sanchis-Guarner, R., Winter, M., and Kenter, J. (2014). UK National Ecosystem Assessment Follow-on. Work Package Report 5: Cultural ecosystem services and indicators. UNEP-WCMC, LWEC, UK.

Corbin, A. (1994). Meereslust. Das Abendland und die Entdeckung der Küste. Frankfurt am Main, Fischer Taschenbuch.

Döring, M., Settekorn, W., and Von Storch, H. (2005). *Küstenbilder, Bilder der Küste. Interdisziplinäre Ansichten, Ansätze und Konzepte*. Hamburg: Hamburg University Press.

Douvere, F., and Ehler, C. (2009). New perspectives on sea use management: Initial findings from European experience with marine spatial planning. *Journal of Environmental Management*, 90(1), pp. 77–88.

Fletcher, R., Baulcomb, C., Hall, C., and Hussain, S. (2014). Revealing marine cultural ecosystem services in the Black Sea. *Marine Policy*, 50, pp. 151–161. doi:http://dx.doi.org/10.1016/j.marpol.2014.05.001.

Gee, K. (2010), Offshore wind power development as affected by seascape values on the German North Sea coast. *Land Use Policy*, 27, pp. 185–194.

Gill, A.B. (2005). Offshore renewable energy: ecological implications of generating electricity in the coastal zone. *Journal of Applied Ecology*, 42(4), pp. 605–615.

Gunderson, L.H. and Holling, C.S. (2002). *Panarchy: understanding transformations in human and natural systems*. Washington, DC: Island Press.

Grüner Kreisverband Dithmarschen. (2009). *Es reicht*. www.gruene-dithmarschen.de/archiv/windenergie-ohne-ende/.

Hasse, J. (2007). "Nordseeküste" – Die touristische Konstruktion besserer Welten. Zur Codierung einer Landschaft [in German]. In: N. Fischer, S. Müller-Wusterwitz and B. Schmidt-Lauber, eds., *Inszenierungen der Küste*. Berlin: Reimer-Verlag, pp. 239–258.

ICES. (2013). Report of the joint HZG/LOICZ/ICES workshop: Mapping cultural dimensions of marine ecosystems (WKCES), 17–21 June 2013, Helmholtz Zentrum Geesthacht, Germany. ICES CM 2013/SSGHIE:07.

Jobstvogt, N., Watson, V., and Kenter, J.O. (2014). Looking below the surface: The cultural ecosystem values of UK marine protected areas. *Ecosystem Services,* 10, pp. 97–110.

Kenter, J.O., Reed, M. S., Irvine, K.N., O'Brien, E., Brady, E., Bryce, R., Christie, M., Church, A., Cooper, N., Davies, A., Hockley, N., Fazey, I., Jobstvogt, N., Molloy, C., Orchard-Webb, J., Ravenscroft, N., Ryan, M., and Watson, V. (2014). UK National Ecosystem Assessment Follow-on. Work Package Report 6: Shared, Plural and Cultural Values of Ecosystems. UNEP-WCMC, LWEC, UK.

Kinzig, A. P., Ryan, P., Etienne, M., Allison, H., Elmqvist, T., and Walker, B.H. (2006). Resilience and regime shifts: Assessing cascading effects. *Ecology and Society*, 11(1), pp. 20. Available at http://www.ecologyandsociety.org/vol11/iss1/art20/.

Kumar, P. (2010). *The economics of ecosystems and biodiversity – ecological and economic foundations*. London: Earthscan.

Lange, M., Burkhard, B., Garthe, S., Gee, K., Kannen, A., Lenhart, H., and Windhorst, W. (2010). Analyzing coastal and marine changes: offshore wind farming as a case study. *LOICZ Research & Studies* No. 36. GKSS Research Centre, Geesthacht, Germany.

Lenhart, H., Ahrendt, K., Burkhard, B., Garthe, S., Gloe, D., Kühn, W., Mendel, B., Nerge, P., Opitz, S., and Schmidt, A. (2010). Analyzing coastal and marine changes: Offshore wind farming as a case study. [pdf] pp. 52–96. Available at www.researchgate.net/profile/Marcus_Lange/publication/233932674_Analyzing_Coastal_and_Marine_Changes_Offshore_Wind_Farming_as_a_Case_Study/links/0fcfd50d1829574f8e000000/Analyzing-Coastal-and-Marine-Changes-Offshore-Wind-Farming-as-a-Case-Study.pdf

Liquete, C., Zulian, G., Delgado, I., Stips, A., and Maes, J. (2013). Assessment of coastal protection as an ecosystem service in Europe. *Ecological Indicators*, 30, pp. 205–217.

LUC. (2015). National Seascape Assessment for Wales. NRW Evidence Report 80.

Luisetti, T., Turner, R.K., Jickells, T., Andrews, J., Elliott, M., Schaafsma, M., Beaumont, N., Malcolm, S., Burdon, D., Adams, C., and Watts, W. (2014). Coastal zone ecosystem services: From science to values and decision making; a case study. *Science of the Total Environment*, 493, pp. 682–693.

Martinez, G., Orbach, M., Frick, F., Donargo, A., Ducklow, K., and Morison, N. (2014). The cultural context of climate change adaptation: Cases from the U.S. East Coast and the German Baltic Sea coast. In: Martinez, G., Fröhle, P., and Meier, H.-J., ed., *Social Dimensions of Climate Change Adaptation in Coastal Regions*. Washington, DC: World Bank Group, pp. 85–100.

Millennium Ecosystem Assessment (Millennium Ecosystem Assessment). (2005). *Ecosystems and human well-being: synthesis report*. Washington, DC: Island Press.

Milcu, A. I., Hanspach, J., Abson, D., and Fischer, J. (2013). Cultural ecosystem services: A literature review and prospects for future research. *Ecology and Society*, 18(3), pp. 44. doi:/10.5751/ES-05790-180344.

Nunneri, C., Lenhart, H., Burkhard, B., and Windhorst, W. (2008). Ecological risk as a tool for evaluating the effects of offshore wind farm construction in the North Sea. *Regional Environmental Change*, 8, pp. 31–43.

Oil and Gas UK. (2014). *Economic report 2014*. Available at http://oilandgasuk.co.uk/wp-content/uploads/2015/05/EC041.pdf

Oxford Maritime. (2015). *The economic impact of the UK Maritime Services Sector*. Report commissioned by Maritime UK, May 2015.

Petersen, J. K., and Malm, T. (2006). Offshore windmill farms: threats to or possibilities for the marine environment. *Ambio*, 35(2), pp. 75–80.

Punt, M. J., Groeneveld, R. A., Van Ierland, E. C., and Stel, J. H. (2009). Spatial planning of offshore wind farms: A windfall to marine environmental protection? *Ecological Economics*, 69, pp. 93–103.

Ratter, B.M.W., and Gee, K. (2012). Heimat - a German concept of regional perception and identity as a basis for coastal management in the Wadden Sea. *Ocean and Coastal Management*, 68, pp. 127–137.

Redman, C.L., and Kinzig, A.P. (2003). Resilience of past landscapes: Resilience theory, society, and the Longue Durée. *Conservation Ecology*, 7(1), p. 14. Available at www.consecol.org/vol7/iss1/art14

Ruiz-Frau, A., Hinz, H., Edwards-Jones, G., and Kaiser, M.J. (2013). Spatially explicit economic assessment of cultural ecosystem services: Non-extractive recreational uses of the coastal environment related to marine biodiversity. *Marine Policy*, 38, pp. 90–98.

Statistisches Amt Für Hamburg Und Schleswig-Holstein. (2005). Statistischer Bericht A. I. 1S, Bevölkerungsentwicklung in den Gemeinden Schleswig-Holsteins.

Suesser, D., Doering, M., and Ratter, B.M.W. (2017). Harvesting energy: Place and local entrepreneurship in community-based renewable energy transition. *Energy Policy*, 101, pp. 332–341.

Sun and Wind Energy. (2017). *Expansion figures for offshore wind power in Germany*. Available at www.sunwindenergy.com/offshore-wind-energy/expansion-figures-offshore-wind-power-germany

Turner, K., Schaafsma, M., Elliott, M., Burdon, D., Atkins, J., Jickells, T., Tett, P., Mee, L., Van Leeuwen, S., Barnard, S., Luisetti, T., Paltriguera, L., Palmieri, G., and Andrews, J. (2014). UK National Ecosystem Assessment Follow On. Work Package Report 4: Coastal and marine ecosystem services: principles and practice. UNEP-WCMC, LWEC, UK.

UK National Ecosystem Assessment. (2010). Chapter 1: Introduction to the UK National Ecosystem Assessment, Coordinating Lead Authors: Claire Brown and Matt Walpole. UNEP-WCMA, Cambridge.

Vogel, M. (2005). Akzeptanz von Windparks in touristisch bedeutsamen Gemeinden der deutschen Nordseeküstenregion. Eine empirische Untersuchung, durchgeführt vom Studiengang Cruise Industry Management unter der Leitung von Prof. Dr. Michael Vogel. Hochschule Bremerhaven, Institut für Maritimen Tourismus, 8. Dezember 2005.

Walker, B. H., Anderies, J. M., Kinzig, A. P., and Ryan, P. (2007). *Exploring resilience in social-ecological systems*. CSIRO, Collingwood, Victoria, Australia.

Wilhelmsson, D., Malm, T., and Öhman, M.C. (2006). The influence of offshore windpower on demersal fish. *ICES Journal of Marine Science*, 63, pp. 775–784. doi:10.1016/j.icesjms.2006.02.001.

Chapter 8

Environmental implications of offshore energy

Andrew B. Gill, Silvana N.R. Birchenough,
Alice R. Jones, Adrian Judd, Simon Jude,
Ana Payo-Payo and Ben Wilson

Introduction

Activities associated with offshore energy extraction and generation have implications for the surrounding marine environment (Gill 2005; Inger et al. 2009; Boehlert and Gill 2010; Miller et al. 2013), and while our understanding of those implications has improved, particularly for marine megafauna (i.e., large mammals such as whales, dolphins and seals, and large fish such as sharks and some bony fishes), much remains unknown about how offshore energy activities affect marine ecosystems (Inger et al. 2009; Boehlert and Gill 2010). This uncertainty can limit the capacity to environmental changes associated with offshore energy development bring marine spatial planning processes, subsequently reducing confidence in within planning decisions which promote an energy mix to combat climate change.

Most evidence of how offshore energy activities affect the marine environment comes from studies of direct changes or responses (termed 'effects') that can be quantified for only a few species or habitats (known as environmental 'receptors') (Boehlert and Gill 2010). There is also a preponderance of negative effects reported, such as the potential threats of temporary displacement or mortality to megafauna (Simmonds and Brown 2010). Often, the motivation to examine these hypothesised effects is a requirement under an environmental impact assessment (Judd et al. 2015; Willsteed et al. 2017). One important and generally unanswered question is whether the effects observed are ecologically meaningful, such as population-level change resulting from reduced fitness of individual animals (King et al. 2015).

To determine if an ecological change resulting from the development and operation of offshore energy has occurred requires considering both direct (e.g., mortality) and indirect effects (e.g., changes to reproductive output associated with food web change), as well as determining pathways of effects and the spatial and temporal scales over which changes occur (Gill 2005; Boehlert and Gill 2010; Miller et al. 2013). This chapter examines the implications for the environmental receptors of offshore energy and highlight those aspects most in need of addressing for effective marine spatial planning.

Offshore energy planning policy and legislation regarding environmental impacts

Many countries legislate assessments of the environmental effects of offshore energy extraction and generation within the marine environment (e.g., European Directives on Environmental Impact Assessment and Strategic Environmental Assessment; Environmental Protection Laws in China). Within this national legislation, there is often a complex array of regulations covering the exploration for and production of offshore energy resources (e.g., oil, gas, wind, wave, tidal, biomass) and assessing their potential environmental effects (Judd et al. 2015). This complexity of intertwined regulation makes it difficult to provide a detailed and lucid summary of the legislation at national, regional or global scales. The Ocean Energy Systems' reports and website (www.ocean-energy-systems.org) provide summaries of the offshore energy and environment policy and regulatory landscape of 20 member countries across the globe in relation to ocean energy, and therefore represent an indispensable resource. Rather than attempt to describe in detail the legislative requirements of each country involved in offshore energy, we provide an overview of the common requirements, with reference to specific legislation for illustration.

The need for and extent of offshore energy exploration and production are driven by domestic policies on energy demands, energy supply (e.g., national energy resources, energy import requirements, energy export potential), and energy mix (e.g., coal, oil, gas, wind, wave, tidal, nuclear, biomass or solar). Most countries have some form of offshore energy policy (Wright 2014). Common features of these policies that are relevant in the context of planning the integration of offshore energy into the marine environment are the requirements to (1) obtain permits or licences from government authorities to do exploratory surveys to determine the location and value of energy resources, (2) get permission from the seabed owner (state or private) to explore for and produce energy, (3) obtain permits/licences from government to construct infrastructure to extract energy resources (e.g., oil or gas platforms, offshore wind farms, offshore energy converters, subsea power cables), (4) obtain permits/licences from government to operate the energy-generation facility and connect to the domestic transmission grid, (5) obtain permits/licences from government for the use, treatment and discharge of chemicals, (6) do environmental impact assessments, and (7) do some form of environmental monitoring to test the predictions in the environmental impact assessment and/or the consequences of construction and operation of the facility.

For several countries, energy policies are informed at a higher level by strategic environmental assessments that aim to strike a balance between promoting economic development of offshore energy resources and effective environmental protection. One aim here is to achieve this balance through a regional rather than a site-based approach, taking into account multiple demands on the marine environment (see details within WavEC 2015). The outputs of strategic environmental assessments feed into national and local marine spatial planning decisions on current and future opportunities for exploiting marine resources. Where there is commercial interest to exploit the energy reserves that the strategic environmental assessments have indicated could be developed sustainably, the successful applicant is required to do a detailed, site-based environmental impact assessment following a leasing round.

Assessing the environmental impacts of offshore energy

Most offshore energy developers are required through some form of national legislation to do a formal environmental impact assessment that will inform permit or licence decision making (Wright 2014). The purpose of an environmental impact assessment is to identify the potential impacts on the environment from proposed development, determine the biological relevance of those impacts, and to propose options to avoid or reduce the largest impacts. Such an assessment provides an opportunity for the developer to engage with the regulator, other sea users and the public throughout the inception and planning phases of the proposed works, and is intrinsically linked to marine spatial planning for the area.

Environment and ecological effects of offshore energy

When considering the potential changes to the marine environment associated with offshore energy, such as within the environmental impact assessment process, it is imperative to understand the interacting components of the system (Lindeboom et al. 2015). Traditionally, environmental changes have been associated with physical (e.g., built structure or seabed) or chemical (e.g., fluid spill) aspects. The focus has also been on direct and acute effects (Boehlert and Gill 2010); however, many changes can occur indirectly and potentially over long periods (i.e., chronic effects; Gill 2005). These changes are linked to factors influencing ecological change, but can be more substantial than acute effects because they impart ecosystem changes (which can result in alteration of the benefits that humans receive from the ecosystem such as fish for consumption, i.e., change to ecosystem services) in the longer term (i.e. decades; Gill 2005; Inger et al. 2009; Miller et al. 2013).

Determining the outcome of environmental interactions

A useful way to encapsulate environmental changes that occur, regardless of the energy source, is by systematically describing the system and its causal links. Borrowing from the environmental risk-assessment literature, the source-pathway-receptor concept (e.g., Narayan et al. 2012) can be applied to break down the interactions into identifiable sources that cause some change (e.g., underwater noise), the pathways by which the sources have an effect (e.g., sound propagation in water), and identifying likely receivers showing some change (e.g., marine mammals sensitive to underwater sound). Causes and effects of changes associated with offshore energy can then be highlighted (Table 8.1).

If the change is some deviation from normal activity, such as diversion of animals from migration routes (e.g., anguillid eels following subsea cables; Westerberg and Lagenfelt 2008) or short-term displacement behaviour (e.g., harbour porpoise moving away during pile-driving; Thomsen 2010), then this is defined as a 'response' or an 'effect'. However, such a change might not necessarily affect the ecosystem because it could be temporary and fall within the tolerance range of the species in question. However, if multiple effects occur such that there is some overriding or permanent burden to the species (e.g., cumulative effect) or they reduce some component of the population's fitness (e.g., interfering with mating; Christiansen and Lusseau 2015), then a biological impact can result (*sensu* Boehlert and Gill 2010). These impacts are central to the consideration of implications to the marine environment from offshore energy extraction.

Types of energy extraction

Energy from marine resources is required either directly to generate power (e.g., offshore wind farm generating electricity) or indirectly for use as a fuel source (e.g., oil for transport). To consider the environmental changes that can occur from energy extraction, understanding the elements of the extraction process is required. It is also important to identify which of those elements could cause a change in an environment receptor and the pathway of change (i.e., understanding source-pathway-receptor relationship; Table 8.1).

Non-renewable energy sources

Many environmental changes caused by the oil and gas sector occur via effects on particular receptor animal groups (Table 8.1). Demand for and consumption of oil and gas (and derived products) continue to rise steadily; hence, those effects on the marine environment are expected to continue (World Ocean Review 2014). Perhaps more importantly, in some regions there are known energy resources within mineral deposits (e.g., manganese nodules; Table 8.1) that cannot yet be extracted with current technology; in reality, however, it is just a matter of time before the technology is developed or adapted to access and extract these non-renewable resources (Glover and Smith 2003).

Exploitable oil and gas reserves are found only in particular regions (e.g., oil—Gulf of Mexico, Persian Gulf and Australasia; gas—North Sea, India and Bangladesh), and this has limited their production to some extent. However, new reserves are being found (e.g., Chapter 14), and known resources that are as yet unexploitable are predicted to become available through environmental change (e.g., drilling for gas under the retreating Arctic ice), or accessible through technological advancements (e.g., exploiting deeper waters with floating structures). Thus, an increasing offshore extraction of oil and gas is expected (World Ocean Review 2014).

Table 8.1 Summary of offshore energy sources, the main attributes linked to potential sources of environmental stressors, and likely receptors following a source-pathway-receptor approach to identifying key components

Type	Main attributes	Environmental effects	Main receptor groups
Oil and Gas	Drilling platforms/vessels, seabed or floating Pumping stations at sea/seabed Pipelines	Direct physical disturbance of seabed Localised suspension of sediments Noise disturbance Chemical pollution of water	Mammals Fish Birds Macroinvertebrates Benthos
Liquefied natural gas	Movement by shipping (rather than pipelines) Drilling platforms/vessels—seabed or floating Pumping stations at sea/seabed	Direct physical disturbance of seabed Localised suspension of sediments Noise disturbance Potential chemical contamination if ship wrecked	Mammals Fish Birds Benthos
Methane hydrate	Deeper seabed areas associated with past large-scale plankton productivity Continental shelf margins and deep adjacent waters (but not open ocean/deep sea)	Direct physical disturbance of seabed Localised suspension of sediments Unstable when released from seabed sediments through mining carbon compounds similar to other fossil fuels Increased CO_2, reduced O_2 and acidification if mining not strictly controlled and managed	Benthos Pelagic plankton

Type	Main attributes	Environmental effects	Main receptor groups
Offshore wind	Focussed in northern Europe, but activity across world, moving from shallow to offshore environments (floating wind turbines)	Direct physical disturbance of seabed	Mammals
		Localised suspension of sediments	Fish
		Noise disturbance	Birds
		Cable deployment and electromagnetic fields (operation only)	Benthos
		Collision risk	
		De facto no-fishing zones	
Wave	Research and development stage with initial small-scale array deployment, but potential across seas and coastal zones	Entanglement in mooring cables	Mammals
		Noise disturbance	Fish
		Cable deployment and electromagnetic fields (operation only)	Benthos
Tidal stream/ Ocean currents	Location-specific for some tidal stream devices	Collision risk	Mammals
	Larger deployment for ocean currents	Cable deployment and electromagnetic fields (operation only)	Fish
		Altered hydrodynamics affecting biophysical properties and processes	Birds (i.e., divers)
			Macroinvertebrates
Tidal barrage/ Lagoons	Zoning off coastal areas to regulate water movement	Construction includes dam/barrier – potential direct and indirect effects on receptors	Mammals
			Fish
			Birds
			Benthos

(Continued)

Type	Main attributes	Environmental effects	Main receptor groups
Ocean thermal energy conversion (otec)	Temperature difference between warm surface waters and colder deeper waters	Change in local environment temperature	Benthos
	Only applicable in warmer climates where temperature difference of water bodies is great enough to generate extractable power	Movement of nutrients in lower water column to upper water	Plankton
Osmotic power	Osmotic differences between fresh and marine water	Change in salinity	Benthos
Algal biofuels	Culture of marine algae (macro + micro) for extraction of compounds used for alternative fuel	Reduction of light penetration	Fish
	potential co-location with other renewable energy devices	Increased organic production in water column and higher organic deposition to seabed	Benthos
			Invertebrates Plankton and planktivores

Perhaps the largest currently untapped, carbon-based resource is methane hydrates: water-trapped molecules of methane that occur in stable, low-temperature areas of the seabed (Chong et al. 2016). The source of the methane is microorganism metabolism of the remains of marine plankton (decomposition) within the seabed sediments. Hence, the areas of primary methane-hydrate deposits are in the most productive parts of the ocean (e.g., upwelling areas) at the edge of continental shelves and adjacent seabeds (Chong et al. 2016). Currently, industry is only testing the capacity to extract these sources, so the likelihood of full-scale mining in the near future is low. However, given that the potential resource is so vast, there could be a large economic incentive to develop viable mining techniques quickly. The biggest apparent issue is that once mined from the seabed (i.e., disturbed), methane will be released (Table 8.1), further exacerbating climate change and its consequences for marine life via temperature rise and acidification (Denman et al. 2011). Decisions to allow mining of these deposits should thus consider the potential increases to existing and future environmental pressures (Table 8.1).

Renewable energy sources

Offshore wind power represents the largest renewable energy development in the marine environment globally, with northern Europe being the focus of activity over the past few decades (91% of world's offshore installed wind power in Europe 2015; Council, G. W. E. Annual report 2015). At present, offshore wind farms can be located anywhere where there is sufficient wind resource and suitable seabed for construction of the wind turbine arrays and cable burial. Most offshore wind turbines deployed around the world are steel monopoles that are particularly suitable for soft-bottom substrates (e.g., sand, sandy mud; Figure 8.1a). Other foundations, such as jackets, gravity-based and suction caisson, have been used where seabed characteristics are hard or alternative foundations are needed (Figure 8.1a). There is now increasing interest in the technology of floating wind turbines, which can have a weighted platform on or near the surface and different mooring-line designs anchoring the turbines to the seabed (Figure 8.1b). Floating turbines will enable the sector to expand its coverage of the seas farther offshore and potentially reduce the pressures associated with coastal planning of offshore energy developments. Depending on the different types of foundations employed, each will have a different environmental footprint (Figure 8.1a and b) and concomitant effect on the seafloor and associated communities.

For the wave and tidal stream sector, technology development is in its early stages so that typical deployments and device types are yet to be determined. Some devices will be fixed to the seabed, while others will use mooring lines and be in mid-water or at the surface. The environmental interactions associated with wave and tidal devices are not well understood because few devices have been deployed to date. Some of the interactions are predicted to

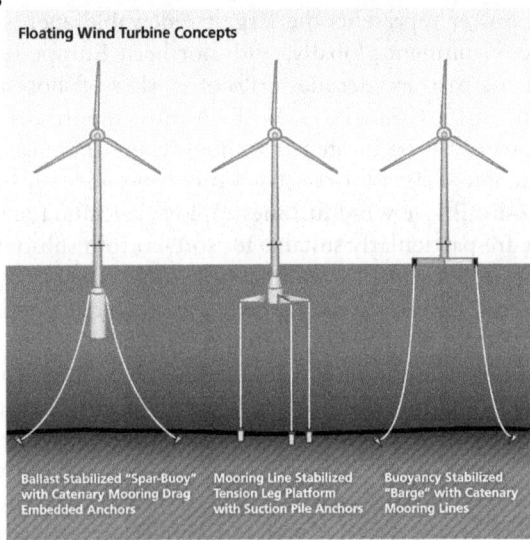

a

Monopile   Tri-Pod   Jacket   Suction Caisson   Gravity Base

b

Floating Wind Turbine Concepts

Ballast Stabilized "Spar-Buoy"   Mooring Line Stabilized   Buoyancy Stabilized
with Catenary Mooring Drag   Tension Leg Platform   "Barge" with Catenary
Embedded Anchors   with Suction Pile Anchors   Mooring Lines

Figure 8.1 (a) Types of offshore wind turbine-foundation structures currently in use in shallow waters (< 50 m depth). (b) Examples of floating offshore wind turbines showing subsurface structures and mooring-line options.

Source: IPCC 2012; Figure 7.19 from Wiser, R., Z. Yang, M. Hand, O. Hohmeyer, D. Infield, P. H. Jensen, V. Nikolaev, M. O'Malley, G. Sinden, A. Zervos, 2011: Wind Energy. In IPCC Special Report on Renewable Energy Sources and Climate Change Mitigation [O. Edenhofer, R. Pichs-Madruga, Y. Sokona, K. Seyboth, P. Matschoss, S. Kadner, T. Zwickel, P. Eickemeier, G. Hansen, S. Schlomer, C. von Stechow (eds)], Cambridge University Press, Cambridge, United Kingdom and New York, NY, USA.

equate to those of offshore wind (i.e., arrays of devices emitting sound and subsea cables emitting electromagnetic fields), but there are some device-specific considerations, such as risk of collision with rotating underwater tidal turbines for marine mammals or fishes (Table 8.1).

Algal biofuel has some potential as a partial alternative to fossil fuels; however, more research and development is required if it is to become a serious competitor for existing fuels (non-renewable) and be broadly available (Hannon et al. 2010). Hence, environmental interactions have not been seriously considered thus far. However, large-scale algal culture could have plausible ecosystem consequences associated with organic enrichment leading to changes in productivity and potentially excessive nutrient richness, known as eutrophication (Table 8.1).

Tidal barrages and lagoons use the natural tidal regime and the height difference related to low and high water to generate electricity hydraulically (Burrows et al. 2009; Pethick et al. 2009). They effectively obstruct part of the coast (often in an estuary) and regulate the water movement and hence, the natural processes at the site. There are few examples in operation, but they remain a potential energy source in a few places where there is sufficient tidal range (e.g., Bay of Fundy, Canada; Cardiff Bay, Wales). Little is known about the likely changes to the marine ecosystem that would result from tidal barrages or lagoons, but they are generally considered to have potentially negative impacts on the unprotected feeding areas of migratory and resident birds (Frid et al. 2012). Negative consequences might result via changing water inundation patterns and processes, which is predicted to alter local benthic habitats and species assemblages (Frid et al. 2012). Tidal barrages and lagoons also represent potential migration barriers to some subsurface species, such as salmonids—effects that can also arise from other marine renewable technologies (Burrows et al. 2009; Pethick et al. 2009; Frid et al. 2012).

The offshore renewable energy sector is one of the most studied renewable technologies in terms of their implications for the environment. While the industry is still evolving, there is some evidence to indicate effects arising at the scale of both devices (wind turbines, wave and tidal stream devices) and site (offshore wind arrays). However, those examining potential effects need to take account of the different spatial and temporal scales at which individual animals or populations might be affected. Furthermore, scale will contribute to any assessment of cumulative effects and will be essential to both decision making and public acceptance of offshore renewable energy development.

Interactions between offshore energy and receptors

When examining the potential interactions between offshore energy and environmental receptors, the four different phases of activity need to be specifically considered: exploration, construction, operation, and decommissioning. These phases have both similar and different identifiable stressors and pathways, and so the specific spatial extents and different time scales related to each phase must be examined. This is important to quantify the biological importance of any particular impact (Gill 2005; Boehlert and Gill 2010; Miller et al. 2013), which can be either direct (e.g., injury to individual animals) or indirect (e.g., food or habitat availability changed leading to population redistribution).

Chemical pollution [all phases]

Extracting mineral resources to produce energy has been synonymous with the risk of chemical pollution, usually in the form of localised oil spills, the consequences of which can be considerable and last for many years (see Table 8.2). Oil and gas installations arguably pose the most severe risk of large-scale chemical pollution (for an overview, see Geraci 2012; Fingas 2014). The most recent example is the Deepwater Horizon oil well blowout in the Gulf of Mexico (2010), which caused the largest oil spill in the history of offshore energy production (eight times the Exxon Valdez oil spill; McNutt et al. 2012; Haney et al. 2014a) and occurred in an area of high marine biodiversity; 28 species of marine mammals have been recorded in the affected area, including 5 species listed as threatened by the International Union for Conservation of Nature (Campagna et al. 2011). The magnitude and severity of the Deepwater Horizon incident and the many resulting studies of oil pollution focussed on the Gulf of Mexico provide findings that are regarded as the best insight available into both acute and chronic impacts. Acute impacts include mortality of marine mammals (Williams et al. 2011) and birds, with an estimate of 9×10^5 bird deaths directly caused by the Deepwater Horizon oil spill (Table 8.2; Haney et al. 2014a). This is to date the largest bird mortality by far caused by an oil spill (Table 8.2). The estimates of bird and mammal mortality in Williams et al. (2011) and Haney et al. (2014a) do not take account of indirect or chronic population effects; hence, the total mortality caused by the Deepwater Horizon oil spill is likely to be even higher (Haney et al. 2014b).

Chronic effects of the exposure to crude oil and associated toxic volatile compounds have been investigated for some species. For example, individual bottlenose dolphin *Tursiops truncatus* fitness decreased as a result of chronic exposure to oil spill-related chemical compounds, in some cases leading to individual death years after the initial pollution event (Schwacke et al. 2014; Venn-Watson et al. 2015). Similar findings have also been reported in the North Sea, where chronic exposure to organochlorines associated with discharges from offshore petroleum installations and industry-related shipping activity reduced the local population of Baltic grey seals *Halichoerus grypus* (Jenssen 1996). Predatory seabirds are particularly susceptible to

Table 8.2 Worldwide oil spills from platform (regular font) and *shipping* (italics) accidents, with bird mortality estimated in 10^5 individual birds

Name	Location, year	Birds ($\times 10^5$)	Reference
Deepwater Horizon	Gulf of Mexico, 2010	9	Haney et al. (2014a, 2014b)
Tricolor	English Channel, 2004	1.0	Camphuysen and Leopold, (2004)
Selendang Ayu	Alaska, 2004	1.4	Byrd and Reynolds (2007)
Prestige	Spain, 2002	2.0	Munilla et al. (2011)
Erika	France, 1999	1.2–3.0	Camphuysen et al. (2005)
Exxon Valdez	Alaska, 1989	1.0–6.9	Piatt and Ford (1996)
Stylis	North Sea, 1981	2.0–3.0	Camphuysen et al. (2005)

direct impacts such as the loss of waterproofing of oiled feathers, which reduces insulation and buoyancy (Leighton 1993). Birds trying to remove the oil from their feathers often ingest oil and they can also bioaccumulate toxins by ingesting contaminated prey, both of which make the birds vulnerable to secondary health problems that reduce survival and breeding success (Leighton 1993).

Crude oil pollutants can impact different life history stages of a receptor animal. There is growing evidence of dose-dependent physiological effects on cardiac function of larval fish of several predatory species (Incardona et al. 2014) and contraction in the cardiac muscle cells of adult Bluefin tuna (*Thunnus thynnus*; Brette et al. 2014). Haddock (*Melanogrammus aeglefinus*) and Atlantic cod (*Gadus morhua*) eggs and larvae in two areas in the North Sea with extensive oil production have greater mortality and more deformities than in areas without (Balk et al. 2011). These impacts add to existing pressures. Bluefin tuna (*T. thynnus*) adult spawning habitat in the Gulf of Mexico, for example, have been degraded as a result of the Deepwater Horizon oil spill, adding to existing pressures of climate change and fishing on the population (Muhling et al. 2012; Hazen et al. 2016).

No matter how carefully designed the oil and gas installations might be, there is always a risk of leak during construction, operation or manufacturing (Wiese et al. 2001), and drilling waste and water produced from the drilling process often have elevated pollutant concentrations (Balk et al. 2011). Therefore, oil from spills or waste is common, and cleaning up is challenging. *In situ* evaluation reports suggest that the main environmental impact of oil spills is determined by the timing and location rather than by the extent of the spillage (Russell 2005). MSP alongside permitting requirement can provide the means to ensure risks of chemical pollution are managed.

Chemical pollution in the context of renewable energy will be site specific and often concerns the resuspension of sediments via dredging, placement of construction vessels, or cable laying. Such activity can disturb sediments and raise turbidity and potentially smother habitats, or the eggs or larvae of benthic species. Sediment disturbance might also release sediment-bound chemicals, some of which are a legacy of former petrochemical and textile industries (Fox et al. 1999). Other potential pollution sources are liquids or materials used for energy devices (e.g., paints, hydraulic fluids, or fuel from maintenance vessels), which could be expected from any activity that re-lies on introduced structures (e.g., oil and gas platforms). Some research has suggested that organic enrichment from marine algae culture might lead to eutrophication, which in turn could also cause chemical pollution (Hannon et al. 2010). However, such effects depend on the nature of the algal culture, the hydrodynamics of the area and any other sources that trigger eutrophica-tion (e.g., discharges from coastal urbanised areas, river outflows).

Energy emissions

Energy emissions come in the form of acoustic, electromagnetic or heat en-ergy. The emission of energy into the marine environment affects marine organisms if it impairs their ability to detect sounds or stimuli of interest, disrupts their behaviour or causes direct physical harm to the sensory sys-tem. The topic is not well understood, with most studies focussing on sound effects on marine mammals and a few sensitive fish species (Southall et al. 2007; Hawkins and Popper 2014). Other animals are not considered suffi-ciently in terms of their susceptibility to sound produced from offshore en-ergy development (Slabbekoorn et al. 2010), and understanding of how other energy emissions affect species is still rudimentary (Boehlert and Gill 2010; Gill et al. 2014).

Noise [all phases]

Underwater noise comes from the vibration of a structure that is transmit-ted through the water or the sediment (e.g., operating wind turbine). The predominantly low-frequency sound can travel for tens or hundreds of kilo-metres if the source is large enough (Thomsen et al. 2006). Most acoustic emissions are produced in the construction phase (e.g. during pile driving of wind turbine foundations' Figure 8.1; Blackwell et al. 2004a). Such acute noise negatively affects marine mammals via direct physical damage to the auditory system (Gordon et al. 2003) and also has the potential to harm other marine organisms (fish and invertebrates; Hawkins and Popper 2014). Short-term exclusion of species from areas around offshore energy sites can therefore occur during construction, but this might not necessarily lead to long-term displacement (e.g., as observed in some marine mammals; Thompson et al. 2013).

Prior to construction, geophysical exploration of the seabed at prospective offshore energy sites is commonly done using seismic surveys that involve ship-based air-gun arrays, which produce pulses of energy reflected by the seabed. Seismic surveys might be audible to as much as 4000 km away (Nieukirk et al. 2012), but are unlikely to harm sensitive marine mammals unless they are close to where direct physical damage to the auditory system can occur (Nowacek et al. 2013). Other animals have not yet been considered sufficiently in this regard. At close ranges, there is evidence that both seismic surveys and pile-driving noise can mask marine mammal communication (Richardson and Thomson 1995; Tyack and Janik 2013; Cerchio et al. 2014; Blackwell et al. 2015). Seismic surveys have also been linked to disorientation, stranding and altered dive profiles in marine mammals, leading to decompression sickness in some species (Gordon et al. 2003).

Operational noises from offshore energy installations are generally of lower frequency and intensity than noises emitted during site exploration and construction, and appear to be of less concern to biodiversity (Truebano et al. 2013). Additionally, non-injurious acoustic emissions from operating devices (particularly tidal stream and wave) could potentially help animals notice the devices and therefore avoid colliding with them (Wilson and Carter 2013). However, the operational phase is the longest part of the lifecycle of an installation, and there is still a potential threat to sensitive marine animals from ongoing exposure to low-intensity operational noise (e.g., Koschinski et al. 2003; Blackwell et al. 2004b; Madsen et al. 2006; Slabbekoorn et al. 2010). Chronic exposure is of particular concern when viewed in the context of cumulative impacts from other human-generated noise sources, such as shipping, fishing, research and military operations (Richardson and Thomson 1995; Wright and Weilgart 2011).

Noise from vessel traffic during construction contributes to cumulative noise exposure, although its impact depends on how long construction lasts (months to years depending on the size of the development; e.g., Slabbekoorn et al. 2010). While noise intensity from offshore energy devices is lower during operation than construction, when set in the context of cumulative impacts from other noise sources, these inputs are potentially detrimental (Slabbekoorn et al. 2010; Hawkins and Popper 2014). Cumulative noise exposure can exceed the threshold expected to cause permanent auditory damage in grey seals (Hastie et al. 2015), and the cumulative impacts of noise have been raised as a potential worldwide issue for fish and invertebrates (Slabbekoorn et al. 2010). Marine spatial planners should therefore ideally consider other local sources of noise and cumulative exposure when assessing the potential impacts of offshore energy developments.

Electromagnetic fields [operation]

Electromagnetic fields are emitted from subsea cables during the operational phase of transmitting electricity (or associated with electrical components along seabed pipelines). Until the advent of marine renewable energy developments and offshore power transmission networks, electromagnetic fields were not considered as a potential source of environmental disturbance. However, recent research has identified that the scale and type of electromagnetic field emission depend on whether the cables transmit direct or alternating current, with magnetic fields the dominant emission for the former and magnetic and electric fields generated by the latter (Gill et al. 2014).

Whether these types of electromagnetic fields affect animals that are able to detect either electric or magnetic fields is currently unknown (Gill et al. 2014), but there are potential effects in terms of navigation and/or orientation to electromagnetic fields (Öhman et al. 2007; Westerberg and Lagenfelt 2008). Most cetaceans and turtles are able to perceive the Earth's magnetic field, so there is potential for them also to be able to detect anthropogenic magnetic fields superimposed on the geomagnetic field (Gill et al. 2014). In addition, many marine fish and invertebrate species possess magnetic or electroreceptive senses, and therefore as with sound, marine animals could be harmed indirectly via changes in the behaviour, distribution or abundance of their prey. There is evidence of short-term altered movement paths in salmonids, sturgeon (Acipenseridae), European eels (Anguillidae) (Öhman et al. 2007; Westerberg and Lagenfelt 2008; Gill et al. 2012a) and elasmobranchs (Öhman et al. 2007; Gill et al. 2014) from electromagnetic fields associated with power and subsea interconnector cables between countries. With the increase in offshore energy generation and the networks of cables being suggested for deployment, marine spatial planners will also have to consider electromagnetic fields further within the process.

Heat [operation]

Little is known about the impact of heat emissions on marine biota, but most research that has been done concerns heat transfer from oil and gas pipes to the surrounding water. Heat loss from pipelines can facilitate the formation of gas hydrates and waxes that slow the rate of flow of the material being piped (Nysveen et al. 2005). Direct electric heating is often applied to reduce this (another potential source of electromagnetic field emissions). Also, subsea cables produce heat when transmitting electricity that is emitted over short distances into the seabed where they are normally buried (Pilgrim et al. 2013). Whether localised heating of the seabed has any implications for benthic infauna or epifauna is unknown.

One important component of any energy emissions is that they are not device- or industry-specific. For example, noise or electromagnetic fields from any source will add to the overall emissions environment. This has been acknowledged by the fact that environmental legislation, for example the European Marine Strategy Framework Directive requires consideration of site specific energy emissions to ensure vulnerable habitats are avoided and identified impacts minimised; marine spatial planning and permitting play a significant role.

Collision and barriers [operation]

There is potential for marine animals to collide with offshore energy structures whether the structures are submerged (e.g., wind turbine towers, tidal turbines and mooring cables) or above the surface of the water (e.g., floating wave energy devices, offshore wind turbines and oil platforms). Risk of physical injury through collision or entanglement would be more likely during periods of poor visibility, high water turbidity or foul weather (Wilson et al. 2006). One area of concern is the risk of marine animals that inhabit high-tidal-stream locations being injured by the moving blades of tidal turbines. For example, harbour porpoises (*Phocoena phocoena*) and fishes (Viehman et al. 2014) aggregate in dynamic marine environments (Jones et al. 2014), which could increase the risk of them using areas that overlap with tidal energy arrays. Models determining the potential and frequency of collisions (assuming neither avoidance nor attraction and that strikes are fatal) suggested that interaction rates could be frequent enough to reduce local populations of animals, but rare enough that they are challenging to observe directly (Wilson et al. 2006).

The main concern above water is that seabirds follow visual cues such as vertical structures or lighting to navigate the oceans, and artificial offshore structures attract seabirds (Wiese et al. 2001). Collision occurs when birds fail to avoid offshore structures, causing severe trauma and death (Drewitt and Langston 2006; Ronconi et al. 2015). Behaviour, weather conditions or installation characteristics determine the magnitude of collision risk in birds (Furness et al. 2013; Langton et al. 2011). Severe weather conditions or nocturnal activity increases collision risks (Ronconi et al. 2015), and marine structures aggregate fish that are attractive to foraging birds and mammals (Castro et al. 2004; Wilhelmsson et al. 2006), potentially increasing collision risk. In addition to the physical risk from the structures themselves, there is also the risk of ship strike around energy installations as a result of the increased vessel traffic associated with construction, maintenance and decommissioning (Henkel et al. 2014). Such collisions could potentially increase pollution from leaking chemicals and reduce safety for other offshore users.

Barriers [all phases]

Barriers, habitat loss and displacement effects, together with collision with submerged, floating or aerial structures, are the most common risks posed by offshore energy installations (Masden et al. 2009; Langton et al. 2011; Gill et al. 2012b; Ronconi et al. 2015). These installations, especially wind farms, can act as visual barriers to some species such as eiders (*Somateria mollissima*) that are reluctant to approach installations (Masden et al. 2009). Subsea cable electromagnetic field emissions might act as barriers or at least divert migratory fishes such as Atlantic salmon (*Salmo salar*) or European eels (*Anguilla anguilla*) (Gill et al. 2012a). Consequently, offshore installations might impede access to potential feeding and resting areas, forcing deviation from natural migratory routes and delaying migration, which could increase the susceptibility to predation and the risk of disorientation (birds, Ronconi et al. 2015; fish, Gill et al. 2012a). There is limited research examining how barriers affect receptor population dynamics, but it appears that while barriers or displacement can reduce breeding performance, they generally do not increase bird mortality (Ronconi et al. 2015).

Incineration by lit objects [operation]

Offshore oil and gas installations often burn the gas surplus at sea and create vertical flares. Seabirds are attracted to artificial lighting and flares around offshore platforms, and birds flying close by are sometimes incinerated (Ronconi et al. 2015). The casualties are particularly common in small songbirds (Passeriformes) and storm petrels (Hydrobatidae; Wiese et al. 2001). Some evidence suggests that reducing the height of the flare might decrease the number of affected birds (Russell 2005).

Species assemblages

Effects on the seabed community [all phases]

There are different types of physical disturbance that can occur during the different phases of an offshore energy project. Each stage (pre-construction, construction, operation and decommission) can directly affect seafloor communities, but the outcome depends on the habitat type and community characteristics (e.g., sensitivity, resilience, ability to recover). We describe the individual effects below and consider the potential for cumulative effects.

INTERTIDAL AND SUBTIDAL AREAS

Construction during the installation of cables or pipelines in nearshore areas between the extraction site and land can last from several months to years depending on the type of development and the number of cables or pipes. There is limited understanding of the immediate changes directly imposed on seabed communities from this type of construction, although the footprint of activity associated with cable or pipeline laying is only a narrow corridor of a few metres along the seabed and intertidal zone. Similar to prolonged dredging and/or sediment removal, species and habitats have variable recovery (Barrio Froján et al. 2008). Habitat loss and destruction of substrata or temporary displacement of sediment have different implications for different habitat types and benthic communities. The severity of these effects is related to the hydrodynamics of the local environment, and the habitat or sediment types in question. The sensitivity (recoverability and vulnerability) of specific habitats or species is perhaps more important for assessing the potential cumulative effects of construction in nearshore areas.

OFFSHORE SEABED AND OFFSHORE STRUCTURES

Installation of a structure itself can kill benthic fauna (non-mobile, sessile or low-mobility fauna) directly below the foundation, or displace more mobile benthic epifauna (Wilhelmsson and Malm 2008). In addition, resuspended sediment can smother the surrounding sea floor. Depending on the site, there are site-specific and localised effects on the benthic communities inhabiting the area of the foundation and/or structure (Wilhelmsson and Malm 2008).

Construction invariably disturbs the sediment and modifies habitats. Community structure, dynamics and environmental conditions (sediment type, hydrodynamics and wave environment) vary spatially. For high-energy environments, typical of much of the southern North Sea for example, disturbances are likely to be temporary and small compared to heavy-gear trawl fishing (Lindeboom et al. 2011). In contrast, areas with long-lived species, such as the clam *Arctica islandica,* are more likely to suffer longer-term impacts (Butler et al. 2010).

During the operational phase, new offshore structures and/or foundations change local ecosystem characteristics by modifying structural and functional ecological attributes. Structures provide a surface for colonisation and can act as artificial reefs over decadal scales (Boehlert and Gill 2010). Colonisation by a range of benthic species (sessile: ascidians, serpulids, hydroids, barnacles, mussels, algae and crustaceans) can occur over several years (Langhamer et al. 2009; Wilhelmsson and Malm 2008; Whomersley and Picken 2003) depending on the ecology and biology of the specific colonising species (Langhamer et al. 2009; Petersen and Malm 2006; Whomersley and Picken 2003). Studies of communities on artificial substrata suggest that there are clear differences compared to natural substrata because the former are less diverse (Langhamer et al. 2009; Wilhelmsson and Malm 2008). Introducing solid structures on soft sediments alters biodiversity and in some cases, modifies the entire ecosystem (Inger et al. 2009). Larger structures with high surface area offer more space for colonisation, and rough surfaces such as untreated concrete are more easily colonised, although concrete structures are often treated with silane/silicone to reduce 'fouling' (colonisation) (Petersen and Malm 2006). The ecological relevance of changing species assemblages and community structures is also scale-dependent.

SCOURING

Scouring (removal of seafloor sediment by hydrodynamic forces) is high in some areas and is affected by foundation design (Lindeboom et al. 2011). Scouring increases turbidity and suspended sediments around the structure, thus removing substrata and changing habitat morphology (Rees et al. 2006). Alteration of highly mobile sediments can change local species composition and/or abundance, and recently scoured areas are colonised by 'opportunistic' benthic species (i.e., fast-growing and short-lived; Coates et al. 2014). The tolerance of benthic species to disturbed areas depends on the site's characteristics, food availability and competition from other benthic organisms (Birchenough and Frid 2009). The effects of scouring are influenced by local hydrography, bathymetry, sediment type and benthos in and around the zone (Coates et al. 2014). There is currently little known about this aspect of offshore energy, although research is currently underway (e.g., in the United Kingdom, Bell et al. 2014). Small-scale changes in flow regime can occur around all types of energy installations as water flows around the structures or through tidal turbines (Cazenave et al. 2014). At tidal stream and wave installations, hydrodynamic energy is removed, which potentially influences the local and regional biophysical habitat. This could impact marine animals directly by altering the hydrodynamic properties of habitats on which they rely (Scott et al. 2010; Jones et al. 2014) or indirectly by changing prey and benthic coloniser composition and abundance (Hendrick et al. 2016).

Artificial reef effects [operation]

Worldwide, artificial reefs are constructed and deployed in coastal waters to manage fisheries, mitigate human-caused damage, protect and rehabilitate particular habitats, or to increase the value of an area for recreation (e.g., Jensen 2002; Claudet and Pelletier 2004; Seaman 2007). Oil and gas structures attract marine life (e.g., Whomersley and Picken 2003; Schroeder and Love 2004; Kaiser and Pulsipher 2005), and offshore renewable energy developments are proving to do likewise (e.g., Wilhelmsson and Malm 2008; Reubens et al. 2013; Coates et al. 2014). These artificial reefs provide new habitats for many sessile and mobile colonising species. The device structures represent different types of substratum, shapes and submersion time, but have generally lower diversity relative to natural reef habitats (e.g., Wilhelmsson and Malm 2008; Ashley et al. 2014).

For offshore wind farms in Europe, several benthic fish species occur at higher densities around turbines compared to surrounding waters and seabed types (e.g., Wilhelmsson et al. 2006; Bergström et al. 2013). A meta-analysis of demersal and sedentary fish suggested that often only a few benthic and semipelagic, hard bottom-associated species respond in this manner (Andersson and Öhman 2010), while the densities of some soft sediment-associated species commonly remain unchanged or decrease (Ashley et al. 2014). More mobile pelagic (e.g., horse mackerel *Trachurus trachurus*) and demersal fish species such as cod *Gadus morhua* and pouting *Trisopterus luscus* have been recorded in shoals closely associated with wind turbines (Bergström et al. 2013; Reubens et al. 2013), with cod having high site fidelity over several months around wind turbines (De Troch et al. 2013).

It is clear that offshore energy structures provide habitat for several colonising species, and that they can aggregate fauna and alter local community structure. Hence, biomass can increase across different species and thus change energy transfer and increase the complexity of trophic linkages (Gill 2005). Local biodiversity can increase through the colonisation of typical reef-dwelling species in areas otherwise largely devoid of hard-bottom habitats. Whether the colonisation process of the offshore energy structure is *in situ* production or simply the product of attraction of species from elsewhere is currently unknown (Lindeboom et al. 2015). Although mobile species (such as fishes) can reproduce and grow around the turbines, this might only represent a temporary aggregation. For sedentary species such as mussels and barnacles, an increase in biomass has been observed at offshore windfarms (Andersson and Öhman 2010; Lindeboom et al. 2011). For such colonisation by species normally associated with hard substrata, there might be potential for increased population sizes on these artificial structures.

Food web effects [operation]

Offshore energy structures offer favourable substrata for colonising invertebrates, with studies on wind turbines in operation suggesting that the community is dominated by filter-feeding taxa such as barnacles and blue mussels, or by filtering assemblages constituting anemones, hydroids and solitary sea squirts (e.g., Wilhelmsson and Malm 2008; Maar et al. 2009; Ashley et al. 2014; Coates et al. 2014). Biota on the structures represent a potential food resource for predatory species (e.g., cod *Gadus morhua*; Reubens et al. 2013). Fish and sessile organisms associated with turbines can increase benthic productivity around the turbines through the deposition of organic material such as faecal matter, organic litter and dead organisms (Coates et al. 2014). This can in turn attract benthos-feeding, soft sediment-associated species (Wilhelmsson et al. 2006). Regardless of the energy industry or technology type, the potential ecosystem impacts can change the food web and trophic interactions between predators and prey, and hence have implications for the processes that drive population dynamics and community composition (e.g., predator-prey dynamics, trophic linkage, abundance, body size, population demographics). While the evidence of any positive effects of offshore energy installations remains scarce, it appears that their structures can enhance foraging opportunity as a consequence of fish aggregation and also act as stepping stones between foraging areas, thus allowing seabirds to exploit previously non-existent or inaccessible food resources (Castro et al. 2004; Wilhelmsson et al. 2006; Ronconi et al. 2015). The potential redistribution and enhanced dispersal opportunities appear to be more likely with offshore renewable energy developments because each installation is close enough to link a greater number of aggregation sites.

There is some evidence from tracking of marine mammals that both grey and harbour seals forage selectively around anthropogenic structures such as subsea pipelines and within windfarms (Russell et al. 2014). In addition, passive acoustic surveys have shown that gas rigs in the North Sea provide foraging habitat for porpoises, made possible because the structures act to aggregate fishes (Todd et al. 2009). These examples indicate the potential for offshore energy developments to have indirect, positive impacts on marine mammals, but this attraction might also increase the risk of negative interactions with offshore energy structures (such as collision, entanglement or noise exposure).

Human-made structures of any type in the marine environment can change the biological community, which has implications for the integrity of biodiversity. For example, the resultant community that interacts with wind farm structures will be different from the community before construction (i.e., an increase in hard-bottom species). At this stage, we are not in a position to determine whether this is ecologically better or worse than the community that existed previously. Furthermore, there is the opportunity for invasive/non-native species to take advantage of the new habitats (Kerckhof et al. 2010). It is likely that these will be the main elements of future discussion and research.

Decommissioning energy structures

Decommissioning obsolete offshore energy infrastructure (whether for fossil fuels or renewable energy) is an expensive problem that is becoming increasingly challenging as options such as abandonment, *in situ* toppling, and deep-sea dumping become less socially acceptable (Fowler et al. 2014). Deciding what to do with the thousands of aging offshore oil and gas structures is currently a pressing issue for many nations (Fowler et al. 2014). The most obvious parts to deal with are the fixed structures associated with platforms above and below the sea surface (such as jackets, top-sides, storage tanks), but the wider infrastructure also includes seabed pipelines, wellheads, cabling and accumulations of (often contaminated) drill cuttings (Schroeder and Love 2004). While large machinery was used to construct these structures, removing them is more challenging still because they will have been modified, degraded, contaminated or buried during their operational lives, and thus will have changed beyond original specifications.

Removing structures could have several ecological repercussions, above water as well as below. Therefore, understanding the colonisation and use of offshore energy structures by marine animals during the construction and operational phases is crucial for providing an appropriate environmental context to use when making decisions regarding decommissioning. For marine animals, decommissioning involves similar issues to construction, but with the addition of blasting risks associated with demolition and liberation of contaminants accumulated over decades of operation (Schroeder and Love 2004). Removing renewable energy structures (with the exception of tidal barrages and major scour protection) is likely to be less challenging because of the smaller size of each device, their modular nature, and designs that routinely require lifting for servicing or replacement. However, a notable difference concerns the nature of the resource extracted. Oil and gas comes from reserves below the seabed, and thus the impacts on the marine environment are an incidental consequence of the process of extraction. However, wave and tidal stream energy (and to some extent, wind) is part of the hydrodynamic environment (Gill 2005), and so removing the energy devices can alter the physical conditions at the site, such as flow and sedimentation regimes and wave patterns. Thus, operating devices will alter conditions well beyond the footprint of actual developments, because they are directly removing hydrodynamic energy from the environment (Neill et al. 2009). Given that the lifetime of renewable energy infrastructure is several decades, these altered conditions would most likely represent a new ecosystem state. For example, at tidal stream and wave installations, hydrodynamic energy is harvested (i.e., removed), and this can influence the local and regional biophysical habitat. This can in turn impact marine animals directly by altering key hydrodynamic habitats (Scott et al. 2010; Jones et al. 2014) or indirectly via changes in the occurrence and abundance of prey species (Embling et al. 2012).

Decommissioning offshore energy structures would rapidly change existing ecosystem conditions by removing the hard substratum and the colonising fauna associated with the structure(s). Specifically, there would be an immediate reduction in biodiversity and abundance of the hard-substratum benthic community (Bacchiocchi and Airoldi 2003; Langhamer et al. 2009; Langhamer and Wilhelmsson 2009). Nearby soft-sediment species would also be released from predation pressure coming from the mobile species associated with the structure. While these changes are most pertinent for benthic fauna and habitats (Miller et al. 2013), there would be changes to trophic pathways that could impinge on other marine predators.

Spatial and temporal considerations

Many countries have marine policies designed to achieve clean, healthy, productive and biologically diverse oceans and seas (Wright 2014). Such policies need to govern activities at a site to take into account the variable spatial and temporal scales over which changes occur, such that integrated planning for regulation and enforcement is possible. When considering time scales, large, offshore energy development can take years to construct and operate for decades. When assessing the impact of offshore energy on marine biota, the duration of each activity potentially causing change, and determining whether the development area is part of a seasonal migratory route, feeding ground, nursery or mating area must be considered. This approach is required to avoid negatively affecting a biologically important proportion of a population potentially at risk at certain times of the year.

One of the main challenges facing offshore energy developers, regulators and conservation agencies is how best to assess the cumulative effects of major offshore developments (Judd et al. 2015; Willsteed et al. 2017). In many cases, cumulative effects will not be restricted to within a dedicated marine spatial planning area (intra-zone), but might occur between zones (inter-zone) such that developers and regulators would need to work together. There is not only a need to assess projects individually, but also to focus more on wider coordination between projects. This type of strategy would help to characterise some of the additional effects, such as increasing turbidity, displacement of species, or resuspension of sediment resulting from areas where there is a co-location of activities (e.g., an aggregate extraction site next to an offshore windfarm). At a broader geographic scale, neighbouring countries will often face similar challenges; to facilitate the assessment of trans-boundary cumulative effects, exchanges via early dialogue and discussions will give managers and regulators the opportunity to integrate knowledge from current and future offshore energy developments (Willsteed et al., 2017).

Cumulative-impact approaches look beyond the direct effects of the development under consideration and take a more holistic view, including any other potentially harmful activities within the same area (or in other parts of the impacted species' ranges; Willsteed et al. 2017). Ideally, cumulative, population-level analyses would have access to detailed information regarding a species' response to different types of disturbance, as well as baseline population data (e.g., abundance, vital rates, viability) and information describing habitats and distributions (Wright and Kyhn 2015). In reality, these data are rarely available for most marine animals (but see Pirotta et al. 2014). In the absence of detailed empirical data, King et al. (2015) described an interim framework for assessing the population consequences of disturbance to marine mammals, which combines available empirical data with expert elicitation (Harwood et al. 2014). That framework provides a promising mechanism for identifying and preventing individual- and population-level impacts of offshore energy development on many receptor animals, and can be a useful tool for future collaborative, multi-sector marine spatial planning.

The influence of climate change (including ocean acidification) on marine species and ecosystems poses a further challenge to understanding the changes associated with offshore energy development. The likely effects of climate change will be detected in species and ecosystems in different ways, and these can be positive or negative depending on the location and the species considered (Mora et al. 2013; Birchenough et al. 2015; Elliott et al. 2015). Some of the changes already arising from temperature shifts and altered pH have marked effects on the growth and development of some benthic species (Birchenough et al., 2015; Elliott et al., 2015). Cumulative assessments of the temporal and spatial aspects of climate change and acidification in addition to offshore energy developments (among other human activities) will be imperative to predict how ecosystems might shift, something that will be essential to ongoing marine spatial planning.

Future considerations for the marine environment and offshore energy

Following the recent United Nations Framework on Climate Change COP-21/22 Agreement, nations are urged to develop strategies that will limit rising global temperatures by reducing their greenhouse-gas emissions to the mid-21st Century (United Nations Framework Convention on Climate Change, 2015). As such, many countries are expected to set targets to reduce carbon dioxide emissions by moving away from non-renewable and towards renewable energy, which brings the spotlight onto the marine environment and its renewable energy potential.

The prospects of environmental changes continue to hinder the development and deployment of offshore renewable energy systems around the world (Copping et al. 2013; http://tethys.pnnl.gov/technology-type/offshore-wind). Because global environmental policy tends to focus on global-scale outcomes (e.g., emissions reductions), local-scale degradation at a site is generally not considered explicitly. Any plans to deploy offshore energy infrastructure to mitigate climate change can therefore experience local, site-specific impediments generally highlighted during environmental impact assessment. Acknowledging and considering the local environmental changes within the context of more strategic environmental plans could include a more balanced assessment of positive and negative impacts that will benefit from integrated policy and planning. Hence, marine spatial planning is being increasingly recognised as a critical approach to ensure the efficient and equitable use of maritime space, and sustainable exploitation of offshore energy resources (Wright 2014). For example, across Europe, this has resulted in the initiation of national marine plans to assist member states to meet 'good environmental status' required under the Marine Strategy Framework Directive (Willsteed et al. 2017). To achieve implementation of the legislative and policy requirements outlined therein, the potential cumulative effects arising from large-scale developments and activities within a marine spatial planning area are required.

Some of the existing and future effects of offshore energy exploitation are likely to be undetected given the difficulty of monitoring marine species. However, there are taxa such as seabirds or fishes that are sensitive to offshore energy installation and operation, and they could prove to be ideal indicators of marine ecosystem health (Furness and Greenwood 1993). Therefore, the environmental effects of offshore energy measured using such indicators could foster the development of general guidelines for marine spatial planning. Identifying impacts, applying appropriate mitigation, and monitoring of offshore energy developments form an important part of the decision-making process, and any rulings have to be based on robust scientific evidence obtained by appropriate and targeted monitoring to determine ecologically relevant changes (Lindeboom et al. 2015). Some nations have guidelines for minimising elements such as construction-related noise (Compton et al. 2008), recommending, for example, that construction work is timed to take account of any seasonally sensitive receptors such as marine mammals or spawning fishes, or by physically mitigating impacts by using bubble curtains and/or solid casings to decrease the propagation of pile-driving noise (Nehls et al. 2007).

Environmental risk and uncertainty

Uncertainty remains high when assessing the environmental implications of offshore energy, which inevitably makes specific decisions risky. The oil and gas sector of offshore energy has historically focussed on the environmental risk of major pollution events, with risk assessment following the Offshore Sector Health and Safety approach that applies the 'ALARP' principle ('As Low As Reasonably Possible') (Aven and Vinnem 2005). This principle relies on a robust assessment of realistic risk, and not simply the worst-case scenarios.

Appropriate risk assessments are therefore still sparse and generally supplemented or substituted with environmental impact assessments (Gill et al. 2012b) that provide a structured approach for identifying, investigating and assessing impacts; hence, they are not unlike risk frameworks. However, risk frameworks apply a logical sequence of steps to calculate the risk posed by an activity done in a repeatable, transparent manner. This enables decision makers to address issues that pose the greatest unmanaged risk. Guidance for avoiding and mitigating threats to some receptors such as marine mammals is available, and this should be incorporated into spatial planning frameworks for marine renewable energy development (e.g., Southall et al. 2007; Dolman and Simmonds 2010; Petruny et al. 2014). However, ongoing research means that best practice will continue to be updated as new data become available (Tougaard et al. 2015). Site-specific attributes should also be considered alongside general recommendations, both for reducing the risk to local populations of marine animals, and to reduce unnecessary mitigation costs to industry at low-risk sites (Tarpgaard et al. 2015).

Some knowledge gaps can be viewed as insurmountable; hence, expert judgement is needed (O'Hagan et al. 2006). However, expert judgement still requires scientifically supported foundations; otherwise, decision makers are obliged to invoke the precautionary principle or to ignore it. Neither approach is suitable for meeting national or international legislative requirements, or minimising or mitigating environmental impacts (Judd et al. 2015).

The focus thus moves to organisms and processes that have greater information (and by association, greater certainty). There is clearly an imbalance between the available evidence and remaining uncertainty when it comes to robust marine spatial planning, so better monitoring, modelling and experimentation are still required.

Conclusions

Offshore energy development is driven by a need to meet rising energy demand, to ensure energy security, and to assist limiting anthropogenic climate change. Development and expansion of offshore renewable energy installations are likely to be a priority for many coastal nations over the coming decades. Consequently, marine spatial planning will necessarily have energy as a central consideration among all the other anthropogenic activities taking place offshore. This marine spatial planning processes will have to consider the environmental implication of planning choices adequately to ensure that the future use of the marine environment is sustainable. Improved understanding of the environmental impacts will lead to more robust marine spatial planning decisions.

Highlights

- Development of offshore energy changes the marine environment.
- Direct and indirect effects of offshore energy occur from many different sources, pathways and receptors.
- While some negative effects are evident, the extent to which they threaten marine species or ecosystem integrity most often remains unclear.
- While there remains much uncertainty, existing knowledge and future targeted monitoring will assist marine spatial planning.
- An adaptive approach to marine spatial planning as knowledge improves will ensure an energy mix embracing offshore renewable energy development with environmental interactions appropriately considered.

References

Andersson, M.H. and Öhman, M.C. (2010). Fish and sessile assemblages associated with wind-turbine constructions in the Baltic Sea. *Marine and Freshwater Research*, 61(6), pp. 642–650. doi:10.1071/MF09117.

Ashley, M.C., Mangi, S.C. and Rodwell, L.D. (2014). The potential of offshore windfarms to act as marine protected areas–A systematic review of current evidence. *Marine Policy*, 45, pp. 301–309.

Aven, T. and Vinnem, J.E. (2005). On the use of risk acceptance criteria in the offshore oil and gas industry. *Reliability Engineering & System Safety*, 90(1), pp. 15–24.

Bacchiocchi, F. & Airoldi, L. (2003). Distribution and dynamics of epibiota on hard structures for coastal protection. *Estuarine, Coastal and Shelf Science*, 56(5), pp. 1157–1166.

Balk, L., Hylland, K., Hansson, T., Berntssen, M.H.G., Beyer, J., Jonsson, G., Melbye, A., Grung, M., Tortensen, B.E., Børseth, J.F. and Skarphedinsdottir, H. (2011). Biomarkers in natural fish populations indicate adverse biological effects of offshore oil production. *PLoS One*, 6(5), p. e19735. doi:10.1371/journal. pone.0019735.

Bell, P.S., McCann, D.L., Scott, B.E., Williamson, B.J., Waggitt, J.J., Ashton, I., Johanning, L., Blondel, P., Creech, A., Ingram, D. and Norris, J. (2014). Flow and Benthic ECology 4D – FLOWBEC – An overview. *Proceedings of the 2nd International Conference on Environmental Interactions of Marine Renewable Energy Technologies (EIMR2014), 28 April – 02 May 2014, Stornoway, Isle of Lewis, Outer Hebrides, Scotland*. Stornoway, Isle of Lewis, EIMR Conference Team: EIMR2014–2885.

Bergström, L., Sundqvist, F. and Bergström, U. (2013). Effects of an offshore wind farm on temporal and spatial patterns in the demersal fish community. *Marine Ecology Progress Series*, 485, pp. 199–210.

Birchenough, S.N. and Frid, C.L. (2009). Macrobenthic succession following the cessation of sewage sludge disposal. *Journal of Sea Research*, 62(4), pp. 258–267.

Birchenough, S.N., Reiss, H., Degraer, S., Mieszkowska, N., Borja, Á., Buhl-Mortensen, L., Braeckman, U., Craeymeersch, J., De Mesel, I., Kerckhof, F. and Kröncke, I. (2015). Climate change and marine benthos: A review of existing research and future directions. *WIREs Clim Change*, 6(2), pp. 203–223. doi:10.1002/wcc.330.

Blackwell, S.B., Greene Jr., C.R. and Richardson, W.J. (2004a). Drilling and operational sounds from an oil production island in the ice-covered Beaufort Sea. *The Journal of the Acoustical Society of America*, 116(5), pp. 3199–3211.

Blackwell, S.B., Lawson, J.W. and Williams, M.T. (2004b). Tolerance by ringed seals (*Phoca hispida*) to impact pipe-driving and construction sounds at an oil production island. *The Journal of the Acoustical Society of America*, 115(5), pp. 2346–2357.

Blackwell, S.B., Nations, C.S., McDonald, T.L., Thode, A.M., Mathias, D., Kim, K.H., Greene Jr, C.R. and Macrander, A.M. (2015). Effects of airgun sounds on bowhead whale calling rates: Evidence for two behavioral thresholds. *PLoS ONE*, 10(6), p. e0125720.

Boehlert, G.W. and Gill, A.B. (2010). Environmental and ecological effects of ocean renewable energy development – A current synthesis. *Oceanography*, 23, pp. 68–81.

Brette, F., Machado, B., Cros, C., Incardona, J.P., Scholz, N.L. and Block, B.A. (2014). Crude oil impairs cardiac excitation-contraction coupling in fish. *Science*, 343(6172), pp. 772–776.

Burrows, R., Yates, N.C., Hedges, T.S., Li, M., Zhou, J.G., Chen, D.Y., Walkington, I.A., Wolf, J., Holt, J. and Proctor, R. (2009). Tidal energy potential in UK waters. *Proceedings of the ICE-Maritime Engineering*, 162(4), pp. 155–164.

Butler, P.G., Richardson, C.A., Scourse, J.D., Wanamaker, A.D., Shammon, T.M. and Bennell, J.D. (2010). Marine climate in the Irish Sea: Analysis of a 489-year marine master chronology derived from growth increments in the shell of the clam *Arctica islandic. Quaternary Science Reviews,* 29(13–14), pp. 1614–1632.

Byrd, G.V. and Reynolds, H.L. (2007). Results of a drift experiment to estimate seabird carcass deposition on beaches at Unalaska Island, Alaska, in the vicinity of the wreck of the M/V (Preassessment data report No. 4). U.S. Fish and Wildlife Service, Maritime National Wildlife Refuge, Homer, Alaska, USA.

Campagna, C., Short, F.T., Polidoro, B.A., McManus, R., Collette, B.B., Pilcher, N.J., Sadovy de Mitcheson, Y., Stuart, S.N. and Carpenter, K.E. (2011). Gulf of Mexico oil blowout increases risks to globally threatened species. *BioScience,* 61(5), pp. 393–397.

Camphuysen, C.J., Chardine, J., Frederiksen, M. and Nunes, M. (2005). Review of the impacts of recent major oil spills on seabirds. *Report of the ICES Working Group on Seabird Ecology.* Texel, Netherlands. www.ices.dk/sites/pub/Publication%20 Reports/Expert%20Group%20Report/lrc/2005/wgse/wgse05.pdf

Camphuysen, C.J. and Leopold, M.F. (2004). The tricolor oil spill: Characteristics of seabirds found oiled in The Netherlands. *Atlantic Seabirds.* 6(3), pp. 109–128.

Castro, J.J., Santiago, J.A. and Santana-Ortega, A.T.(2004). A general theory on fish aggregation to floating objects: An alternative to the meeting point hypothesis. *Reviews in Fish Biology and Fisheries,* 11, pp. 255–277. doi:10.1023/A:1020302414472.

Cazenave, P. W., Torres, R. and Allen, J.I. (2014). Offshore renewable energy device impacts on seasonally stratified seas around the UK: An unstructured modelling approach. *Partnership for Research in Marine Renewable Energy (PRIMARE) 1st Annual conference.* June 2014. Plymouth.

Cerchio, S., Strindberg, S., Collins, T., Bennett, C. and Rosenbaum, H. (2014). Seismic surveys negatively affect humpback whale singing activity off Northern Angola. *PLoS One,* 9(3), p. e86464.

Christiansen, F. and Lusseau, D. (2015). Linking behavior to vital rates to measure the effects of non-lethal disturbance on wildlife. *Conservation Letters,* 8(6), pp. 424–431.

Chong, Z.R., Yang, S.H.B., Babu, P., Linga, P. and Li, X.S. (2016). Review of natural gas hydrates as an energy resource: Prospects and challenges. *Applied Energy,* 162, pp. 1633–1652.

Clark, N.A., (2006). Tidal barrages and birds. *Ibis,* 148, pp. 152–157. doi:10.1111/j.1474-919X.2006.00519.x.

Claudet, J. and Pelletier, D. (2004). Marine protected areas and artificial reefs: A review of the interactions between management and scientific studies. *Aquatic Living Resources,* 17(2), pp. 129–138.

Coates, D.A., Deschutter, Y., Vincx, M. and Vanaverbeke, J. (2014). Enrichment and shifts in macrobenthic assemblages in an offshore wind farm area in the Belgian part of the North Sea. *Marine Environmental Research,* 95, pp. 1–12.

Compton, R., Goodwin, L., Handy, R. and Abbott, V. (2008). A critical examination of worldwide guidelines for minimising the disturbance to marine mammals during seismic surveys. *Marine Policy,* 32(3), pp. 255–262.

Copping, A., Smith, C., Hanna, L., Battey, H., Whiting, J., Reed, M., Brown-Saracino, J., Gilman, P. and Massaua, M. (2013). Tethys: Developing a commons for understanding environmental effects of ocean renewable energy. *International Journal of Offshore Energy*, 3, pp. 41–51.

Council, G. W. E. Annual report (2015) [online] Available at www. gwec.net/wp-content/uploads/2012/06. GLOBAL_INSTALLED_WIND_POWER_CAPACITY_MW_–_Regional_Distribution. jpg [Accessed on 13 October 2015].

Denman, K., Christian, J. R., Steiner, N., Pörtner, H. O., and Nojiri, Y. (2011). Potential impacts of future ocean acidification on marine ecosystems and fisheries: Current knowledge and recommendations for future research. *ICES Journal of Marine Science*, 68(6), pp. 1019–1029.

De Troch, M., Reubens, J. T., Heirman, E., Degraer, S., & Vincx, M. (2013). Energy profiling of demersal fish: A case-study in wind farm artificial reefs. *Marine Environmental Research*, 92, pp. 224–233.

Dolman, S. and Simmonds, M. (2010). Towards best environmental practice for cetacean conservation in developing Scotland's marine renewable energy. *Marine Policy*, 34(5), pp. 1021–1027.

Drewitt, A.L. and Langston, R.H.W. (2006). Assessing the impacts of wind farms on birds. *Ibis*, 148, pp. 29–42. doi:10.1111/j.1474–919X.2006.00516.x.

Elliott, M., Borja, Á., McQuatters-Gollop, A., Mazik, K., Birchenough, S., Andersen, J.H., Painting, S. and Peck, M. (2015). Force majeure: Will climate change affect our ability to attain Good Environmental Status for marine biodiversity? *Marine Pollution Bulletin*, 95(1), pp. 7–27. doi:10.1016/j.marpolbul.2015.03.015.

Embling, C.B., Illian, J., Armstrong, E., van der Kooij, J., Sharples, J., Camphuysen, K.C.J. and Scott, B.E. (2012). Investigating fine-scale spatio-temporal predator–prey patterns in dynamic marine ecosystems: A functional data analysis approach. *Journal of Applied Ecology*, 49(2), pp. 481–492.

Fingas, M. (2014). *Handbook of oil spill science and technology*. Burlinton: John Wiley & Sons.

Fowler, A.M., Macreadie, P.I., Jones, D.O.B. and Booth, D.J. (2014). A multi-criteria decision approach to decommissioning of offshore oil and gas infrastructure. *Ocean & Coastal Management*, 87, pp. 20–29.

Fox, W.M., Johnson, M.S., Jones, S.R., Leah, R.T. and Copplestone, D. (1999). The use of sediment cores from stable and developing salt marshes to reconstruct historical contamination profiles in the Mersey Estuary, UK. *Marine Environmental Research*, 47(4), pp. 311–329.

Frid, C., Andonegi, E., Depestele, J., Judd, A., Rihan, D., Rogers, S.I. and Kenchington, E. (2012). The environmental interactions of tidal and wave energy generation devices. *Environmental Impact Assessment Review*, 32(1), pp. 133–139.

Froján, C.R.B., Boyd, S.E., Cooper, K.M., Eggleton, J.D. and Ware, S. (2008). Long-term benthic responses to sustained disturbance by aggregate extraction in an area off the east coast of the United Kingdom. *Estuarine, Coastal and Shelf Science*, 79(2), pp. 204–212.

Furness, R.W. and Greenwood, J.J.D. (1993). *Birds as Monitors of Environmental Change*. Dordrecht: Springer.

Furness, R.W., Wade, H.M. and Masden, E.A. (2013). Assessing vulnerability of marine bird populations to offshore wind farms. *Journal of Environmental Management*, 119, pp. 56–66. doi:10.1016/j.jenvman.2013.01.025.

Furness, R.W., Wade, H.M., Robbins, A.M. and Masden, E.A. (2012). Assessing the sensitivity of seabird populations to adverse effects from tidal stream turbines and wave energy devices. *ICES Journal of Marine Science Journal du Conseil*, 69(8), pp. 1466–1479. doi:10.1093/icesjms/fss131.

Geraci, J. (2012). Sea mammals and oil: confronting the risks. San Diego, CA: Elsevier.

Gill, A. B. (2005). Offshore renewable energy: Ecological implications of generating electricity in the coastal zone. *Journal of Applied Ecology*, 42(4), pp. 605–615.

Gill, A. B., Bartlett, M., and Thomsen, F. (2012a). Potential interactions between diadromous fishes of UK conservation importance and the electromagnetic fields and subsea noise from marine renewable energy developments. *Journal of Fish Biology*, 81(2), pp. 664–695.

Gill, A.B., Jude, S., et al. (2012b). Review of environmental risk and uncertainty for supporting policy development and decision making for the marine renewable energy sector. NERC MREKEP Report. Cranfield University.

Gill, A.B., Gloyne-Philips, I., Kimber, J. and Sigray, P. (2014). Marine renewable energy, electromagnetic (EM) fields and EM-sensitive animals. In: M. A. Shields and A. I. Payne (ed.), *Marine renewable energy technology and environmental interactions*. Netherlands: Springer, pp. 61–79.

Glover, A.G. and Smith, C.R. (2003). The deep-sea floor ecosystem: Current status and prospects of anthropogenic change by the year 2025. *Environmental Conservation*, 30(3), pp. 219–241.

Gordon, J., Gillespie, D., Potter, J., Frantzis, A., Simmonds, M.P., Swift, R. and Thompson, D. (2003). A review of the effects of seismic surveys on marine mammals. *Marine Technology Society Journal*, 37(4), pp. 16–34.

Halpern, B.S. (2003). The impact of marine reserves: Do reserves work and does reserve size matter? *Ecological applications*, 13(sp1), pp. 117–137.

Haney, J.C., Geiger, H.J. and Short, J.W. (2014a). Bird mortality from the Deepwater Horizon oil spill. I. Exposure probability in the offshore Gulf of Mexico. *Marine Ecology Progress Series*, 513, pp. 225–237. doi:10.3354/meps10991.

Haney, J.C., Geiger, H.J. and Short, J.W. (2014b). Bird mortality from the Deepwater Horizon oil spill. II. Carcass sampling and exposure probability in the coastal Gulf of Mexico. *Marine Ecology Progress Series*, 513, pp. 239–252. doi:10.3354/meps10839.

Hannon, M., Gimpel, J., Tran, M., Rasala, B. and Mayfield, S. (2010). Biofuels from algae: Challenges and potential. *Biofuels*, 1(5), pp. 763–784.

Harwood, J., King, S., Schick, R., Donovan, C. and Booth, C. (2014). A draft protocol for implementing the interim population consequences of disturbance (PCOD) approach: Assessing the effects of UK offshore renewable energy developments on marine mammal populations. SMRU Marine Report to the Crown Estate. *Scottish Marine and Freshwater Science*, 5(2).

Hastie, G.D., Gillespie, D.M., Gordon, J.C., Macaulay, J.D., McConnell, B.J. and Sparling, C.E. (2014). Tracking technologies for quantifying marine mammal interactions with tidal turbines: Pitfalls and possibilities. In: M. A. Shields and A. I. Payne (ed.), *Marine renewable energy technology and environmental interactions*. Netherlands: Springer, pp. 127–139.

Hastie, G.D., Russell, D.J., McConnell, B., Moss, S., Thompson, D. and Janik, V.M. (2015). Sound exposure in harbour seals during the installation of an offshore wind farm: Predictions of auditory damage. *Journal of Applied Ecology*, 52(3), pp. 631–640.

Hawkins, A.D. and Popper, A.N. (2014). Assessing the impacts of underwater sounds on fishes and other forms of marine life. *Acoustics Today,* 10(2), pp. 30–41.

Hazen, E.L., Carlisle, A.B., Wilson, S.G., Ganong, J.E., Castleton, M.R., Schallert, R.J., Stokesbury, M.J., Bograd, S.J. and Block, B.A. (2016). Quantifying overlap between the Deepwater Horizon oil spill and predicted bluefin tuna spawning habitat in the Gulf of Mexico. *Nature: Scientific Reports,* 6, Article number: 33824.

Hendrick, V.J., Hutchison, Z.L. and Last, K.S. (2016). Sediment burial intolerance of marine macroinvertebrates. *PLoS One,* 11(2), pp. e0149114. doi:10.1371/journal.pone.0149114.

Henkel, S.K., Suryan, R.M. and Lagerquist, B.A. (2014). Marine renewable energy and environmental interactions: Baseline assessments of seabirds, marine mammals, sea turtles and benthic communities on the Oregon shelf. In: M.A. Shields and A.I.L. Payne (ed.), *Marine renewable energy technology and environmental interactions.* Netherlands: Springer, pp. 93–110.

Incardona, J.P., Gardner, L.D., Linbo, T.L., Brown, T.L., Esbaugh, A.J., Mager, E.M., Stieglitz, J.D., French, B.L., Labenia, J.S., Laetz, C.A., Tagal, M., Sloan, C.A., Elizur, A., Benetti, D.D., Grosell, M., Block, B.A. and Scholz, N.L. (2014). Deepwater horizon crude oil impacts the developing hearts of large predatory pelagic fish. *Proceedings of the National Academy of Sciences of the United States of America,* 111(15), pp. E1510–E1518.

Inger, R., Attrill, M.J., Bearhop, S., Broderick, A.C., James Grecian, W., Hodgson, D.J., Mills, C., Sheehan, E., Votier, S.C., Witt, M.J. and Godley, B.J. (2009). Marine renewable energy: potential benefits to biodiversity? An urgent call for research. *Journal of Applied Ecology,* 46(6), pp. 1145–1153.

Jensen, A. (2002). Artificial reefs of Europe: perspective and future. *ICES Journal of Marine Science: Journal du Conseil,* 59(suppl), pp. S3–S13.

Jenssen, B.M. (1996). An overview of exposure to, and effects of, petroleum oil and organochlorine pollution in grey seals (*Halichoerus grypus*). *Science of the Total Environment,* 186(1–2), pp. 109–118.

Jones, A.R., Hosegood, P., Wynn, R.B., De Boer, M.N., Butler-Cowdry, S. and Embling, C.B. (2014). Fine-scale hydrodynamics influence the spatio-temporal distribution of harbour porpoises at a coastal hotspot. *Progress in Oceanography,* 128, pp. 30–48.

Judd, A.D., Backhaus, T. and Goodsir, F. (2015). An effective set of principles for practical implementation of marine cumulative effects assessment. *Environmental Science & Policy,* 54, pp. 254–262.

Kaiser, M.J., and Pulsipher, A.G. (2005). Rigs-to-reef programs in the Gulf of Mexico. *Ocean Development & International Law,* 36(2), pp. 119–134.

Kerckhof, F., Rumes, B., Jacques, T., Degraer, S. and Norro, A. (2010). Early development of the subtidal marine biofouling on a concrete offshore windmill foundation on the Thornton Bank (southern North Sea): First monitoring results. *Underwater Technology,* 29(3), pp. 137–149.

King, S.L., Schick, R.S., Donovan, C., Booth, C.G., Burgman, M., Thomas, L. and Harwood, J. (2015). An interim framework for assessing the population consequences of disturbance. *Methods in Ecology and Evolution,* 6: 1150–1158.

Koschinski, S., Culik, B.M., Henriksen, O.D., Tregenza, N., Ellis, G., Jansen, C. and Kathe, G. (2003). Behavioural reactions of free-ranging porpoises and seals to the noise of a simulated 2 MW windpower generator. *Marine Ecology Progress Series*, 265, pp. 263–273.

Langhamer, O., Wilhelmsson, D. and Engström, J. (2009). Artificial reef effect and fouling impacts on offshore wave power foundations and buoys–A pilot study. *Estuarine, Coastal and Shelf Science*, 82(3), pp. 426–432.

Langhamer, O. and Wilhelmsson, D. (2009). Colonisation of fish and crabs of wave energy foundations and the effects of manufactured holes–a field experiment. *Marine Environmental Research*, 68(4), pp. 151–157.

Langton, R., Davies, I.M. and Scott, B.E. (2011). Seabird conservation and tidal stream and wave power generation: Information needs for predicting and managing potential impacts. *Marine Policy*, 35, pp. 623–630. doi:10.1016/j.marpol. 2011.02.002.

Leighton, F.A. (1993). The toxicity of petroleum oils to birds. *Environmental Reviews*, 1, pp. 92–103. doi:10.1139/a93–008.

Lindeboom, H.J., Kouwenhoven, H.J., Bergman, M.J.N., Bouma, S., Brasseur, S.M. J.M., Daan, R. and Scheidat, M. (2011). Short-term ecological effects of an offshore wind farm in the Dutch coastal zone; a compilation. *Environmental Research Letters*, 6(3), p. 035101.

Lindeboom, H., Degraer, S., Dannheim, J., Gill, A. B. and Wilhelmsson, D. (2015). Offshore wind park monitoring programmes, lessons learned and recommendations for the future. *Hydrobiologia*, 756(1), pp. 169–180.

Maar, M., Bolding, K., Petersen, J.K., Hansen, J.L. and Timmermann, K. (2009). Local effects of blue mussels around turbine foundations in an ecosystem model of Nysted off-shore wind farm, Denmark. *Journal of Sea Research*, 62(2), pp. 159–174.

Madsen, P.T., Wahlberg, M., Tougaard, J., Lucke, K. and Tyack, P. (2006). Wind turbine underwater noise and marine mammals: Implications of current knowledge and data needs. *Marine Ecology Progress Series*, 309, pp. 279–295.

Masden, E.A., Haydon, D.T., Fox, A.D., Furness, R.W., Bullman, R. and Desholm, M. (2009). Barriers to movement: Impacts of wind farms on migrating birds. *ICES Journal of Marine Science: Journal du Conseil*, 66, pp. 746–753. doi:10.1093/icesjms/fsp031.

McNutt, M.K., Camilli, R., Crone, T.J., Guthrie, G.D., Hsieh, P.A., Ryerson, T.B., Savas, O. and Shaffer, F. (2012). Review of flow rate estimates of the Deepwater Horizon oil spill. *Proceedings of the National Academy of Science*, 109(50), pp. 20260–20267. doi:10.1073/pnas.1112139108.

Miller, R.G., Hutchison, Z.L., Macleod, A.K., Burrows, M.T., Cook, E.J., Last, K.S. and Wilson, B. (2013). Marine renewable energy development: Assessing the Benthic Footprint at multiple scales. *Frontiers in Ecology and the Environment*, 11(8), pp. 433–440.

Mora, C., Frazier, A.G., Longman, R.J., Dacks, R.S., Walton, M.M., Tong, E.J., Sanchez, J.J., Kaiser, L.R., Stender, Y.O., Anderson, J.M. and Ambrosino, C.M. (2013). The projected timing of climate departure from recent variability. *Nature*, 502(7470), pp. 183–187.

Muhling, B.A., Roffer, M.A., Lamkin, J.T., Ingram, G.W., Upton, M.A., Gawlikowski, G., Muller-Karger, F., Habtes, S. and Richards, W.J. (2012). Overlap between Atlantic bluefin tuna spawning grounds and observed Deepwater

Horizon surface oil in the northern Gulf of Mexico. *Marine Pollution Bulletin,* 64(4), pp. 679–687.

Munilla, I., Arcos, J.M., Oro, D., Álvarez, D., Leyenda, P.M. and Velando, A. (2011). Mass mortality of seabirds in the aftermath spill. *Ecosphere,* 2(7), pp. 1–14.

Narayan, S., Hanson, S., Nicholls, R.J., Clarke, D., Willems, P., Ntegeka, V. and Monbaliu, J. (2012). A holistic model for coastal flooding using system diagrams and the Source-Pathway-Receptor (SPR) concept. *Natural Hazards and Earth System Science,* 12(5), pp. 1431–1439.

Nehls, G., Betke, K., Eckelmann, S. and Ros, M. (2007). Assessment and costs of potential engineering solutions for the mitigation of the impacts of underwater noise arising from the construction of offshore windfarms. BioConsult SH report, Husum, Germany. On behalf of COWRIE Ltd.

Neill, S.P., Litt, E.J., Couch, S.J. and Davies, A.G. (2009). The impact of tidal stream turbines on large-scale sediment dynamics. *Renewable Energy,* 34(12), pp. 2803–2812.

Nieukirk, S.L., Mellinger, D.K., Moore, S.E., Klinck, K., Dziak, R.P. and Goslin, J. (2012). Sounds from airguns and fin whales recorded in the mid-Atlantic Ocean, 1999–2009. *Journal of the Acoustical Society of America,* 131(2), pp. 1102–1112.

Nowacek, D.P., Bröker, K., Donovan, G., Gailey, G., Racca, R., Reeves, R.R., Vedenev, A.I., Weller, D.W. and Southall, B.L. (2013). Responsible practices for minimizing and monitoring environmental impacts of marine seismic surveys with an emphasis on marine mammals. *Aquatic Mammals,* 39(4), pp. 356–377.

Nysveen, A., Kulbotten, H., Lervik., Bomes, A.H. and Hoyer-Hansen, M. (2005). Direct electrical heating of subsea pipelines – Technology development and operating experience. In: *Petroleum and Chemical Industry Conference, 2005. Industry Applications Society 52nd Annual,* pp. 177–187, 12–14 September 2005 doi:10.1109/PCICON.2005.1524553.

O'Hagan, A., Caitlin, E., Buck, C.E., Daneshkhah, A., Eiser, J.R., Garthwaite, P.H., Jenkinson, D.J., Oakley, J.E. and Rakow, T. (2006). *Eliciting Distributions – Uncertainty and Imprecision, in Uncertain Judgements: Eliciting Experts' Probabilities.* Chichester: John Wiley and Sons, Ltd.

Öhman, M.C., Sigray, P. and Westerberg, H. (2007). Offshore windmills and the effects of electromagnetic fields on fish. *AMBIO: A Journal of the Human Environment,* 36(8), pp. 630–633.

Petersen, J. K. and Malm, T. (2006). Offshore windmill farms: Threats to or possibilities for the marine environment. *AMBIO: A Journal of the Human Environment,* 35(2), pp. 75–80.

Pethick, J.S., Morris, R.K., and Evans, D.H. (2009). Nature conservation implications of a Severn tidal barrage–A preliminary assessment of geomorphological change. *Journal for Nature Conservation,* 17(4), pp. 183–198.

Petruny, L. M., Wright, A.J. and Smith, C.E. (2014). Getting it right for the North Atlantic right whale (Eubalaenaglacialis): A last opportunity for effective marine spatial planning? *Marine Pollution Bulletin,* 85(1), pp. 24–32.

Piatt, J.F. and Ford, R.G. (1996). How many seabirds were killed by the Exxon Valdez oil spill? In: Proc Exxon Valdez Oil Spill Symposium. *Presented at the American Fish Society Symposium, Vol. 18,* pp. 712–719. https://alaska.usgs.gov/

science/biology/seabirds_foragefish/products/publications/How_many_Sb_
killed_by_Spill.pdf

Pilgrim, J., Catmull, S., Chippendale, R., Tyreman, R. and Lewin, P. (2013). Off-shore wind farm export cable current rating optimization. In: *Proceedings of EWEA Offshore Wind Conference, Frankfurt, Germany. https://eprints.soton.ac.uk/361105/*

Pirotta, E., New, L., Harwood, J. and Lusseau, D. (2014). Activities, motivations and disturbance: An agent-based model of bottlenose dolphin behavioral dynamics and interactions with tourism in Doubtful Sound, New Zealand. *Ecological Modelling*, 282, pp. 44–58.

Rees, H.L., Pendle, M.A., Limpenny, D.S., Mason, C.E., Boyd, S.E., Birchenough, S. and Vivian, C.M.G. (2006). Benthic responses to organic enrichment and climatic events in the western North Sea. *Journal of the Marine Biological Association of the United Kingdom*, 86(1), pp. 1–18.

Reubens, J. T., Braeckman, U., Vanaverbeke, J., Van Colen, C., Degraer, S. and Vincx, M. (2013). Aggregation at windmill artificial reefs: CPUE of Atlantic cod (*Gadus morhua*) and pouting (*Trisopterus luscus*) at different habitats in the Belgian part of the North Sea. *Fisheries Research*, 139, pp. 28–34.

Richardson, W.J. and Thomson, D.H. (1995). *Marine Mammals and Noise*. San Diego, CA: Academic Press.

Ronconi, R.A., Allard, K.A. and Taylor, P.D. (2015). Bird interactions with offshore oil and gas platforms: Review of impacts and monitoring techniques. *Journal of Environmental Management*, 147, pp. 34–45. doi:10.1016/j.jenvman.2014.07.031.

Russell, D.J., Brasseur, S.M., Thompson, D., Hastie, G.D., Janik, V.M., Aarts, G., McClintock, B.T., Matthiopoulos, J., Moss, S.E. and McConnell, B. (2014). Marine mammals trace anthropogenic structures at sea. *Current Biology*, 24(14), pp. R638–R639.

Russell, R.W. (2005). Interactions between migrating birds and offshore oil and gas platforms in the Northern Gulf of Mexico. (Final Report No. OCS Study MMS2005–009). *U.S. Dept. of the Interior, Minerals Management Service, New Orleans, LA*.

Schroeder, D.M. and Love, S.M. (2004). Ecological and political issues surrounding decommissioning of offshore oil facilities in the Southern California Bight. *Ocean & Coastal Management*, 47(1–2), pp. 21–48.

Schwacke, L.H., Smith, C.R., Townsend, F.I., Wells, R.S., Hart, L.B., Balmer, B.C., Collier, T.K., De Guise, S., Fry, M.M., Guillette Jr, L.J. and Lamb, S.V. (2014). Health of Common Bottlenose Dolphins (*Tursiops truncatus*) in Barataria Bay, Louisiana, Following the Deepwater Horizon Oil Spill. *Environmental Science & Technology*, 48(1), pp. 93–103.

Schwemmer, P., Mendel, B., Sonntag, N., Dierschke, V. and Garthe, S. (2010). Effects of ship traffic on seabirds in offshore waters: implications for marine conservation and spatial planning. *Ecological Applications*, 21, pp. 1851–1860. doi:10.1890/10–0615.1.

Scott, B.E., Sharples, J., Ross, O.N., Wang, J., Pierce, G.J. and Camphuysen, C.J. (2010). Sub-surface hotspots in shallow seas: fine-scale limited locations of top predator foraging habitat indicated by tidal mixing and sub-surface chlorophyll. *Marine Ecology Progress Series*, 408, pp. 207–226.

Seaman, W. (2007). Artificial habitats and the restoration of degraded marine ecosystems and fisheries. In J. Ryland and G. Relini (ed.), *Biodiversity in enclosed seas and artificial marine habitats*. Netherlands: Springer, pp. 143–155.

Simmonds, M.P. and Brown, V.C. (2010). Is there a conflict between cetacean conservation and marine renewable-energy developments? *Wildlife Research,* 37(8), pp. 688–694.

Southall, B.L., Bowles, A.E., Ellison, W.T., Finneran, J.J., Gentry, R.L., Greene Jr, C.R., Kastak, D., Ketten, D.R., Miller, J.H., Nachtigall, P.E. and Richardson, W.J. (2007). Marine mammal noise exposure criteria: Initial scientific recommendations (2007). *Aquatic Mammals,* 33(4), pp. 273–275.

Slabbekoorn, H., Bouton, N., van Opzeeland, I., Coers, A., ten Cate, C. and Popper, A.N. (2010). A noisy spring: The impact of globally rising underwater sound levels on fish. *Trends in Ecology & Evolution,* 25(7), pp. 419–427.

Smyth, K., Christie, N., Burdon, D., Atkins, J.P., Barnes, R. and Elliott, M. (2015). Renewables-to-reefs?–Decommissioning options for the offshore wind power industry. *Marine pollution Bulletin,* 90(1), pp. 247–258.

Tarpgaard, E., Mikaelsen, M., et al. (2015). Why a site-specific evaluation of the need for noise mitigation must replace the general evaluations. *EWEA Offshore.* Copenhagen: PO.ID 205.

Thompson, P.M., Hastie, G.D., Nedwell, J., Barham, R., Brookes, K.L., Cordes, L.S., Bailey, H. and McLean, N. (2013). Framework for assessing impacts of pile-driving noise from offshore wind farm construction on a harbour seal population. *Environmental Impact Assessment Review,* 43, pp. 73–85.

Thomsen, F., Lüdemann, K., Kafemann, R. and Piper, W. (2006). Effects of offshore wind farm noise on marine mammals and fish. Biola, Hamburg, Germany, COWRIE Ltd.

Thomsen, F. (2010). Sound impacts. In: J. Huddleston (ed.), *Understanding the environmental impacts of offshore windfarms information.* Oxford: COWRIE.

Todd, V.L., Pearse, W.D., Tregenza, N.C., Lepper, P.A. and Todd, I.B. (2009). Diel echolocation activity of harbour porpoises (*Phocoena phocoena*) around North Sea offshore gas installations. *ICES Journal of Marine Science,* 66(4), pp. 734–745.

Tougaard, J., Carstensen, J., Teilmann, J., Skov, H. and Rasmussen, P. (2009). Pile driving zone of responsiveness extends beyond 20 km for harbor porpoises (*Phocoena phocoena L.*). *The Journal of the Acoustical Society of America,* 126(1), pp. 11–14.

Tougaard, J., Wright, A.J. and Madsen, P.T. (2015). Cetacean noise criteria revisited in the light of proposed exposure limits for harbour porpoises. *Marine Pollution Bulletin,* 90(1), pp. 196–208.

Truebano, M., Embling, C., Witt, M., Godley, B.J., Attrill, M. (2013). The potential impacts of marine renewable energy on marine mammals. In: *Marine Renewables Biodiversity and Fisheries.* Ed: Attrill, M. Friends of the Earth, Marine Institute Plymouth University, pp. 1–29.

Tyack, P. and Janik, V. (2013). Effects of noise on acoustic signal production in marine mammals. In: H. Brumm (ed.), *Animal communication and noise.* Berlin: Springer, pp. 251–271.

United Nations Framework Convention on Climate Change. (2015). Adoption of the Paris Agreement, FCCC/CP/L.9/Rev.1.

Venn-Watson, S., Colegrove, K.M., Litz, J., Kinsel, M., Terio, K., Saliki, J., Fire, S., Carmichael, R., Chevis, C., Hatchett, W. and Pitchford, J. (2015). Adrenal gland and lung lesions in gulf of Mexico common bottlenose dolphins (Tursiops truncatus) found dead following the deepwater horizon oil spill. *PLoS One*, 10(5), pp. e0126538.

Viehman, H.A., Zydlewski, G.B., McCleave, J.D. and Staines, G.J. (2014). Using hydroacoustics to understand fish presence and vertical distribution in a tidally dynamic region targeted for energy extraction. *Estuaries and Coasts*, 38(1), pp. 215–226.

WavEC (2015). Consenting processes for ocean energy on OES member countries. A report prepared by WavEC for the OES under *ANNEX I -Review, Exchange and Dissemination of Information on Ocean Energy Systems*, p. 57.

Weilgart, L. (2012). Are there technological alternatives to air guns for oil and gas exploration to reduce potential noise impacts on cetaceans? In: A.N. Popper and A. Hawkins (ed.), *Effects of noise on aquatic life*. New York: Springer Press, pp. 605–607.

Westerberg H. and Lagenfelt, I. (2008). Sub-sea power cables and the migration behaviour of the European eel. *Fisheries Management and Ecology*, 15, pp. 369–375.

Whomersley, P. and Picken, G.B. (2003). Long-term dynamics of fouling communities found on offshore installations in the North Sea. *Journal of the Marine Biological Association of the UK*, 83(5), pp. 897–901.

Wiese, F.K., Montevecchi, W.A., Davoren, G.K., Huettmann, F., Diamond, A.W. and Linke, J. (2001). Seabirds at risk around offshore oil platforms in the north-west Atlantic. *Marine Pollution Bulletin*, 42(12), pp. 1285–1290.

Wilhelmsson, D. and Malm, T. (2008). Fouling assemblages on offshore wind power plants and adjacent substrata. *Estuarine, Coastal and Shelf Science*, 79(3), pp. 459–466.

Wilhelmsson, D., Malm, T. and Öhman, M.C. (2006). The influence of offshore windpower on demersal fish. *ICES Journal of Marine Science: Journal du Conseil*, 63, pp. 775–784. doi:10.1016/j.icesjms.2006.02.001.

Williams, R., Gero, S., Bejder, L., Calambokidis, J., Kraus, S.D., Lusseau, D., Read, A.J. and Robbins, J. (2011). Underestimating the damage: interpreting cetacean carcass recoveries in the context of the Deepwater Horizon/BP incident. *Conservation Letters*, 4(3), pp. 228–233.

Willsteed, E., Gill, A.B., Birchenough, S.N. and Jude, S. (2017). Assessing the cumulative environmental effects of marine renewable energy developments: establishing common ground. *Science of the Total Environment*, 577, pp. 19–32.

Wilson, B., Batty, R., et al. (2006). Collision risks between marine renewable energy devices and mammals, fish and diving birds: Report to the Scottish Executive.

Wilson, B. and Carter, C. (2013). The use of acoustic devices to warn marine mammals of tidal-stream energy devices. Edinburgh, Scottish Government Report: 33.

Witt, M.J., Sheehan, E.V., Bearhop, S., Broderick, A.C., Conley, D.C., Cotterell, S.P., Crow, E., Grecian, W.J., Halsband, C., Hodgson, D.J. and Hosegood, P. (2012). Assessing wave energy effects on biodiversity: The Wave Hub experience. *Philosophical Transactions of Royal Society London, Series A: Mathematical Physical and Engineering Sciences*, 370(1959), pp. 502–529. doi:10.1098/rsta.2011.0265.

World Ocean Review. (2014). World Ocean Review 1: Living with the oceans – a report on the state of the world's oceans. [Online] Available at http://world-oceanreview.com/en/comments/feed/wor-1/energy/fossil-fuels/2/ [Accessed: 24 August 2015].

Wright, A. J. and Kyhn, L.A. (2015). Practical management of cumulative anthropogenic impacts with working marine examples. *Conservation Biology,* 29(2), pp. 333–340.

Wright, A.J. and Weilgart, L. (2011). Assessing cumulative impacts of underwater noise with other stressors on marine mammals. *The Journal of the Acoustical Society of America,* 129(4), pp. 2394–2394.

Wright, G. (2014). Strengthening the role of science in marine governance through environmental impact assessment: A case study of the marine renewable energy industry. *Ocean & Coastal Management,* 99, pp. 23–30.

Chapter 9

Meaningful stakeholder participation in marine spatial planning with offshore energy

Katherine L. Yates

Introduction

As our oceans get busier, and competition for resource access grows, the need for more effective and integrated management of the marine environment becomes increasingly pressing (Crowder et al., 2006; Douvere et al., 2007). Marine spatial planning is seen as the main tool to resolve conflict between stakeholders, integrate multiple sectors, and rationalise the multifaceted complexities of marine management (Douvere et al., 2007; Ehler and Douvere, 2009, Chapter 1). Marine spatial planning is inherently a participatory process in which stakeholders play a fundamental role (Ehler & Douvere, 2009; Calado et al., 2010).

The importance of stakeholder participation in marine spatial planning is recognised in international, regional, and national policies (Commission of the European Community, 2008; Flannery & Cinnéide, 2008; Calado et al., 2010), and in some places is now required by law (Kerr et al., 2014). Indeed, stakeholders are included to some capacity in almost all marine spatial planning process (Collie et al., 2013); however, the scope and extent of stakeholder involvement differs greatly among countries (Collie et al., 2013), and is often culturally influenced (Elher, personal communication, Ocean Visions Consulting, 2017).

Incorporating stakeholders is important for many reasons, not least because they can enhance the quality of resultant decisions (Brody, 2003; Reed, 2008) and improve the support for marine spatial plans (Pomeroy & Douvere, 2008). However, the impacts of incorporating stakeholders on the quality of, and the support for, planning outcomes are highly influenced by the process by which those plans were derived (Reed, 2008). Unfortunately, many stakeholders do not feel satisfied with participation processes (Pita et al., 2011; Yates, 2014; Reilly, Hagan, & Dalton, 2016), and many planning endeavours have failed due to poor stakeholder engagement (e.g., Weible, 2008; Anderson, 2013).

Many marine spatial planning processes were initiated as a result of expanding offshore energy developments, particularly marine renewables (Jay, 2010, Chapters 1 and 12). The combined effects of rising demands for energy, desires to curtail carbon emissions, and limited development options on land have seen a large rise in offshore wind and an increasing interest in other types of marine renewable energy sources (Pelc & Fujita, 2002; Jay, 2010). Discovery of new fossil fuel deposits also opens up the possible expansion of non-renewable, offshore energy extraction (Chapter 14). Offshore energy competes for space with existing, often longstanding uses/ stakeholders, and the resultant conflicts have been a substantial driver to rationalise management and move to more integrated governance. In this sense, offshore energy is both a driver of and a stakeholder in marine spatial planning.

In this chapter, I consider why stakeholders should be the heart of marine spatial planning, and I explore the reality of stakeholder participation, which is invariable, tricky, costly, often ineffective, and sometimes just fails. Following a discussion on successful incorporation of stakeholders into marine spatial planning, in which I consider the added complexities of offshore energy, I conclude by looking at some of the tools that can be used to enable more effective and transparent stakeholder participation.

Stakeholders at the heart of marine spatial planning

Marine spatial planning problems are complex, multifaceted, and often contentious. To optimise the allocation of space and meet objectives, which usually span social, economic, and ecological aims, it is essential that as complete as possible an understanding of the planning problem is obtained. To do this, it is fundamental that stakeholders, a vital source of knowledge that can improve both the planning process and the resultant plans, are involved at all stages of marine spatial planning (Commission of the European Community, 2008).

Stakeholder involvement allows planners to appreciate the wider context of a given planning scenario, and targeting a broad range of expert and 'non-expert' input is important to gain as much of the multifaceted nature of this context as is feasible (Middendorf & Busch, 1997; Reed, 2008; Fletcher et al., 2013). Different stakeholder groups have different priorities (Clarke, Thurstan, & Yates, 2016), and even differ on how they think the marine spatial planning process should be done (Gopnik et al., 2012). Stakeholders are best placed to explain their own priorities and to describe the conflicts they experience with other stakeholders (Yates, 2014), and they can also provide insight into wider social and political values that would not be gained through adopting an exclusively scientific approach (Middendorf & Busch, 1997). Data gaps are common in marine spatial planning, and stakeholders can provide valuable data that might not have otherwise been available, such as the distribution of current use of marine space (Yates & Schoeman, 2013). These data can then be incorporated into the decision-making process, enhancing the planners' ability to find the most efficient planning options (Yates & Schoeman, 2014). Stakeholders can also identify and develop opportunities, such as co-location (Chapters 10, 11, and 13) that might not have occurred to a planner.

With stakeholders, planners can explore and get a more holistic understanding of the potential impacts of different planning options. It is not only the planner(s) that gain(s) from this improved understanding; different stakeholders can also develop a greater appreciation of each other's needs, priorities, and aspirations. Seen by some as a learning and iterative procedure that is as important as the final plan, a collaborative, stakeholder-centred marine spatial planning process should allow stakeholders to learn about and recognise the legitimacy of each other's viewpoints (Keen & Mahanty, 2006). Indeed, stakeholders involved in a marine spatial planning pilot study in Dorset (United Kingdom) identified the opportunity to "... network with people you wouldn't normal come across" as a strength of the process, and they attributed their improved knowledge of marine spatial planning and issues to the "... broad selection of people involved" (Fletcher et al., 2013).

Being involved in marine spatial planning improves stakeholders' understanding of the complexities involved and provides planners and managers the opportunity to explain the need for change, including any underpinning legislation or policy drivers that influence the planning process (Fletcher et al., 2013). Increased understanding of both the need for management measures and the process by which they developed should lead to greater acceptance and compliance (Innes, 1996). Thus, as well as stakeholders contributing to the planning process, involving them can also be seen as an important capacity-building exercise (Nutters & Pinto da Silva, 2012; Anderson, 2013).

A stakeholder-centred planning process also increases the perception of ownership and inclusiveness of any resultant plans, which in turn should increase the ease of implementation of the final plan (Cinnéide, 2007). This sense of ownership of the planning process combined with an improved understanding of other viewpoints is thought to lead stakeholders away from competitive-interest bargaining and towards consensus building (Healey, 2006), which in turn should lead to greater support and reduced inter-stakeholder conflict (Flannery & Cinnéide, 2008). Indeed, some suggest that while the plan is the ultimate goal, it is the participatory process and the increased capacity and support that results from it that is in fact the most valuable outcome (Flannery & Cinnéide, 2008).

Stakeholder attitudes and behaviours can influence both positively and negatively the outcome of management endeavours (Pita et al., 2011; Alexander et al., 2012), and evidence shows that the flipside of successful engagement is also true—inadequate stakeholder engagement can cause the planning process to fail. Even a small group of disengaged stakeholders can railroad a project; for example, in South Australia, failure to address the concerns of the community led to the emergence of a small protest group and ultimately, the abandonment of a windfarm development (Anderson, 2013).

Of course, involving stakeholders is not a small or straightforward task; on the contrary, it is inherently complicated, costly, and potentially risky. Bringing different groups of stakeholders together could be the best way to foster greater understanding of the issues and improved appreciation of other viewpoints, but it does mean bringing together a lot of opposing and potentially hostile points of view. Discussing conflict can be a means to resolve it, but these discussions need multiple meetings, substantial time, and careful, experienced facilitation (Reed, 2008). The more inclusive the process and the more stakeholders involved, the more time and monetary costs are likely to be required.

Some argue that stakeholder participation can become 'talking shops' that achieve little and delay decisive action (Bojórquez-Tapia et al., 2004; Vedwan et al., 2008). Good leadership and clear aims for the participation process are important to help avoid this, but it is essential to acknowledge from the beginning that stakeholder participation will take time, cost money, and is likely get messy along the way. With this acknowledgement, adequate resources to support stakeholder participation must also be allocated (Degnbol & Wilson, 2008; Fletcher et al., 2013). Consideration of resource need should extend beyond government departments to include stakeholder groups who will need to allocate sufficient staff and members' time to engage fully in the entire planning process (Fletcher et al., 2013).

Stakeholder participation does not take place in a power vacuum (Reed, 2008). Among stakeholder groups, and indeed within a given stakeholder group, there will be those that are, or are perceived to be, more powerful than others, and there will be those that have been previously marginalised. The offshore energy industry is a prime example of a 'powerful' stakeholder with a lot of resources that can intimidate small, coastal communities (Baker, Roberts, & Shaw, 2004). Participation processes, if not carefully organised, can reinforce existing group dynamics and perceptions rather than empowering the marginalised. Careful consideration of equity of representation, as well as equity of impact, has to be made when developing participation processes. Finding ways to balance power, and the perception of power, is essential for allowing all stakeholders to contribute. It is also important to recognise that stakeholder groups are themselves heterogeneous and representatives sent to engage in marine spatial planning might also not be truly representative of the stakeholder groups they purport to represent (Yates, 2014). For example, the fishing community, which is arguably the stakeholder group that will be most directly impacted by the expansion of offshore energy, has diverse spatial priorities depending on target species, gear type, boat size, and home port (Yates & Schoeman, 2013). Participation processes need to enable incorporation of this detail within stakeholder subgroups to allow as complete as possible a picture of the *user-user* and *user-nature* conflicts that need to be addressed during the planning processes.

There is also a need to ensure that stakeholders have the support they need to be able to participate (Fletcher et al., 2013). Marine spatial planning is complicated, and aspects can be highly technical. The credibility of participation processes has in the past been criticised on the basis that stakeholders do not have sufficient expertise to engage meaningfully (e.g., Fischer and Young, 2007). While this might seem a reasonable concern, it does display a certain arrogance of traditional 'expertise' and does not acknowledge that one of the main benefits of the processes is mutually increased understanding of the issues contained within marine spatial planning. Fishers might not understand the intricacies of offshore wind-farm structures, but nor do engineers understand the intricacies of fishing, and neither group is likely to appreciate fully the impact of expanding or contracting their activities on the socio-economic situation of the wider coastal community. Moreover, traditional science is increasingly recognising that stakeholders, particularly fishers, can contribute greatly to our understanding of the marine environment (Hind, 2012). Marine spatial planning will inevitably require many different types of 'experts', and those experts will need to develop skills to communicate between sectors. Only by bringing stakeholders together will planners be able to start to piece together the complex reality of a planning situation. Arguably, only with all those different experts contributing will the most innovative solutions be found.

Meaningful participation

Marine spatial planning is most commonly led by government agencies, and most processes involve stakeholders; however, participation varies substantially in the methods employed and the amount of stakeholder involvement in decision making (Collie et al., 2013). For participation to be meaningful, stakeholders have to have some influence on the outcome of planning (Flannery & Cinnéide, 2008). The most meaningful participation will involve stakeholders right from the phase of objective setting (Ehler & Douvere, 2009). Of course, this does not always happen, and there are many different models that stakeholder participation can follow.

Participation can be divided into five main categories depending on degree of participation ('Participation Spectrum'; International Association of Public Participation 2004; see Figure 1). At the most minimal end of the spectrum, there is one-way communication through information provision. This is a heavily top-down approach, with the government agencies doing the planning and the stakeholders simply being told about it. This the simple provision of information is not meaningful participation. Consultation, a participation method commonly employed by government agencies in Europe, is somewhat more involved, where planners seek stakeholder views about proposed plans. This is still predominantly top-down. Examples include the German Exclusive Economic Zone Plan and Norwegian Barents Sea Plan, where government agencies primarily developed the plan and stakeholders subsequently participated in a public review process (Collie et al., 2013). The extent to which stakeholder views are incorporated in the planning decisions, and the impact consultation has on the final plan, vary greatly. Importantly, while consultation demonstrates a commitment to receive stakeholder feedback, it does not commit to any specific use of that feedback. Whether consultation can constitute meaningful participation will depend on the specifics of the consultation process and the extent to which stakeholder views influence outcomes. It is inevitable, however, that even the most meaningful consultation will be limited in terms of benefits of stakeholder participation, including the extent to which stakeholders feel engaged in the process and the subsequent ownership over the resulting plans. Indeed, experience of consultation with fishers around offshore renewable energy developments indicates that consultation is inadequate, with stakeholders going so far as to say it is just a 'box-ticking' exercise (Kerr et al., 2014; Reilly, Hagan, & Dalton, 2016).

At the other end of the scale, there are bottom-up approaches where planning starts with maximum stakeholder participation (Figure 9.1). In these partnerships, participation means that stakeholders are part of setting the agenda and visioning before planning even begins, and the final decision-making is normally shared across the partnership. In the California Marine Life Protection Act, for example, stakeholders played active roles in developing goals, synthesising data, assessing impacts, and submitting

Increasing stakeholder impact on planning decisions

Inform	Consult	Involve	Collaborate	Empower
Provide information to assist stakeholders in understanding the planning process, problems, opportunities and/or solutions. One directional. No pathway to impact the decisions.	Provide stakeholders the opportunity to feedback on planning options and/or decisions. Stakeholders concerns and aspirations listened to. Possibly inform stakeholders how their input influenced decisions.	Work directly with stakeholders throughout the process to ensure that concerns and aspirations are constantly understood and considered. Feedback provided on how stakeholder input influenced decisions.	Partner with stakeholders in all aspects of planning, including the development of alternatives and the identification of preferred solutions. Advice and recommendations of stakeholders incorporated to maximum extent possible.	Final decision in the hands of stakeholders.
Relatively rare	More common in Europe	More common in America and Australia		Relatively rare
China, Marine Functional Zoning planning process	Norway, Barents Sea Marine Plan Germany, Exclusive Economic Zone Plan	USA, California Marine Life Protection Act Australia, Great Barrier Reef Marine Park Zoning Plan		St. Kitts and Nevis Marine Spatial Plan
Top-down				Bottom-up

Figure 9.1 The spectrum of stakeholder participation.
Source: Adapted from the International Association for Public Participation core values with marine spatial planning examples taken from Collie et al. (2013).

suggested plans (Weible, 2008; Gleason et al., 2010; Collie et al., 2013). In some cases, participation can go so far as to empower stakeholders with the final decision making, such as in St. Kitts and Nevis, where the stakeholder participants had equal status with the government throughout the planning process (Collie et al., 2013).

Bottom-up approaches should inherently give stakeholders more input to the planning decision, and thus have a much greater opportunity to be meaningful. However, meaningful participation also requires that participation is effective. Stakeholders empowered to make planning decisions will need more support, including mediation, facilitation, and technical and scientific expertise. More meaningful participation approaches are inherently more expensive and can take longer, so more funding must be allocated at the outset to ensure that the process can continue to completion. Some argue that neither entirely top-down (inform) nor bottom-up (empower) approaches can work by themselves, and that some combination should be used (Fleming & Jones, 2012).

All stakeholders or some

All individuals, groups, and organisations that are in some way affected, involved, or interested in marine spatial planning can be considered 'stakeholders' (Pomeroy & Douvere, 2008). To what extent all of these different stakeholders can and should be incorporated in marine spatial planning is debatable. While stakeholder participation is beneficial, and even fundamental to successful marine spatial planning, it is also costly in both time and money. As the number of stakeholders involved increases, it is likely that those costs will also increase. Resources will always be finite, so decisions will have to be made about stakeholder selection. Some stakeholders, such as fishers and offshore energy companies, depend directly on access to the marine environment for their livelihoods. In the case of fishers, their jobs and connection to the sea are also central aspects of their social well-being and cultural heritage (Britton, 2012). It has been argued that stakeholders like fishers with direct dependencies and historic connections are more entitled to a role in marine spatial planning, and that stakeholders should be categorised into 'primary', 'secondary', and 'tertiary' in terms of their entitlement to participate (Pomeroy & Douvere, 2008). Others argue that this 'entitlement' approach makes the process elitist and contradicts the good practice in stakeholder participation developed in terrestrial planning (Ritchie & Ellis, 2016). Instead, they argue that marine spatial planning should begin from a position where all the views of all stakeholders are central to the process (Ritchie & Ellis, 2016). This opinion is echoed by others who identify that including the wider stakeholder community as an important element of marine spatial planning and the ecosystem approach it embodies (Fleming & Jones, 2012). Involving all stakeholders also means that everyone's concerns can be addressed when they arise, which is generally easier than dealing with them later (Fletcher et al., 2013). But involving all stakeholders greatly limits the options of how they can participate; there are unavoidable constraints on how many people can fit around a table and have an effective discussion, and the more stakeholders there are, the fewer the resources available for supporting each individual's participation. Unavoidably, the more inclusive a stakeholder participation process, the less meaningful it can be for each individual.

Methods such as spatial access priority mapping (Yates & Schoeman, 2013) and projects such as ScotMap (Kafas et al., 2014) enable participation of many individual stakeholders to move from the more traditional consultation following initial planning, to actual involvement in plan development. Nevertheless, even when all opinions are sought, in reality a few individuals will have to be selected to have a more active role in the planning process. For example, in a marine spatial planning project in the Firth of Clyde, Scotland, there was a steering group comprised of a 'core' membership of representatives from marine industries and regulatory bodies in the area, which was complimented with a wider public consultation (Flannery, 2012). This might seem like a sensible approach, but it does mean that the core membership was afforded a much more meaningful mode of participation, with a greater opportunity to influence the planning process than the general public. That is not necessarily a problem, but it is a choice and should be a conscious, explicit one.

Neither do all stakeholders want everyone else to be involved. For example, some fishers in the United Kingdom consider that the general public should not be involved as stakeholders in marine spatial planning because they believe that the general public has limited knowledge of the marine environment and that they are overly influenced by conservation organisations (Jones, 2008; Fleming & Jones, 2012). However, public involvement as stakeholders is part of many marine spatial planning policies and legislation, including the United Kingdom Marine Bill (The House of Commons, 2009). Thus, mechanisms to enable the wider public to participate in a meaningful and transparent way need to be found, and choices about how the public's contribution is incorporated need to be explicit and openly communicated.

Barriers to meaningful participation

Integration

Integrated governance and management of the marine environment is a long-standing international aspiration, identified as an important action within the United Nations Convention of the Law of the Sea (Douvere & Ehler, 2009), which predates marine spatial planning by several decades (Chapter 5). Marine spatial planning is now seen as the main tool with which to achieve integration (Crowder et al., 2006; Douvere et al., 2007; Ehler & Douvere, 2009; Kidd & Ellis, 2017). But there are many aspects to integration, some of which offer substantial barriers to meaningful stakeholder participation.

First, there is a need to get stakeholders to agree that integrating existing sectoral management is valuable and worth the substantial effort and investment required (Ritchie & Ellis, 2016). Pre-existing conflict among stakeholders, including government agencies that are reluctant to share power and collaborate with other agencies, can make adopting an integrated approach difficult (Guentte & Alder, 2007; Yates et al., 2013).

Another issue is lack of relevant knowledge and data, which is one of the most commonly highlighted issues for developing integrated management (Douvere & Ehler, 2011). While stakeholders represent a huge resource in terms of both data and knowledge, there is an acknowledged need to find robust ways to integrate information from stakeholders with other forms of data (Alexander et al., 2012). There is an expanding literature, for example, on the knowledge of fishers and how it should be used to inform marine management (Hind, 2015). However, there are technical and cultural challenges to integrating that fisher knowledge, which is often qualitative and generally in non-standard formats that are incongruous with more traditional, highly systematic, and quantitative scientific data. These two types of stakeholders—scientists and fishers—have different expertise and often only limited exposure to each other's expertise. Fisher knowledge can be perceived by scientists to be 'sub-standard' compared to scientific data, and this attitude can be a substantial barrier to the integration of fishers knowledge (Soto, 2006). Not only is integration of stakeholder information important for filling knowledge gaps, it is also important for legitimacy and stakeholder buy-in. Stakeholders understandably want to see their input used. Failure to integrate relevant data could also be used by those dissatisfied with the planning process to try and discredit any decisions as inadequate or incomplete (Collie et al., 2013).

Communication

Clear, consistent, targeted, and timely communication with stakeholders is fundamental for effective participation, and so it must be at the heart of marine spatial planning (Nutters & Pinto da Silva, 2012; Kannen, 2014). Good communication is essential for establishing and maintaining trust among stakeholders and is central to encouraging collaboration (Dougill et al., 2006). Internal communication within stakeholder groups also needs to be strong, or the message and wider engagement can be diluted (Fletcher et al., 2013). Communicating the meaning of participation clearly and the extent to which stakeholder will influence decisions is important to ensuring consistency between expectation and outcomes (Nutters & Pinto da Silva, 2012). In particular, roles and objectives should be clearly defined to avoid misunderstandings (Portman, 2009). In the case of the Lyme Bay (United Kingdom) planning process, the failure to communicate desired outcomes clearly was considered a major factor in the breakdown of trust (Fleming & Jones, 2012). Stakeholders also need to understand clearly what is expected from them throughout the process to enable them to provide the desired information, feedback, and input (Portman, 2009).

Communication strategies should also ensure that content is understandable to the target audience (Portman, 2009). This is not just about translating the technical terminology to the lay person, but also communication between different types of technological sectors. Developing a glossary of terms to facilitate integration of highly specialised knowledge held within different sectors and by different stakeholders is one suggestion for enhancing communication between stakeholder groups (OSPAR Commission, 2009). Communication methods employed should also acknowledge that different stakeholder groups access different sources of information and that diverse communication channels are required, especially if one wishes to encourage more inclusive participation. For example, industry-targeted publications, such as the *Fishing News* (fishingnew.co.uk), can be an effective way to reach the wider stakeholder group within an industry sector to invite their input. Where possible, exploiting trusted, pre-existing communication channels is advantageous (Fletcher et al., 2013), but by its very nature, marine spatial planning will requires the creation of new lines of communication among stakeholders (Nutters & Pinto da Silva, 2012).

Transparency

Developing transparency in marine spatial plans is central for ensuring ongoing support and is highlighted in many policy documents and guidelines (Gilliland & Laffoley, 2008; Pomeroy & Douvere, 2008; Gopnik et al., 2012), including occurring as one of the 10 'key principles' in the European Community's Roadmap for marine planning (Commission of the European Community, 2008). Transparency underpins the accountability and legitimacy of both the planning process and the resultant plan (Calado et al., 2010), and increased transparency of marine management is one of potential benefits assigned to undertaking marine spatial planning (Ehler, 2008). Stakeholders themselves identify a transparent process as an important aspect of success (Gopnik et al., 2012). Stakeholders need to know the 'rules of the game' if they are to engage fully with processes. Transparency is also central to the development and maintenance of trust between stakeholders, which is essential for developing collaborations and for progressing the process in the direction of consensus building (Reed, 2008; Kerr et al., 2014). Some go so far as to suggest that without a transparent process, it is unlikely that stakeholder discussions will lead to a positive outcome (Fleming & Jones, 2012).

To ensure transparency, stakeholder roles and the decision-making process need to be clear, and resultant decisions need to be communicated and justified to stakeholders (COM, 2010). Stakeholders need to know in advance what will happen to the information they provide, how it will used, and who will have access to it. For example, fishers report fears that information they share on their most important fishing grounds might be used by competitors or by marine conservation interests as an argument for designating marine protected areas (Degnbol & Wilson, 2008, Chapter 15). This issue also arises for offshore energy companies who understandably might be reluctant to share commercially sensitive information if doing so is not protected by some form of confidentially agreement. However, this raises the complexity of designing a transparent process. If that commercially sensitive information is pertinent to the planning process and could impact a planning decision, then creating enough trust through ensuring confidentiality is important. yet, in doing so, the ability to be transparent in how a planning decision was achieved is limited. Thus, two different kinds of transparency have to be managed: internal and external (Degnbol & Wilson, 2008). Internally, ways have to be found for stakeholders to be transparent with each other; external transparency—with wider stakeholders and ultimately, the public—requires that all components of marine spatial planning must be accountable. Therefore, internal transparency and/or external transparency must be limited, and careful attention must be paid to balancing the two types (Degnbol & Wilson, 2008).

Planning with offshore energy might be different

Many marine spatial planning processes have been driven by the expansion of offshore energy, particularly in Europe (Jay, 2010, Chapter 1), while others have been sparked by a desire to enhance marine conservation through the establishment of marine protected areas or marine zoning (State of California, 1999; Day, 2002). Reviewing how planning processes are different when they involve offshore energy as a stakeholder and (often) also a driver compared to processes that do not, is important for considering how to promote meaningful, transparent participation.

One of the main differences is the amount of resources that offshore energy companies have compared to other stakeholders. Their capacity to have dedicated teams to engage in planning usually far exceeds that of other groups of stakeholders. Fishers who wish to participate have to do so at a combined personal loss of (at least) a day's earnings and any associated travel, which can present a substantial barrier to participation, especially when combined with a perception that their participation will not impact anything (Yates, 2014). Even the organisations that represent fishers are becoming overwhelmed by the burden of these new participatory approaches, and thus may eschew involvement even when they feel it is important (Nutters & Pinto da Silva, 2012).

The other main difference between marine spatial planning processes that include offshore energy (and sometimes those that involve designating marine protected area) is that offshore energy is itself a driver of marine spatial planning, as well as a stakeholder within it. As such, offshore energy as a stakeholder might be, or at least perceived to be, in a more powerful position than other stakeholders (Fleming & Jones, 2012). Offshore energy, and particularly renewables, often has an explicit mandate for expansion driven by national and international commitments to reduce greenhouse-gas emissions. Many countries have targets for renewables energy developments (e.g., NI Assembly, 2013) which are generally established at a national level and through processes separate to those for marine spatial planning. Any marine spatial planning with offshore energy will have to take into account these targets, with some inevitability of resource allocation (space) for offshore energy, regardless of objections or conflicts that come up during the planning process. Indeed, there are many examples where marine spatial planning has been initiated because of strategic objectives to expand offshore energy. The main objective of the Bay of Biscay (Spain) Marine Energy Project was allocating an area for developing wave-energy devices (Jones et al., 2013). Likewise, the Pentland Firth and Orkney Waters Pilot Regional Marine Spatial Plan (United Kingdom) was driven by an objective to promote wave and tidal energy (Johnson et al., 2016), and promoting oil and gas production drove the Barents Sea Integrated Management Plan in Norway (Olsen et al., 2014). In all of these cases, other sectors and objectives were considered, but the main driver was opening up space for the development of offshore energy.

As such, offshore energy starts off on the front foot, knowing that through the process they will gain access; it is how much access and exactly where that still needs to be debated. The opposite is true for other stakeholders, like fishers, who know they will lose access to resources and who start out fighting to minimise loss (for more, see Chapter 6). It could be argued that marine spatial planning that aims to meet international commitments under the Convention on Biological Diversity (United Nations, 1992) by designating 10% of the marine environment under some form of protection presents a similar situation (a pre-existing strategic goal that impacts planning objectives). However, while conservation can be a driver like offshore energy, it is not in itself a 'stakeholder'. Conservation interests can be held by and represented through various stakeholders, but conservation gain (improved ecological integrity and resilience) is a common asset shared across all stakeholders (Chapter 3). Within marine spatial planning with conservation objectives, there is also far more room for debate and compromise. There can be a requirement to designate 10% of the area as protected, but what restrictions (if any) there are within those protected areas is not fixed. In contrast, allocation of area for offshore energy normally excludes other users (although see Chapter 10). As such, marine spatial planning that involves offshore energy as a stakeholder and as a driver varies from a process that does not in three ways: (1) the positions from which stakeholder start, (2) the balance of power, and (3) the options available.

Some argue that experiences to date with marine spatial planning involving offshore energy show that the process is driven mainly by the objectives of offshore energy to such an extent that it could be considered 'strategic sectoral planning' (Jones et al., 2016). This concern is echoed by fishers. A fishing industry representative participating in the Lyme Bay planning process in southern England said

> … my concern is that they talk about integrated planning, but we are seeing wind farm planning and MCZs [a type of marine protected area] going ahead regardless, so there is not integrated planning. We are seeing the individual industries being planned separately. It is a piecemeal approach of closing areas to mobile gear. We need a more strategic approach.
> (Fleming & Jones, 2012)

During a planning workshop for the Mull of Kintyre in Scotland, the fishing association representative pointed out that "… the negotiations are based upon an assumption that tidal energy would take precedence over other stakeholders" (Chapter 15).

If achieving strategic offshore energy objectives is the overriding priority of the process, then trade-offs and compromises for *user-user* and *user-nature* conflicts have to align ultimately with those offshore energy objectives, meaning that some conflicts will inevitably go unresolved (Jones et al., 2013; Johnson et al., 2016). In these situations, early stakeholder engagement and meaningful participation are even more important to maximise the possibility of buy-in to a process that will inevitably leave some groups feeling (initially, at least) marginalised. Building trust among stakeholders can help develop collaborations that might lead to innovations, such as co-location (Chapters 10, 11, and 13) that can minimise the impact of the inevitability of resource loss to offshore energy. Transparency about these overarching objectives is essential for developing and maintaining that trust.

Moving forward

Tools that allow collection of robust stakeholder data from many individual stakeholders can enable more inclusive planning, balance power, and encourage the use of stakeholder data in decision making, especially if the data are collected in a way that facilitates integration with other datasets (Yates & Schoeman, 2013). Participatory mapping, which had been widely used for gathering local ecological knowledge (see Huntington, 2000), is one way of capturing spatial stakeholder data. To be most effective for marine spatial planning, participatory mapping needs to be quantitative and, in the likely event of only a subset of the stakeholder community making a map, it needs to be scalable to represent the whole community.

Spatial access priority mapping used in Northern Ireland with the fishing community is one example of a participatory mapping method that can gather quantitative, spatial data from any stakeholder group that can be weighted and scaled to enable the final result to model the entire community (Yates & Schoeman, 2013). In brief, the method requires the stakeholders to identify all their priority areas, as many and as big as they like, and to indicate any difference in priority between the individual areas. Then, their priority is divided over the total area(s) chosen to give each unit area a value to that stakeholder. The resultant maps are intuitive (Figure 9.2), instantly highlighting areas of high conflict potential if assigned for offshore energy (or any other use that prohibited fishing). Importantly, the method is appropriate across sectors and the data are quantitative, so they can be incorporated into trade-offs, spatial analysis, or into decision-support tools like Marxan (marxan.org) or Zonation (github.com/cbig) that help to find optimal zoning configurations (Klein et al., 2008; Yates et al., 2015). The method itself is simple and easy to communicate, and stakeholders can readily see the impacts of their data on planning outcomes (e.g., Yates & Schoeman, 2014). These participatory mapping approaches can require a substantial investment, on the part of both the planning agency and the

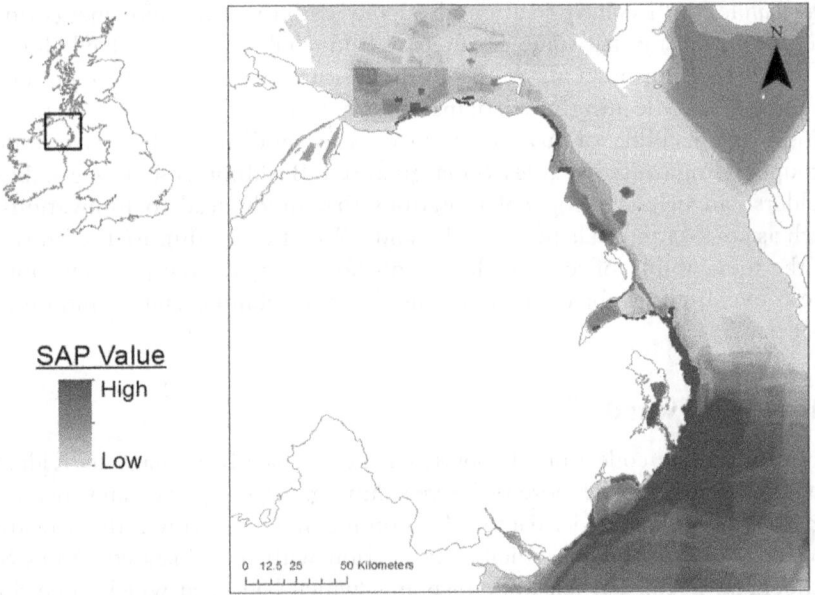

Figure 9.2 Participatory mapping of stakeholder values (SAP = spatial access priority) to inform marine spatial planning. Maps were made with the Northern Irish fishing community between 2011 and 2012 and provide a quantitative, spatial representation of stakeholder priorities. For full details, see Yates and Schoeman (2013).

stakeholders involved, especially if they are aiming for inclusive participation. To realise the potential value of these engagement approaches, there has to be a firm commitment from the outset to include the data, integrating them with other sources, and a transparent pathway to inform decision making.

Decision-support tools for spatial planning can be used effectively to show stakeholders the impacts of different planning alternatives, and can facilitate discussions around compromises and trade-offs (Merrifield et al., 2013, Chapter 15). Decision-support tools come in many forms, some of which might be too technical for wide participation, but there are many others that some support stakeholders can use (and have used) to enable and inform discussions (Stelzenmüller et al., 2013). For example, Chapter 15 outlines in detail how 'Geodesign' tools were used to facilitate the negotiation of spatial trade-offs around tidal energy development off the Mull of Kintyre, Scotland. The tools enabled the collection of data from stakeholders, the review of data by them, and then the exploration of possible planning solutions based on that data. The development of interactive decision-support tools, as well as the ability of stakeholders to submit their own plans for consideration, is credited with the success of the California Marine Life Protection Act planning process (Weible, 2008; Gleason et al., 2010). Most decision-support tools are custom-made for a given planning problem (Stelzenm et al., 2013), but there are free, widely applicable options. Functionality will be limited, but there is no reason why some sort of interactive planning tool cannot be made available (with support) for use with stakeholders.

When used effectively, trade-off analysis (Chapter 2) can also increase transparency because it requires clarity in the data, targets, goals, and issues, shows the impacts of different options, and makes the decision-making process more explicit (Gleason et al. 2010; Collie et al., 2013; Yates et al., 2015). Stakeholders can see how different options are weighed against each other and the impacts of different choices, so trade-off analysis can help stakeholders to accept the logic of decisions they might not like from a personal point of view. Trade-off analysis, participatory mapping, and spatial optimisation tools (like Marxan and Zonation) can all be used in conjunction to develop and explore the implications of different zoning plans (Yates et al., 2015).

Conclusion

Ensuring meaningful, effective stakeholder participation in marine spatial planning with offshore energy requires clear communication throughout a transparent process that balances power and ensures integration of stakeholder input. The (real or perceived) notion that offshore energy is at an inherent advantage makes effective engagement with and empowerment of other stakeholders especially important. In this power-imbalance context, it will be even more essential that participation mechanisms clearly show stakeholders how their input has been used if they are to engage in and trust the process. By using methods that allow robust collection of stakeholder data and then integrating those data into trade-off analysis and decision-support tools, planners create mechanisms for engaging create mechanisms for engaging stakeholders in an inclusive, transparent and more meaningful way. Stakeholders will have the opportunity to inform the processes rather than just be consulted on planning outcomes, and planners can readily communicate the impact stakeholder input had. Practicalities will dictate that only a few 'entitled' stakeholders will be involved in the more meaningful negotiation and decision-making aspects of participation, but this way their decisions will be informed by more inclusive participation.

Highlights

- Effective stakeholder participation is essential for successful marine spatial planning.
- To be effective, stakeholder participation has to be meaningful, with clear pathways to impact decisions.
- Offshore energy is a driver as well as stakeholder in marine spatial planning, potentially affording it more power (real or perceived) than other stakeholders.
- Participation methods must balance power among stakeholders while clearly acknowledging legislative and policy commitments that could unbalance it.
- Tools that allow for transparent collection and incorporation of stakeholder input will ensure more meaningful, robust participation and more defensible decisions.

References

Alexander, K.A., Janssen, R., Arciniegas, G., O'Higgins, T.G., Eikelboom, T. and Wilding, T.A. (2012). Interactive marine spatial planning: Siting tidal energy arrays around the Mull of Kintyre. *PLoS One*, 7(1), p. e30031.

Anderson, C. (2013). The networked minority: How a small group prevailed in a local windfarm conflict. *Energy Policy*, 58, pp. 97–108. doi: 10.1016/j.enpol.2013.02.048.

Baker, M., Roberts, P. and Shaw, R. (2004). *Stakeholder Involvement in Regional Planning. National Report of the Town and Country Planning Association Study supported by the Office of the Deputy Prime Minister*, TCPA London.

Bojórquez-Tapia, L.A., de la Cueva, H., Díaz, S., Melgarejo, D., Alcantar, G., Solares, M.J., Grobet, G. and Cruz-Bello, G. (2004). Environmental conflicts and nature reserves: redesigning Sierra San Pedro Martir National Park, Mexico. *Biological Conservation*, 117(2), pp. 111–126.

Britton, E. (2012). *The 'Having, Doing and Being' of Fishing Well: Assessing the Social Wellbeing of Northern Ireland's Fishing Households in a Changing Coastal Environment*. Ulster University, Northern Ireland.

Brody, S.D. (2003). Measuring the effects of stakeholder participation on the quality of local plans based on the principles of collaborative ecosystem management. *Journal of Planning Education and Research*, 22, pp. 407–419.

Calado, H., Ng, K., Johnson, D., Sousa, L., Phillips, M. and Alves, F. (2010). Marine spatial planning : Lessons learned from the Portuguese debate. *Marine Policy*, 34(6), pp. 1341–1349.

COM. (2010). Communication from the commission to the European parliament, the council, the European economic and social committee and the committee of the regions: Maritime spatial planning in the EU—achievements and future development COM, vol. 771.

Cinnéide, M. (2007). Enhancing the development process in Lithuania: towards truly participative governance. In: S. Giguére, ed., *Baltic Partnerships: Integration, Growth and Local Governance in the Baltic Sea Region*. Paris: OECD, pp. 83–108.

Clarke, B., Thurstan, R. and Yates, K. (2016). Stakeholder perceptions of a marine protected area. *Journal of Coastal Research*, 75, pp. 622–626

Collie, J.S., Beck, M.W., Craig, B., Essington, T.E., Fluharty, D., Rice, J. and Sanchirico, J.N. (2013). Marine spatial planning in practice. *Estuarine, Coastal and Shelf Science*, 117, pp. 1–11. doi: 10.1016/j.ecss.2012.11.010.

Commission of the European Community. (2008). *Roadmap for Maritime Spatial Planning : Achieving Common Principles in the EU, COM(2008)*. [pdf], p. 12. Available at: http://eur-lex.europa.eu/LexUriServ/LexUriServ.do?uri=COM:2008:0791:FIN:EN:PDF.

Crowder, L.B., Osherenko, G., Young, O.R., Airamé, S., Norse, E.A., Baron, N., Day, J.C., Douvere, F., Ehler, C.N., Halpern, B.S. and Langdon, S.J. (2006). Resolving mismatches in US Ocean Governance. *Science*, 313(5787), pp. 617–618.

Day, J.C. (2002). Zoning – lessons from the Great Barrier Reef Marine Park. *Ocean & Coastal Management*, 45(2–3), pp. 139–156.

Degnbol, D and Wilson, D.C. (2008). Spatial planning on the North Sea : A case of cross-scale linkages. *Marine Policy*, 32, pp. 189–200.

Dougill, A.J., Fraser, E.D.G., Holden, J., Hubacek, K., Prell, C., Reed, M.S., Stagl, S. and Stringer, L.C. (2006). Learning from doing participatory rural research: Lessons from the peak district national park. *Journal of Agricultural Economics*, 57(2), pp. 259–275. doi: 10.1111/j.1477–9552.2006.00051.x.

Douvere, F. and Ehler, C.N. (2009). New perspectives on sea use management: initial findings from European experience with marine spatial planning. *Journal of environmental management*, 90(1), pp. 77–88. doi: 10.1016/j.jenvman.2008.07.004.

Douvere, F. and Ehler, C.N. (2011). The importance of monitoring and evaluation in adaptive maritime spatial planning. *Journal of Coastal Conservation*, 15, pp. 305–311.

Douvere, F., Maes, F., Vanhulle, A. and Schrijvers, J. (2007). The role of marine spatial planning in sea use management: The Belgian case. *Marine Policy*, 31(2), pp. 182–191. doi: 10.1016/j.marpol.2006.07.003.

Ehler, C. (2008). Conclusions: Benefits, lessons learned, and future challenges of marine spatial planning. *Marine Policy*, 32(5), pp. 840–843.

Ehler, C. and Douvere, F. (2009a). *Marine Spatial Planning - A Step-by-Step Approach. Intergovernmental Oceanographic Commission and Man and the Biosphere Programme. IOC Manual and Guides No. 53, ICAM Dossier No. 6.*, Paris.

Fischer, A. and Young, J.C. (2007). Understanding mental constructs of biodiversity: Implications for biodiversity management and conservation. *Biological Conservation*, 136(2), pp. 271–282. doi: 10.1016/j.biocon.2006.11.024.

Flannery, W. (2012). A roadmap for marine spatial planning: A critical examination of the European Commission's guiding principles based on their application in the Clyde MSP Pilot Project. *Marine Policy*, 36(1), pp. 265–271.

Flannery, W. and Cinnéide, M.Ó. (2008). Marine spatial planning from the perspective of a small seaside community in Ireland. *Marine Policy*, 32(6), pp. 980–987.

Fleming, D.M. and Jones, P.J.S. (2012). Challenges to achieving greater and fairer stakeholder involvement in marine spatial planning as illustrated by the Lyme Bay scallop dredging closure. *Marine Policy*, 36(2), pp. 370–377. doi: 10.1016/j. marpol.2011.07.006.

Fletcher, S., McKinley, E., Buchan, K.C., Smith, N. and McHugh, K. (2013). Effective practice in marine spatial planning: A participatory evaluation of experience in Southern England. *Marine Policy*, 39, pp. 341–348. doi: 10.1016/j. marpol.2012.09.003.

Gilliland, P.M. and Laffoley, D. (2008). Key elements and steps in the process of developing ecosystem-based marine spatial planning. *Marine Policy*, 32(5), pp. 787–796.

Gleason, M., McCreary, S., Miller-Henson, M., Ugoretz, J., Fox, E., Merrifield, M., McClintock, W., Serpa, P. and Hoffman, K. (2010). Science-based and stakeholder-driven marine protected area network planning: A successful case study from north central California. *Ocean & Coastal Management*, 53(2), pp. 52–68. doi: doi. org/10.1016/j.ocecoaman.2009.12.001.

Gopnik, M., Fieseler, C., Cantral, L., McClellan, K., Pendleton, L. and Crowder, L. (2012). Coming to the table: Early stakeholder engagement in marine spatial planning. *Marine Policy*, 36(5), pp. 1139–1149. doi: 10.1016/j.marpol.2012. 02.012.

Guentte, S. and Alder, J. (2007). Lessons from marine protected areas and integrated ocean management initiatives in Canada. *Coastal Management*, 35, pp. 51–78.

Healey, P. (2006). *Collaborative Planning : Shaping Places in Fragmented Societies*. 2nd ed. Hampshire: Palgrave Macmillan.

Hind, E.J. (2012). *Last of the Hunters or the Next Scientists? Arguments for and Against the Inclusion of Fishers and Their Knowledge in Mainstream Fisheries Management*. The National University of Ireland, Galway.

Hind, E.J. (2015). Knowledge research: A challenge to established fisheries science. *ICES Journal of Marine Science*, 72, pp. 341–358.

Huntington, H. (2000). Using traditional ecological knowledge in science: Methods and applications. *Ecological Applications*, 10(5), pp. 1270–1274.

Innes, J.E. (1996). Planning through consensus building: A new view of the comprehensive planning ideal. *Journal of the American Planning Association*, 62, pp. 460–472.

Jay, S. (2010). Planners to the rescue: Spatial planning facilitating the development of offshore wind energy. *Marine Pollution Bulletin*, 60(4), pp. 493–499. doi: 10.1016/j. marpolbul.2009.11.010.

Johnson, K., Kerr, S. and Side, J. (2016). The Pentland Firth and Orkney Waters and Scotland – Planning Europe's Atlantic gateway. *Marine Policy*, 71, pp. 285–292.

Jones, P., Qiu, W. and Lieberknecht, L. (2013). *MESMA Work Package 6 (Governance) Typology of Conflicts in MESMA Case Studies*. http://mesma.org/

Jones, P.J.S. (2008). Fishing industry and related perspectives on the issues raised by no-take marine protected area proposals. *Marine Policy*, 32(4), pp. 749–758. Available at: http://linkinghub.elsevier.com/retrieve/pii/S0308597X08000055 (Accessed August 9, 2013).

Jones, P.J.S., Lieberknecht, L.M. and Qiu, W. (2016). Marine spatial planning in reality : Introduction to case studies and discussion of findings. *Marine Policy*, 71, pp. 256–264.

Kafas, A., McLay, A., Chimienti, M. and Gubbins, M. (2014). ScotMap inshore fisheries mapping in Scotland: Recording fishermen's use of the Sea. *Scottish Marine and Freshwater Science*, 5(17), p. 32. doi: 10.4789/1554-1.

Kannen, A. (2014). Challenges for marine spatial planning in the context of multiple sea uses, policy arenas and actors based on experiences from the German North Sea. *Regional Environmental Change*, 14(6), pp. 2139–2150.

Keen, M. and Mahanty, S. (2006). Learning in sustainable natural resource management: challenges and opportunities in the Pacific. *Society and Natural Resources*, 19(6), pp. 497–513.

Kerr, S., Watts, L., Colton, J., Conway, F., Hull, A., Johnson, K., Jude, S., Kannen, A., MacDougall, S., McLachlan, C. and Potts, T. (2014). Establishing an agenda for social studies research in marine renewable energy. *Energy Policy*, 67, pp. 694–702. doi: 10.1016/j.enpol.2013.11.063.

Kidd, S. and Ellis, G. (2017). From the land to sea and back again ? Using terrestrial planning to understand the process of marine spatial planning. *Journal of Environmental Policy & Planning*, 14(1), pp. 49–56.

Klein, C.J., Chan, A., Kircher, L., Cundiff, A.J., Gardner, N., Hrovat, Y., Scholz, A., Kendall, B.E. and Airame, S. (2008). Striking a balance between biodiversity conservation and socioeconomic viability in the design of marine protected areas. *Conservation Biology : The Journal of the Society for Conservation Biology*, 22(3), pp. 691–700.

Merrifield, M.S., McClintock, W., Burt, C., Fox, E., Serpa, P., Steinback, C. and Gleason, M. (2013). Ocean & coastal management MarineMap : A web-based platform for collaborative marine protected area planning. *Ocean & Coastal Management*, 74, pp. 67–76.

Middendorf, G. and Busch, L. (1997). Inquiry for the public good: Democratic participation in agricultural research. *Agriculture and Human Values*, 14, pp. 45–57.

NI Assembly. (2013). *ELECTRICITY the Renewables Obligation (Amendment) Order (Northern Ireland) 2013*, Northern Ireland Assembly, Belfast, Northern Ireland.

Nutters, H.M. and Pinto da Silva, P. (2012). Fishery stakeholder engagement and marine spatial planning: Lessons from the Rhode Island Ocean SAMP and the Massachusetts Ocean Management Plan. *Ocean & Coastal Management*, 67, pp. 9–18.

Olsen, E., Fluharty, D., Hoel, A.H., Hostens, K., Maes, F. and Pecceu, E. (2014). Integration at the round table: Marine spatial planning in multi-stakeholder settings. *PLoS One*, 9(10), p. e109964.

Olsen, E., Holen, S., Hoel, A.H., Buhl-Mortensen, L. and Røttingen, I. (2016). How Integrated Ocean governance in the Barents Sea was created by a drive for increased oil production. *Marine Policy*, 71, pp. 293–300.

OSPAR Commission. (2009). *Overview of National Spatial Planning and Control Systems Relevant to the OSPAR Maritime Area*, No. 444, ISBN:978-1-90684084-6.

Pelc, R. and Fujita, R.M. (2002). Renewable energy from the ocean. *Marine Policy*, 26, pp. 471–479.

Pita, C., Pierce, G.J., Theodossiou, I. and Macpherson, K. (2011). An overview of commercial fishers' attitudes towards marine protected areas. *Hydrobiologia*, 670(1), pp. 289–306.

Pomeroy, R. and Douvere, F. (2008). The engagement of stakeholders in the marine spatial planning process. *Marine Policy*, 32(5), pp. 816–822.

Portman, Michael. (2009). Involving the public in the impact assessment of offshore renewable energy facilities. *Marine Policy*, 33(2), pp. 332–338.

Reed, M.S. (2008). Stakeholder participation for environmental management: A literature review. *Biological Conservation*, 141(10), pp. 2417–2431.

Reilly, K., Hagan, A.M.O. and Dalton, G. (2016). Ocean & coastal management moving from consultation to participation : A case study of the involvement of fishermen in decisions relating to marine renewable energy projects on the island of Ireland. *Ocean and Coastal Management*, 134, pp. 30–40.

Ritchie, H. and Ellis, G. (2016). "A system that works for the sea"? Exploring stakeholder engagement in marine spatial planning. *Journal of Environmental Planning and Management*, 6, pp. 701–723

Soto, C.G. (2006). *Socio-Cultural Barriers to Applying Fishers' Knowledge in Fisheries Management: An Evaluation of Literature Cases*. Simon Fraser University, Canada.

State of California. (1999). *Marine Life Protection Act. Fish and Game Code Section 2850–2863*. State of California, USA.

Stelzenmüller, V., Lee, J., South, A., Foden, J. and Rogers, S.I. (2013). Practical tools to support marine spatial planning : A review and some prototype tools. *Marine Policy*, 38, pp. 214–227.

The House of Commons. (2009). *Marine and Coastal Access Bill*. London.

United Nations. (1992). *Convention on Biological Diversity*. Rio de Janeiro, Brazil.

Vedwan, N., Ahmad, S., Miralles-Wilhelm, F., Broad, K., Letson, D. and Podesta, G. (2008). Institutional evolution in Lake Okeechobee Management in Florida: Characteristics, impacts, and limitations. *Water Resources Management*, 22(6), pp. 699–718.

Weible, C.M. (2008). Caught in a maelstrom: Implementing California marine protected areas. *Coastal Management*, 36(4), pp. 350–373.

Yates, K.L. (2014). View from the wheelhouse: Perceptions on marine management from the fishing community and suggestions for improvement. *Marine Policy*, 48, pp. 39–50.

Yates, K.L., Payo Payo, A. and Schoeman, D.S. (2013). International, regional and national commitments meet local implementation: A case study of marine conservation in Northern Ireland. *Marine Policy*, 38, pp. 140–150.

Yates, K.L. and Schoeman, D.S. (2013). Spatial access priority mapping (SAPM) with fishers: A quantitative GIS method for participatory planning. *PLoS One*, 8(7), p. e68424.

Yates, K.L. and Schoeman, D.S. (2014). Incorporating the spatial access priorities of fishers into strategic conservation planning and marine protected area design: Reducing cost and increasing transparency. *ICES Journal of Marine Science*, 72, pp. 587–594.

Yates, K.L., Schoeman, D.S. and Klein, C.J. (2015). Ocean zoning for conservation, fisheries and marine renewable energy: Assessing trade-offs and co-location opportunities. *Journal of Environmental Management*, 152, pp. 201–209.

Chapter 10

Capturing benefits

Opportunities for the co-location of offshore energy and fisheries

Tara Hooper, Matthew Ashley and Melanie Austen

Introduction

The foundations, cables, scour protection, and armouring that comprise offshore energy infrastructure act as artificial reefs: the hard substrata provide settlement surfaces for sessile benthic organisms, increasing food availability and providing shelter for more mobile species. Artificial reefs can increase species richness and diversity, relative abundance, and catch per unit effort compared to control sites, and can serve as nursery areas (e.g., Koeck et al., 2011; Spanier et al., 2011; Santos and Monteiro, 2007). Indeed, fishers have been placing structures on the seabed to attract fish for at least 3,000 years (Fabi et al., 2011), and there is recent evidence of extensive use of artificial reefs by recreational anglers and commercial fishers using pots, traps, and nets (Milon, 1989; Murray and Betz, 1994; Ramos et al., 2006).

This prior experience from other artificial structures on the seabed suggests that offshore energy infrastructure could support recreational and commercial fishing through the co-location of the activities (i.e., direct use by fishers of the immediate area surrounding the foundations and cables). There are issues of compatibility with certain types of fishing gear, with a strong perception among fishermen that the deployment of towed or static nets is unlikely to be practical or safe around offshore energy infrastructure (Mackinson et al., 2006). However, the displacement of certain types of fishing activity from an offshore energy site would create a *de facto* marine protected area, which in itself creates the potential for an alternative, indirect benefit to fisheries. Marine protected areas can augment the exploitable fish stock outside the exclusion zone through the migration of adults and juveniles (an effect known as 'spillover'), as well as improving the wider status of the stock through improved reproductive potential and increased egg production (Roberts et al., 2001; Gell and Roberts, 2003).

Where the co-location of capture fisheries is not appropriate, co-location with aquaculture might present an alternative option to allow food production and energy generation to coexist. There is the potential for aquaculture systems to be attached to offshore wind-farm foundations (enhancing the potential economic viability of offshore aquaculture), and for both the aquaculture and energy partners to reduce their costs through shared operations and maintenance (Buck et al., 2004).

In this chapter, we first consider whether the ecological evidence of artificial reef effects arising from offshore energy infrastructure demonstrates any enhancement of the species of interest to commercial and recreational fishers, before reviewing evidence of the use of energy infrastructure by these interest groups. We focus on the Louisiana coast of Gulf of Mexico and the United Kingdom, considering both the long-established offshore oil and gas industry and the developing offshore wind sector. We also discuss the potential for aquaculture within offshore wind farms, reviewing the opportunities and barriers for this alternative approach to co-location.

Ecological evidence of effects on commercial and recreational species

Offshore oil and gas platforms

Relative abundance of fish species

Most research into the effects of oil and gas infrastructure on fish populations has compared fish assemblages at the platforms to those at control sites, and has shown evidence of increases in species richness, diversity, and catch rates in the Gulf of Mexico (Ajemian et al., 2015), the Mediterranean (Consoli et al., 2013), and the Adriatic (Fabi et al., 2004; Scarcella et al., 2011). However, while there is strong evidence of artificial reef effects, their nature is site-specific (Ajemian et al., 2015), and so it is difficult to generalise about the implications for individual species.

In some instances, the fish assemblages recorded at offshore oil and gas platforms are dominated by non-commercial species, particularly blennies (blenniidae) (Consoli et al., 2013). However, there is some evidence that species of commercial or recreational importance (including cod *Gadus morhua*, seabream *Sparus aurata*, anchovy *Engraulis encrasicolus* and jacks Carangidae) can be highly abundant, even to the extent that they dominate the local fish community (Fabi et al., 2004; Seaman et al., 1989; Helvey, 2002; Fujii, 2015). A large and diverse range of other commercial and game fish has been recorded around offshore oil and gas platforms at varying abundance, including tuna *Thunnus thynnus*, mackerel *Scomber scombrus*, and bass *Dicentrarchus labrax* (Consoli et al., 2013; Scarcella et al., 2011; Fabi et al., 2004).

Studies from the North Sea provide evidence that the spatial extent of this increased abundance is limited to about 200 m from the platforms (Løkkeborg et al., 2002; Valdemarsen, 1979; Olsen and Valdermarsen, 1977), and that aggregation effects are species-specific. While species such as cod aggregate close to platforms (Løkkeborg et al., 2002), the homogenous distribution of sole *Solea*

solea in the area surrounding a platform in the Adriatic suggests that this species is unaffected by the presence of the infrastructure (Scarcella et al., 2011). The wider ecological context also influences the likelihood of individuals aggregating at offshore structures. When the expected pattern of increased abundance close to oil platforms in the North Sea was not observed for cod, haddock *Melanogrammus aeglefinus*, or saithe *Pollachius virens*, high densities of large predators were suggested as a possible cause (Soldal et al., 2002; Løkkeborg et al., 2002). The absence of aggregation effects has also been attributed to location-specific phenomena, such as the complexity or material of the platform structure and the diversity (or lack thereof) of the surrounding substrata (Løkkeborg et al., 2002).

Other population effects

Site fidelity (the tendency to remain in, or return to, a particular location), the relative size of individuals, and the associated role of energy infrastructure as nurseries have also been studied at oil and gas platforms. Tagging suggests considerable variability in patterns of site use between species and individuals (Lowe et al., 2009). Long-term research indicates a high site fidelity for cod and widow rockfish *Sebastes entomelas* (Jørgensen et al., 2002; Lowe et al., 2009). Shorter-term studies (< 1 month) show that blue runner *Caranx crysos* and red snapper *Lutjanus campechanus* can remain associated with platforms for one to two weeks, although the authors draw different conclusions about whether this duration of residence indicates that the individuals demonstrate site fidelity (Brown et al., 2010; McDonough and Cowan, 2007).

The use of oil and gas platforms also varies according to life stage. Individuals found closer to the platforms are generally larger than those farther away for species including cod, saithe, haddock, and rockfish (Soldal et al., 2002; Fujii, 2015; Helvey, 2002; Love et al., 2005). However, the converse situation has also been observed; Løkkeborg et al. (2002) report that cod caught closest to a partially decommissioned platform were smaller than those found at greater distances, and saithe showed different size profiles seasonally and between years (Fujii, 2015). The presence of immature individuals at oil and gas platforms infers a role for the structures as nursery habitat. Juveniles of species including rockfish, jacks, mackerel, and lingcod *Ophiodon elongatus* have been found associated with platforms (Helvey, 2002; Lindquist et al., 2005; Love et al., 1999; Love and Westphal, 1990). The abundance of juveniles can be considerable; the density at which juvenile lingcod were present on mussel mounds around a platform base had not previously been observed on any other natural or artificial structures in southern or central California (Love et al., 1999).

Invertebrate species

Fish are not the only group of fisheries interest that use oil and gas infrastructure—commercially important shellfish have also been recorded, although they are rarely the subject of directed research. Mussels *Mytilus* spp. are often the dominant fouling organism on offshore energy structures (Wolfson et al., 1979; Southgate and Myers, 1985, and references therein), and species including *Nephrops*, lobster *Homarus gammarus*, squid *Loligo vulgaris*, cuttlefish *Sepia officinalis*, octopus *Octopus vulgaris*, and spider crab *Maja squinado* have been associated with gas platforms in the Adriatic Sea (Scarcella et al., 2011; Fabi et al., 2004). These species were not abundant, although the capture methods used were designed to target fish, and so might not have been the most appropriate for assessing invertebrate populations.

Offshore wind farms

There is a smaller body of evidence on the effects of offshore wind farms on species of commercial and recreational importance. The studies that do exist are overwhelmingly from the north-east Atlantic, although this is to be expected as northern Europe had 91% of global offshore wind capacity in 2015 (GWEC, 2016). Offshore wind farm studies tend to support the research from oil and gas platforms, suggesting that certain species aggregate at the infrastructure, which is used at different life stages. There is evidence that species including mussels, brown crab *Cancer pagurus*, pouting *Trisopterus luscus*, and eel *Anguilla anguilla* have increased abundance at offshore wind farm foundations (Hooper and Austen, 2014; Reubens et al., 2013a; Bergström et al., 2013; Wilhelmsson and Malm, 2008). Tagging has revealed that individual cod can spend periods of several months within offshore wind-farm sites, staying close to the turbines (Winter et al., 2010; Reubens et al., 2013b), and the fouling communities on the foundations provide a food source for juvenile cod (Reubens et al., 2014).

Research at offshore wind farms provides further evidence that the artificial reef effects of energy infrastructure are species-specific, and in particular, might have limited benefits for flatfish. As has been found for oil and gas platforms, there is no evidence that the foundations of offshore wind turbines have a particular attraction for sole (Winter et al., 2010), and neither was there evidence for any effects on the density of flounder *Platichthys flesus* (Bergström et al., 2013). For other species, the impacts can be negative; for example, the abundance of dab *Limanda limanda* declined following the construction of an offshore wind farm (Vandendriessche et al., 2014).

Opportunities for recreational fishing

Gulf of Mexico

There are approximately 4,000 active oil and gas platforms currently in the Gulf of Mexico (National Oceanic and Atmospheric Administration, 2013), and those off the Louisiana coast have become a particular focus of recreational angling due to an absence of natural reefs in the area (Stanley and Wilson, 1989). The platforms were the principal destination for 73% of anglers interviewed out of New Orleans (Gordon, 1993), and an average of about five sport-fishing boats per platform per month were recorded visiting a sample of 164 oil and gas structures (Ditton and Auyong, 1984). It is common for anglers to visit multiple platforms, usually up to six in a single trip (Gordon, 1993). Some anglers will travel 80 km or more offshore to access the platforms, particularly those using larger boats (Ditton and Auyong, 1984; Stanley and Wilson, 1989).

The importance of operational oil and gas platforms to sport fishing was the driver for a programme to convert decommissioned rigs to artificial reefs (Stanley and Wilson, 1989). This 'rigs to reefs' approach was seen as a cost-effective fishery enhancement tool that, after piloting in the early 1980s, became government policy in the United States (Reggio, 1987). The conversion process usually involves the *in situ* toppling, partial removal, or relocation of the platform support structure to create fish habitat (Jørgensen, 2012). By 2013, over 330 platform jackets had been converted as part of the Louisiana Artificial Reef Program (LDWF, 2013).

United Kingdom

United Kingdom waters also contain considerable energy infrastructure, primarily in the North Sea and to a lesser extent, the Irish Sea. At the end of 2014, there were 334 oil and gas fields in production (Oil and Gas Authority, 2015) and over 1,400 offshore wind turbines had been installed or were under construction (4COffshore, 2014). The supporting infrastructure has an even greater footprint; it has been estimated that over 45,000 km of pipelines and cables have been laid by the oil and gas industry in the North Sea region (Oil and Gas UK, 2013). Even with this apparent potential, the co-location of recreational fishing and energy infrastructure that occurs in the Gulf of Mexico has not been reported from the United Kingdom. This is despite the considerable size of the recreational angling sector in the United Kingdom, which has been estimated to involve 884,000 people (Armstrong et al., 2012).

In the case of oil and gas, the co-location of recreational fishing is not possible because under Section 23 of the Petroleum Act (1987), vessels are legally prohibited from entering the safety zones that extend 500 m from any part of an installation. The difference in approach between the United States and northern Europe also extends to decommissioned platforms; the dumping or toppling of steel platforms in the North Sea was banned in 1998 by the OSPAR Commission (which governs international co-operation on the protection of the marine environment of the north-east Atlantic) (Jørgensen, 2012).

Offshore wind farm developments are potentially more accessible, in that United Kingdom Electricity regulations (SI No. 2007/1948) provide for a 500 m exclusion zone during construction, but safety zones for operational turbines extend only 50 m from the foundations, and are rarely put in place (Department for Energy and Climate Change, 2011). Sea anglers have reported that while they expect some overall economic benefit as this energy sector develops, they do not foresee a large market demand for charter trips to offshore wind farms (Mackinson et al., 2006). At present, however, there is little information on the scale of recreational use of operational offshore wind farms in United Kingdom waters.

Hooper et al. (2017a) recently undertook an online survey of 199 anglers around the United Kingdom. A quarter of those respondents reported that they had fished within or around an offshore wind farm, while 73% of those who had not expressed a willingness to do so in the future. A factor in the use of offshore wind farms was whether or not they were present within anglers' usual fishing grounds: the structures were not a sufficient draw that anglers would travel specifically to access them. The anglers reported mixed outcomes in terms of overall catch success at turbines, and also how catches of different species had changed following wind farm construction. There were consistent reports for one species: any angler who mentioned tope *Galeorhinus galeus* reported that catches had declined. These reports came from only five anglers, but suggest an issue that should be investigated further.

In a separate study, two charter vessel operators working in the eastern Irish Sea reported an unsuccessful fishing trial within an offshore wind farm, and expressed the opinion that there was only limited potential for the structures to support the larger, adult fish preferred by anglers due to a lack of complex rock habitats (Ashley, 2014). Outside of the United Kingdom, recreational fishing has been observed within the footprint of Thorntonbank offshore wind farm 30 km off the Belgian coast, but the intensity was variable and appeared to decline with time (Vandendriessche et al., 2013b).

Co-location with commercial fishing

Evidence of fishing activity associated with energy infrastructure

As was the case with recreational fishing, evidence of the use of energy infrastructure by commercial fishers is scarce, and is often anecdotal or self-reported by interview respondents. However, commercial rod and line fishing has been observed at oil and gas platforms in the Gulf of Mexico, at a frequency of about 0.5 vessels per platform per month (about 10% of that of sport fishers), often at distances >100 km offshore (Ditton and Auyong, 1984). The oil rigs in the Campos Basin off the Brazilian coast were the main fishing grounds for the pole and line skipjack tuna fisheries during the late 1970s, and fishers regularly flout the current 500 m exclusion zones, as they still consider the rigs to be good fishing grounds (Jablonski, 2008).

North Sea oil rigs are also surrounded by 500 m exclusion zones, and Osmundsen and Tveteras (2003) reported that trawlers tend to fish as close as possible to the boundary of these areas. This could represent a displacement effect, with the boats staying as close as possible to their past fishing grounds from which they are now excluded. An alternative explanation is that fishers targeting the edge of safety zones are benefitting from 'spillover' effects; the prohibition of fishing around the platforms provides a refuge for fish, and the enhanced populations within the protected area might lead to increased fishery benefits outside the exclusion zone. This effect has been documented for marine protected areas more generally (see, for example, Di Lorenzo et al., 2016). In the context of offshore wind farms, the Ekofisk oil field in Norway was not historically a successful fishing ground for white fish (such as cod, haddock, and whiting), but since oil production started there have been reports of high activity of Scottish seiners close to the border of the safety zone (Valdermarsen, 1979).

Vessel monitoring system data also showed a moderate increase in vessels surrounding offshore wind farms in Belgium (Vandendriessche et al., 2013b). This was attributed to displacement as opposed to spillover, because a separate study on the ecology of the fish did not show major differences in the presence of commercial demersal species in the proximity of the wind farms compared to other areas (Vandendriessche et al., 2013b). A single trawlerman who had attempted to fish within an offshore wind farm in the Irish Sea reported that he had caught no marketable fish and so had not tried again (Ashley, 2014).

Crab and lobster potters are often identified as more likely beneficiaries of artificial reef effects than trawlermen (Hooper and Austen, 2014; Alexander et al., 2013a), but <5% of a sample of crab and lobster fishers from the east coast of England reported having set pots within the footprint of an offshore wind farm, and there were reports that increased siltation within the wind farm made potting unworkable (Hooper et al., 2015). However, 44% of the sample reported that they had set pots around the outside of a wind farm, and usually also fished close to the cable (Hooper et al., 2015), suggesting that they were receiving some benefit from the presence of the infrastructure.

Major evidence gaps relating to co-location with offshore wind

Attraction versus production

Relative fish catches can be higher at artificial reefs than control sites (e.g., Santos and Monteiro, 2007), but there is not yet conclusive evidence of whether the individuals are simply attracted from elsewhere or whether the infrastructure contributes to the production of additional biomass. This issue is of concern because if artificial reefs are merely aggregating individuals and the sites become overexploited, this could have a negative effect on the population as a whole and jeopardise the economics of any fishery (Whitmarsh et al., 2008).

There is recent evidence of both attraction and production at artificial reefs (e.g., Cresson et al., 2014; Cenci et al., 2011; Simon et al., 2011), and a site- and species-specific approach is needed to assess this phenomenon because many different factors affect the mechanisms underlying the function of artificial reefs (Spanier et al., 2011). Evidence from energy infrastructure specifically is minimal. An increase in production of fish biomass at offshore wind farms has been shown locally, but evidence of larger-scale production is lacking (Vandendriessche et al., 2013a).

Noise and electromagnetic field effects

Offshore wind farms can have negative consequences for at least one species of commercially exploited fish (Vandendriessche et al., 2014), and there is particular concern about the potential implications of noise and electromagnetic field effects (Gill et al., 2009; Thomsen et al., 2006). Noise from pile driving is likely to be detectable by fish species at a distance of 80 km from the source (Thomsen et al., 2006). The noise from pile driving can affect behaviour in cod (Mueller-Blenkle et al., 2010) and inhibit the startle response of bass, which might impede their ability to avoid predators (Everley et al., 2016). Behavioural responses were also shown by elasmobranchs in response to electromagnetic fields, but these did not take the form of a consistent or predictable pattern (Gill et al., 2009). It remains that research in this area is limited, and further study is necessary to understand more fully the individual- and population-level effects of noise and electromagnetic field effects, and their implications for fisheries.

Economic benefits

The use and economic benefit of artificial reefs is much less well-documented than the ecological effects. There is no direct evidence that the catches and incomes of net fishers using artificial reefs are different (Koeck et al., 2011) or lower (Islam et al., 2014) than those using other local areas, although Whitmarsh et al. (2008) and Ramos et al. (2006) suggested that fishing regularly at artificial reef sites is profitable, and can be more so than in control areas; however, this last study was a modelled assessment and did not compare the fishers' actual incomes.

Understanding potential profitability is essential for developing successful co-location of energy infrastructure and commercial fisheries. The possibility of profitable outcomes has not yet been demonstrated conclusively, and does not seem to have been the focus of much research. One study suggested that the density of cod and saithe around an offshore platform was below that of interest to commercial fishing, but this was based on a single week of sampling that might not have taken place at the optimum distance from the platform (Valdermarsen, 1979). Another study estimated up to 12 tonnes of cod and saithe were present near a decommissioned platform (Soldal et al., 2002), which is approximately equal to the combined live weight of cod landed at a large United Kingdom port by all <10 m vessels during one month in 2015 (Marine Management Organisation, 2015).

There is a tendency for co-location potential to be considered from the perspective of outcomes for individual fishers, rather than the sector as a whole. Fishers, particularly those using mobile gear, have long been concerned that their operations will be impossible within the footprints of offshore wind farms (e.g., Mackinson et al., 2006), and recent evidence demonstrates that such displacement is indeed occurring (Gray et al., 2016). Displacement is likely to have economic consequences, ranging from additional fuel costs to reduced catches (Chapter 6). The same degree of displacement might not necessarily occur across all gear types, and there is the potential for mitigation to reduce impacts. Therefore, examining the possibilities for co-location should be extended beyond crab and lobster fishing; in particular, research should aim to determine how future designs of offshore wind farms could encourage co-location with other gear types. Long-liners, for example, have suggested that increased spacing between turbine foundations would enable them to access offshore wind farm sites (Mackinson et al., 2006). Expanding the opportunity for co-location to other sectors has the potential to offset some of the economic costs of displacement.

In general, a broader approach to economic impacts should be taken to consider, for example, the degree to which co-location benefits (accruing in all likelihood to a relatively small number of fishers) offset the wider costs resulting from displacement. There is also an absence of empirical evidence on the potential economic benefits to fisheries of spillover and related effects resulting from energy installations acting as *de facto* marine protected areas, and, again, how these are traded off against displacement costs.

Barriers to co-location

Scale and accessibility

One issue in potential profitability of co-located fisheries is whether a development is of an adequate size to support profitable fishing. Understanding the issue of the potential scale of the resource is also hampered because empirical research focuses on the major structures (platforms, turbine foundations), and so evidence is lacking on the role of the cables and pipelines that cover a greater area. The footprint of the smaller offshore wind farms might be insufficient to generate much benefit for crab and lobster fishers, because even small operations tend to work at least 50 pots (Hooper et al., 2015). Scale is less of an issue for larger offshore wind farms; however, these are farther offshore and might be inaccessible to many inshore fishers, although they would be within the range of larger boats (Hooper et al., 2015).

Accessibility is also important for potential recreational use, with 71% of anglers in Louisiana agreeing that distance offshore was a consideration in their choice to fish at particular oil and gas platforms (Gordon, 1993). Recreational fishers may be less willing to travel far offshore to reach wind farms because they are unmanned; Gordon (1993) has suggested that the possibility of emergency assistance from platform personnel might be a factor in anglers' willingness to travel farther to fish at active oil and gas platforms than other artificial reef systems.

Commercial fishers and charter vessel skippers have suggested that the potential of offshore wind farms to sustain economic activity could be augmented through the deployment of additional scour protection or specific artificial reef material and the enhancement of shellfish stocks (Ashley, 2014; Blyth-Skyrme, 2010). However, it is not clear how such schemes would be financed in practice (Hooper et al., 2015).

Concerns and risk perception

Research into issues of commercial fishing and offshore renewables has tended to focus on perceptions of future development as opposed to experiences of existing structures. At least some fishers recognise the possibilities of artificial reef and spillover effects as potential benefits (Alexander et al., 2013a, b; Reilly et al., 2015); however, fishers are uncertain whether these possible gains will be realised (Alexander et al., 2013a), and they have concerns about the legality of fishing close to energy infrastructure, the risk of gear entanglement or vessel collision (particularly in the event of engine failure), and the validity of insurance (Alexander et al., 2013a; Reilly et al., 2015; Hooper et al., 2015). Similar concerns about safety and liability are shared by developers of offshore wind farms (Hooper et al., 2015).

These concerns affect whether fishers are able, or willing, to set gear close enough to the structures to benefit from any artificial reef effects. There appear to be different attitudes towards minimum safe distances in different locations, with 74% of Louisiana sport fishers reporting that they tie up to oil and gas platforms at some point during the year (Gordon, 1993). Practice and opinions in the North Sea suggest a high perceived risk. Most offshore wind farm developers would prefer fishing gear to be deployed at least 100 m from their infrastructure (farther than the minimum distance set in statutory regulations), and practices such as fishers "… attempting to tie on to structures, seeking to harvest mussels" are viewed negatively (Hooper et al., 2015). The fishers themselves are also often reticent to approach offshore wind farms too closely. Potters are willing to set gear much closer to natural reefs and structures such as shipwrecks than offshore wind farms, but they cannot use grappling hooks to retrieve gear lost near turbine foundations as they would in other circumstances because of the risk of snagging cables (Hooper et al., 2015).

Aquaculture as an alternative to capture fisheries

If the barriers are such that the co-location of capture fisheries is inappropriate, an alternative might be for the farming of seaweed, shellfish, or finfish through aquaculture to be established within the footprint of energy installations. In parallel with interest in capture fishery co-location, there has been a recent focus in states bordering the North Sea (particularly Germany) on the potential to establish aquaculture facilities within offshore wind farms.

The opportunity

In general, offshore aquaculture is of broad interest because it has enormous economic potential and could perhaps support new livelihoods in rural coastal areas (Buck et al., 2008), while the further development of nearshore aquaculture is hampered due to poor water quality and the presence of protected areas (Buck et al., 2004). Cultivation offshore is also considered to be more sustainable than intensive techniques (Buck et al., 2008), and is expected to have a lower environmental impact as the dynamic conditions reduce the speed at which organic material accumulates below the farm (Jansen et al., 2016).

There have been optimistic assessments of the range of species that could be cultivated offshore (Gimpel et al., 2015). A more general consensus suggests that bivalve aquaculture is the most feasible (particularly for blue mussels *Mytilus edulis*, and in the medium term, Pacific and European oysters *Crassostrea gigas* and *Ostrea edulis*), while possibilities for fish farming are limited (Jansen et al., 2016; Syvret et al., 2013, Linley et al., 2007; Mee, 2006). Conflicting opinions exist for the potential of seaweed aquaculture (Jansen et al., 2016; Syvret et al., 2013; Linley et al., 2007), which could relate to the different markets and infrastructure currently available in different countries.

There is broad support across different stakeholder groups for the co-location of aquaculture and offshore wind farms in principal (Michler-Cieluch and Kodeih, 2008; Wever et al., 2015), and the possibilities range from co-existing in the same area to sharing infrastructure and joint economic input (Buck et al., 2004). Arguments for co-location include the potential for an alternative livelihood for fishers displaced by wind-farm development (Buck et al., 2004), and the improved light in which the offshore wind industry would be viewed following such collaboration with other sectors (Wever et al., 2015). The potential benefits expected mostly relate to specific technical and economic issues; the footprint of offshore wind farms provides appropriately sized areas free of shipping traffic (Buck et al., 2008), and the turbine foundations provide the possibility for attaching aquaculture systems (Buck and Buchholz, 2005). This latter benefit is considered crucial to ensure the economic feasibility of offshore aquaculture (Buck et al., 2004; Michler-Cieluch and Krause, 2008). It has also been suggested that offshore wind developers could benefit because aquaculture gear and even bottom cultivation can help reduce scour and hence the risk of monopile subsidence (Shellfish Association of Great Britain, 2012).

Both industries could also gain from the effective use of skills, equipment, and infrastructure, such as through shared environmental monitoring (SAGB, 2012). The collective development or adaptation of vessel equipment is envisaged as a first point of cooperation between the sectors (Michler-Cieluch et al., 2009), enabling shared operations and maintenance (Buck et al., 2004). There is also the opportunity to develop technical leadership in the field, with potential export possibilities to countries such as China where the aquaculture sector is much larger (Wever et al., 2015).

Economic viability

The main driver for co-location from the industry perspective is cost savings (Krause et al., 2011), and there is little motivation to proceed unless economic benefit is possible for both parties (Michler–Cieluch et al., 2009). There has been some assessment of the economic issues, which has demonstrated that the cultivation of blue mussels within offshore wind farms can be profitable, with sensitivity testing showing this remains the case even with substantial drops in price or production, or increases in fixed and transport costs (Jansen et al., 2016; Buck et al., 2010). Overall positive earnings for the offshore wind developer are also possible in a co-location context, but only where the industry continues to receive existing subsidies; without these, the development would incur financial losses (Griffin et al., 2015).

Co-location brings the potential for cost savings for both sectors in terms of shared infrastructure costs, operations and maintenance activities, and environmental impact assessment and monitoring (Buck et al., 2004, 2010; Krause et al., 2011). Griffin et al. (2015) suggest that shared operations and maintenance activities could result in substantial savings (of millions of Euros) for both industries, although the extent to which these could be realised depends on how much the shared activities result in the increased use of existing capacity (i.e., filling downtime of vessels and personnel) compared to the need for new investment to accommodate new activities. Also, any incidents of damage as a result of one industry's activities impacting on the other will be costly, and so there is a need to offset the costs of additional insurance against the benefits of co-location (Griffin et al., 2015).

These economic assessments have all been done in an academic context and concern hypothetical developments. In practice, there remains a need for semi-commercial trials to provide successful examples of co-location before financiers would be willing to invest (Mee, 2006). The need for additional data from field trials is compounded because each case for co-location would be unique, because the risks vary depending on the conditions at each offshore wind farm and the type of aquaculture (Syvret et al., 2013).

A pilot study from the Irish Sea

One example of a pilot study (driven by the shellfish industry) comes from the North Hoyle offshore wind farm constructed in 2003 that is approximately 9 km from the North Wales coast in about 10 m of water. As reported by SAGB (2012), the pilot study involved three trial beds of mussel spat laid on the seabed within (but near edge of) North Hoyle, following agreement from the developer (RWE) and permission of the regulator (the Countryside Council for Wales). The growth rates of the mussels were high during the trial, but there was unexplained high mortality as they reached harvestable size, which was perhaps due to the delay in laying spat until late summer while negotiations proceeded.

The pilot demonstrated that it was possible for the activities to coexist, with the large aquaculture vessels (more than 40 m in length) able to navigate effectively near the turbines and with no disruption to the operations of the wind farm. Important in the success of the cooperation was the development of working protocols. These were provided in advance to the wind farm developer and contained a detailed description of the proposal, a method statement, a code of practice, an operational plan, an emergency response plan, and a risk assessment, as well as evidence of vessel insurance and permission from the regulator. The schedule of work, including timing and duration of access into the wind farm, was also agreed in advance, and confirmed immediately before operations began.

Continuing challenges and concerns

Such evidence from pilot studies is limited, and there remain considerable obstacles to overcome if co-location is to develop as a successful commercial venture. The challenges faced in realising the potential for co-location have been raised consistently by stakeholders from both the aquaculture and offshore wind sectors and across different countries, suggesting a consensus on the inherent issues to be addressed (Michler-Cieluch and Krause, 2008). Even the concept of co-location is challenging, as there is no precedent for such collaboration and experience in either industry (Krause et al., 2011). However, perhaps the largest obstacle is the lack of data to support the environmental and economic viability of offshore aquaculture in general (Buck et al., 2008). The harsh offshore environment places enormous stress on the materials used leading to technical challenges, and there are also biological constraints in some offshore areas (Buck et al., 2008; Michler-Cieluch and Krause, 2008).

Consultations with offshore wind developers have shown that their principal concerns are to ensure that any aquaculture activity does not interfere with their primary purpose (of energy generation), hinder their operations, or affect the integrity of their infrastructure (Mee, 2006;

Michler-Cieluch and Krause, 2008). Issues of safety, access, potential damage to cables, insurance, and liability have been raised by stakeholders (Mee, 2006; Michler-Cieluch and Krause, 2008), echoing similar concerns expressed over the co-location of capture fisheries (Hooper et al., 2015). Also, aquaculture activities that interact directly with turbine foundations might increase the likelihood of structural failure through corrosion, bio-fouling, and increasing mechanical load, particularly in the case of jacket foundations (Jansen et al., 2016). Addressing these implications for foun-dation design, risk, and additional safety measures will be a substantial factor in successful co-location (Buck et al., 2008). Other factors considered important for successful co-location include the need for trust, transparency, and regular communication between the parties involved (Michler-Cieluch and Krause, 2008). Successful implementation would also require clarity in rights, rules, and responsibilities, as well as overall regulation (Krause et al., 2011; Michler-Cieluch et al., 2009). Without these, conflicts between sec-tors are foreseen despite the apparent strength of any co-location potential (Buck et al., 2004).

Effective regulation

Stakeholders perceive that bureaucratic obstacles are a major issue and expect co-location to complicate the existing consenting process such as environ-mental impact assessments (Michler-Cieluch and Krause, 2008). In addition, issues such as property rights that are required to ensure security of tenure have not been adequately addressed (Buck et al., 2004). Coherent regulation is required that directly addresses specific issues, such as harmonising the lease duration for both industries (Michler-Cieluch et al., 2009). Multiple use of sites was not foreseen when permits were issued for existing offshore wind-farm developments (Buck et al., 2004), and in some cases the condi-tions within current licences are a barrier to co-location. In the United King-dom, wind farm operators are prohibited from making additional income from the site (Mee, 2006), and co-use is illegal within wind farm sites in the Netherlands (Jansen et al., 2016). To permit effective co-location, the policy system will need to shift away from traditional, single-use management to a multiple-use approach, and testing the organisational issues of co-location should be as important within any pilot as the technical side (Michler-Cieluch and Krause, 2008).

Michler-Cieluch and Krause (2008) proposed a two-track approach to resolve these regulatory issues. The first track is process-oriented, addressing the longer-term requirements of knowledge exchange, social change, and building collaboration, while the second is results-oriented and concerns the short-term pursuit of concrete economic and technical solutions. The estab-lishment of co-location will be difficult if regulatory issues remain unresolved (Michler-Cieluch et al., 2009).

Lessons from the United States

The co-location potential for aquaculture has been postulated since the early stages of development of the offshore wind industry (e.g., Buck et al., 2004), but more than a decade later, has yet to see any real progress. Lessons can perhaps be learned from the experience in the United States, where co-location with offshore oil and gas platforms in the Gulf of Mexico has been proposed since the late 1980s (Bridger, 2004; Mee, 2006). This demonstrated first that industry engagement was slow; it took 10 years for the oil and gas industry to accept the possibilities for co-location and begin to invest in offshore aquaculture experiments (Mee, 2006). Ultimately, aquaculture within oil rigs in the United States was successfully piloted (Mee, 2006), and specific provisions related to the co-location of aquaculture with oil and gas platforms were included when National Offshore Aquaculture Act of 2005 (S. 1195 of the 109th Congress) was introduced. However, this bill was not enacted, and opposition remained to the establishment of offshore aquaculture in general. This included the introduction of another bill (H.R. 3534 of the 111th Congress) that had the apparent goal of delaying the development of offshore aquaculture until more comprehensive legislation could be passed (Upton and Buck, 2010).

This landscape of regulatory uncertainty was identified by the United States Administration as the major barrier to developing open-ocean aquaculture, and to commercial investment in the industry (Upton and Buck, 2010), which is a second lesson for co-location prospects elsewhere. The uncertainty has been compounded for the prospects of co-location with energy installations, as subsequent bills put before Congress (the National Sustainable Aquaculture Act of 2009, H.R. 4363 of the 11th Congress, re-introduced in 2011 as H.R. 2373 of the 112th Congress) sought to prohibit entirely the co-location of offshore aquaculture with operational or decommissioned oil and gas rigs. Ultimately, these also failed to become law.

There have been attempts to move forward in the absence of national legislation, but this has also been a protracted process. The Fishery Management Plan for Regulating Offshore Marine Aquaculture in the Gulf of Mexico took effect in September 2009. However, its implementation was delayed by factors including the drafting of the National Aquaculture Policy and a requirement to consider the implications of the Deepwater Horizon oil spill, as well the need for a consultation process on draft documents (National Marine Fisheries Service, 2015). So, it was not until January 2016 that the Final Rule was published providing the framework for applications for offshore aquaculture permits. This again envisages situations in which aquaculture and oil and gas infrastructure could be co-located (NOAA, 2016). However, even now the future for the industry does not look certain, because at least one legal challenge has been lodged since the Final Rule was published (Civil Action No. 2:16-cv-01271).

Environmental impact

As was the case in the United States, concerns about the environmental impact of offshore aquaculture have been raised in Europe (e.g., Wever et al., 2015). These tend to relate to aquaculture generally, but there are some issues specific to co-location that require more investigation. For example, the impact of suspended mussel longlines is expected to be neutral for seabirds (Roycroft et al., 2007), but offshore wind farms do present a collision risk for bird species (Hooper et al., 2017b). Whether this, and associated bird mortality, would increase with the large increase in potential prey presented by a mussel farm is unknown, but has been raised as a concern (SAGB, 2012). Conversely, there is likely to be some disturbance to marine mammals from the noise created by operations and maintenance vessels (Scheidat et al., 2011). Combining vessel activities for offshore wind farms and aquaculture therefore has the potential to reduce the cumulative impact on marine mammals in the area. Also, both offshore wind farms (Hooper et al., 2017b) and suspended bivalve culture sites (McKindsey et al., 2011) can attract non-native species. Whether the net effect of the combination of these activities, compared to their separate development within the same geographical area, would increase or reduce the potential spread of invasive species has not been investigated.

Fulfilling ambitions for co-location

Some ambitions for the multi-use of offshore spaces go beyond bilateral co-location, with visions for multifunctional structures that provide energy, fisheries, leisure, and biodiversity benefits and are conceived, developed, and managed from the outset for all potential beneficiaries, and not just the energy operators (Dafforn et al., 2015, Lacroix and Pioch, 2011). However, a first step is to secure successful co-location with a single sector. Despite considerable challenges, support remains from most stakeholders for the co-location of offshore wind farms and aquaculture as a pragmatic solution to the demands on ocean space (Wever et al., 2015). Recommendations for achieving effective co-location include establishing clear policy frameworks, mechanisms for financial support as a catalyst for the approach, appropriate licensing for environmental protection, and involving different stakeholders to share their views and jointly develop creative solutions (Stuiver et al., 2016).

Conclusion: incorporating co-location potential within marine planning

The role of marine spatial planning in reducing user conflict and maximising economic opportunities is well recognised (e.g., Directive 2014/89/EU for Maritime Spatial Planning and the US Government's National Ocean Policy

Implementation Plan). Within such planning frameworks, there are also examples of explicit reference to the desire for fisheries and other activities to coexist: for example, in the United Kingdom's Marine Policy Statement (the framework for the Marine Plans for England's coastal and offshore waters) and Scotland's National Marine Plan. There is good ecological evidence that energy infrastructure provides some benefits for some species of interest to fishers. In addition, there is a well-established practice of sport (and some commercial) fishing around offshore oil and gas platforms off the coast of the United States, representing one exemplar of co-location. However, there is a different culture and policy framework in North Sea, as evidenced by the exclusion of fishers from the proximity of oil and gas infrastructure. Such statutory restrictions are less common for offshore wind farms, so there is greater potential for co-location with fishing activity, and this is likely to be a focus of current marine planning needs as the sector continues to expand, particularly in Northern Europe. The United States' example also provides lessons for the co-location of aquaculture because legislative uncertainty and delays have severely hampered the potential for the development of the industry.

While co-location might appear attractive in theory, there is not yet strong evidence that the artificial reef effects provided by the infrastructure of offshore wind developments support an exploitable resource that fishers are willing or able to access, and many questions remain regarding the technical feasibility and other practicalities of co-locating offshore aquaculture. Co-location is unlikely to be universally applicable, because its potential will depend on local environmental conditions, the size and location of the wind farm, and the risk perception of fishers (including the aquaculture sector) and developers.

The potential for co-location can be used to guide decisions on optimum planning, while planning policy has an important role in developing strategies for facilitating any such co-location. In particular, if fishing or aquaculture is to be possible within the footprint of an offshore wind farm, the co-location issue should be considered early in the planning stages. Planning tools can be used to consider trade-offs, and support the selection of optimum sites and degree of co-location (Yates et al., 2015). Design issues such as turbine spacing are also important to allow different gears access to the wind-farm footprint, and fishers can adapt their own practices by modifying their gear type or size (Blyth-Skyrme, 2010).

Developing successful co-location is challenging nonetheless because there is often a lack of trust between fishers and energy companies, with inadequate communication, consultation, and collaboration identified as the primary reasons for the poor relationship between the sectors (de Groot et al., 2014; O'Keeffe and Haggett, 2012; Mackinson et al., 2006). There has been recent effort to address this issue through the development of best-practice guidance for interaction between the fishing and offshore energy sectors in both the United Kingdom and the United States (Fishing Liaison with

Offshore Wind and Wet Renewables Group, 2014, 2015; Moura et al., 2015). Better collaboration between fishers and developers at the earliest planning stage will be essential for successful co-location, but there is also the need for more research to (1) consider ecological impacts beyond the wind-farm footprint (particularly in terms of attraction versus production of fish biomass), (2) evaluate the ecological effects of offshore wind-farm installations more directly in fisheries terms, (3) monitor actual behaviour of fishers and anglers, and (4) assess in more detail the technical and economic feasibility, and the costs and benefits to both parties, of establishing offshore aquaculture within wind farm sites. Such additional information is necessary to provide robust evidence for incorporating co-location into future planning decisions.

Highlights

- There is strong ecological evidence of high relative abundance of important recreational and commercial species around offshore energy infrastructure.
- The Gulf of Mexico exemplifies fishery co-location opportunities, but a different culture and policy framework exists in the North East Atlantic.
- Aquaculture might present an alternative co-location option to capture fishers, although offshore techniques remain at the early stages of development.
- Co-location potential is site-specific and depends on environmental conditions, wind farm design, and the risk perception of fishers and developers.
- Stronger evidence of the fishery implications of offshore wind farms is also necessary to support robust decisions using marine spatial planning.

References

4COffshore. (2014). Global offshore wind farms database. www.4coffshore.com/windfarms. Accessed 05 December 2014.

Ajemian, M., Wetz, J., Shipley-Lozano, B., Shively, J. and Stunz, G. (2015). An analysis of artificial reef fish community structure along the Northwestern Gulf of Mexico shelf: Potential impacts of "Rigs-to-Reefs" programs. *PLoS ONE*, 10(5), e0126354.

Alexander, K. A., Potts, T. and Wilding, T. A. (2013a). Marine renewable energy and Scottish west coast fishers: Exploring impacts, opportunities and potential mitigation. *Ocean and Coastal Management*, 75, 1–10.

Alexander, K., Wilding, T. and Jacomina Heymans, J. (2013b). Attitudes of Scottish fishers towards marine renewable energy. *Marine Policy*, 37, 239–244.

Armstrong, M., Brown A., Hargreaves, J., Hyder, K., Pilgrim-Morrison, S., Munday, M., Proctor, S., Roberts, A. and Williamson, K. (2013). *Sea Angling 2012 – A Survey of Recreational Sea Angling Activity and Economic Value in England.* November 2013. Department of the Environment, Food and Rural Affairs, London.

Ashley, M. (2014). *The Implications of Co-Locating Marine Protected Areas around Offshore Wind Farms.* PhD Thesis. University of Plymouth/Plymouth Marine Laboratory.

Bergström, L., Sundqvist, F. and Bergström, U. (2013). Effects of an offshore wind farm on temporal and spatial patterns in the demersal fish community. *Marine Ecology Progress Series,* 485, 199–210.

Blyth-Skyrme, R. E. (2010). *Options and Opportunities for Marine Fisheries Mitigation Associated with Windfarms.* Final report for Collaborative Offshore Wind Research into the Environment contract FISHMITIG09. COWRIE Ltd., London.

Bridger, C. J. (Ed). (2004). *Efforts to Develop a Responsible Offshore Aquaculture Industry in the Gulf of Mexico: A Compendium of Offshore Aquaculture Consortium Research.* Mississippi-Alabama Sea Grant Consortium, Ocean Springs, MA.

Brown, H., Benfield, M. C., Keenan, S. F. and Powers, S. P. (2010). Movement patterns and home ranges of a pelagic carangid fish, *Caranx crysos,* around a petroleum platform complex. *Marine Ecology Progress Series,* 403, 205–218.

Buck, B. H. and Buchholz, C. M. (2005). Response of offshore cultivated *Laminaria saccharina* to hydrodynamic forcing in the North Sea. *Aquaculture,* 250(3), 674–691.

Buck, B. H., Ebeling, M. W. and Michler-Cieluch, T. (2010). Mussel cultivation as a co-use in offshore wind farms: Potential and economic feasibility. *Aquaculture Economics and Management,* 14(4), 255–281.

Buck, B. H., Krause, G., Michler-Cieluch, T., Brenner, M., Buchholz, C. M., Busch, J. A., Fisch, R., Geisen, M. and Zielinski, O. (2008). Meeting the quest for spatial efficiency: Progress and prospects of extensive aquaculture within offshore wind farms. *Helgoland Marine Research,* 62(3), 269.

Buck, B. H., Krause, G. and Rosenthal, H. (2004). Extensive open ocean aquaculture development within wind farms in Germany: The prospect of offshore co-management and legal constraints. *Ocean and Coastal Management,* 47(3), 95–122.

Cenci, E., Pizzolon, M., Chimento, N. and Mazzoldi, C. (2011). The influence of a new artificial structure on fish assemblages of adjacent hard substrata. *Estuarine, Coastal and Shelf Science,* 91(1), 133–149.

Consoli, P., Romeo, T., Ferraro, M., Sarà, G. and Andaloro, F. (2013). Factors affecting fish assemblages associated with gas platforms in the Mediterranean Sea. *Journal of Sea Research,* 77, 45–52.

Cresson, P., Ruitton, S. and Harmelin-Vivien, M. (2014). Artificial reefs do increase secondary biomass production: Mechanisms evidenced by stable isotopes. *Marine Ecology Progress Series,* 509, 15–26.

Dafforn, K. A., Mayer-Pinto, M., Morris, R. L. and Waltham, N. J. (2015). Application of management tools to integrate ecological principles with the design of marine infrastructure. *Journal of Environmental Management,* 158, 61–73.

Department of Energy and Climate Change. (2011). *Applying for Safety Zones around Offshore Renewable Energy Installations, Guidance notes.* November 2011 (revised), p. 19. Department for Energy and Climate Change, London.

Di Lorenzo, M., Claudet, J. and Guidetti, P. (2016). Spillover from marine protected areas to adjacent fisheries has an ecological and a fishery component. *Journal for Nature Conservation,* 32, 62–66.

Ditton, R. B. and Auyong, J. (1984). Fishing offshore platforms central Gulf of Mexico: An analysis of recreational and commercial fishing use at 164 major offshore petroleum structures. *Government Reports, Announcements and Index, National Technical Information Service (NTIS), US Department of Commerce*, 84(21). US Department of the Interior, Minerals Management Service, Washington, DC.

Everley, K. A., Radford, A. N. and Simpson, S. D. (2016). Pile-driving noise impairs antipredator behavior of the European sea bass *Dicentrarchus labrax*. In *The Effects of Noise on Aquatic Life II* (pp. 273–279). Springer, New York.

Fabi, G., Spagnolo, A., Bellan-Santini, D., Charbonnel, E., Çiçek, B. A., García, J. J. G., Jensen A. C., Kallianiotis A. and Santos, M. N. D. (2011). Overview on artificial reefs in Europe. *Brazilian Journal of Oceanography*, 59(SPE1), 155–166.

Fabi, G., Grati, F., Puletti, M. and Scarcella, G. (2004). Effects on fish community induced by installation of two gas platforms in the Adriatic Sea. *Marine Ecology Progress Series*, 273, 187–197.

Fishing Liaison with Offshore Wind and Wet Renewables Group. (2015). *Best Practice Guidance for Offshore Renewables Developments: Recommendations for Fisheries Disruption Settlements and Community Funds*. August 2015. Fishing Liaison with Offshore Wind and Wet Renewables Group (FLOWW), Edinburgh.

Fishing Liaison with Offshore Wind and Wet Renewables Group. (2014). *Best Practice Guidance for Offshore Renewables Developments: Recommendations for Fisheries Liaison*. January 2014. Fishing Liaison with Offshore Wind and Wet Renewables Group (FLOWW), Edinburgh.

Fujii, T. (2015). Temporal variation in environmental conditions and the structure of fish assemblages around an offshore oil platform in the North Sea. *Marine Environmental Research*, 108, 69–82.

Gell, F. R. and Roberts, C. M. (2003). Benefits beyond boundaries: The fishery effects of marine reserves. *Trends in Ecology and Evolution*, 18(9), 448–455.

Gill, A. B., Huang, Y., Gloyne-Philips, I., Metcalfe, J., Quayle, V., Spencer, J. and Wearmouth, V. (2009). *COWRIE 2.0 Electromagnetic Fields (EMF) Phase 2: EMF-Sensitive Fish Response to EM Emissions from Sub-Sea Electricity Cables of the Type Used by the Offshore Renewable Energy Industry*. (Project reference COWRIE-EMF-1-06). Collaborative Offshore Wind Research into the Environment (COWRIE) Ltd.

Gimpel, A., Stelzenmüller, V., Grote, B., Buck, B. H., Floeter, J., Núñez-Riboni, I., Pogoda, B. and Temming, A. (2015). A GIS modelling framework to evaluate marine spatial planning scenarios: Co-location of offshore wind farms and aquaculture in the German EEZ. *Marine Policy*, 55, 102–115.

Gordon, W. R. Jr. (1993). Travel characteristics of marine anglers using oil and gas platforms in the central Gulf of Mexico. *Marine Fisheries Review*, 55(1), 25–31.

Gray, M., Stromberg, P. -L. and Rodmell, D. 2016. *Changes to Fishing Practices around the UK as a Result of the Development of Offshore Windfarms – Phase I*. 121 pp. The Crown Estate, London.

Griffin, R., Buck, B. and Krause, G. (2015). Private incentives for the emergence of co-production of offshore wind energy and mussel aquaculture. *Aquaculture*, 436, 80–89.

de Groot, J., Campbell, M., Ashley, M. and Rodwell, L. (2014). Investigating the co-existence of fisheries and offshore renewable energy in the UK: Identification of a mitigation agenda for fishing effort displacement. *Ocean and Coastal Management*, 102, 7–18.

GWEC. (2016). *Global Wind Report Annual Market Update 2015*. Global Wind Energy Council. www.gwec.net/publications/global-wind-report-2/global-wind-report-2015-annual-market-update/. Accessed 21 July 2016.

Helvey, M. (2002). Are southern California oil and gas platforms essential fish habitat? *ICES Journal of Marine Science*, 59(suppl), S266–S271.

Hooper, T. and Austen, M. (2014). The co-location of offshore windfarms and decapod fisheries in the UK: Constraints and opportunities. *Marine Policy*, 43, 295–300.

Hooper, T., Hattam, C. and Austen, M. (2017a). Recreational use of offshore wind farms: Experiences and opinions of sea anglers in the UK. *Marine Policy*, 78, 55–60.

Hooper, T., Beaumont, N. and Hattam, C. (2017b). The implications of energy systems for ecosystem services: A detailed case study of offshore wind. *Renewable and Sustainable Energy Reviews* 70, 230–241.

Hooper, T., Ashley, M. and Austen, M. (2015). Perceptions of fishers and developers on the co-location of offshore wind farms and decapod fisheries in the UK. *Marine Policy*, 61, 16–22.

Islam, G. M. N., Noh, K. M., Sidique, S. F. and Noh, A. F. M. (2014). Economic impact of artificial reefs: A case study of small scale fishers in Terengganu, Peninsular Malaysia. *Fisheries Research*, 151, 122–129.

Jablonski, S. (2008). The interaction of the oil and gas offshore industry with fisheries in Brazil: The "Stena Tay" experience. *Brazilian Journal of Oceanography*, 56(4), 289–296.

Jansen, H. M., Burg, S., Bolman, B., Jak, R. G., Kamermans, P., Poelman, M. and Stuiver, M. (2016). The feasibility of offshore aquaculture and its potential for multi-use in the North Sea. *Aquaculture International*, 24(3), 735–756.

Jørgensen, D. (2012). OSPAR's exclusion of rigs-to-reefs in the North Sea. *Ocean and Coastal Management*, 58, 57–61.

Jørgensen, T., Løkkeborg, S. and Soldal, A. V. (2002). Residence of fish in the vicinity of a decommissioned oil platform in the North Sea. *ICES Journal of Marine Science*, 59, S288–S293.

Koeck, B., Pastor, J., Larenie, L., Astruch, P., Saragoni, G., Jarraya, M. and Lenfant, P. (2011). Evaluation of impact of artificial reefs on artisanal fisheries: Need for complementary approaches. *Brazilian Journal of Oceanography*, 59(SPE1), 1–11.

Krause, G., Griffin, R. M. and Buck, B. H. (2011). Perceived concerns and advocated organisational structures of ownership supporting 'Offshore Wind Farm-Mariculture Integration' (pp. 203–218). In Krause, G. (Ed). *From Turbine to Wind Farms - Technical Requirements and Spin-Off Products*, INTECH, Rijeka, Croatia.

Lacroix, D. and Pioch, S. (2011). The multi-use in wind farm projects: More conflicts or a win–win opportunity? *Aquatic Living Resources*, 24(2), 129–135.

Lindquist, D. C., Shaw, R. F. and Hernandez, F. J. (2005). Distribution patterns of larval and juvenile fishes at offshore petroleum platforms in the north–central Gulf of Mexico. *Estuarine, Coastal and Shelf Science*, 62(4), 655–665.

Linley, E. A. S., Wilding, T. A., Black, K., Hawkins, A. J. S. and Mangi, S. (2007). *Review of the Reef Effects of Offshore Wind Farm Structures and their Potential for*

Enhancement and Mitigation. Report to the Department for Business, Enterprise and Regulatory Reform. London.

Løkkeborg, S., Humborstad, O. B., Jørgensen, T. and Soldal, A. V. (2002). Spatio-temporal variations in gillnet catch rates in the vicinity of North Sea oil platforms. *ICES Journal of Marine Science*, 59(suppl), S294–S299.

Louisiana Department of Wildlife and Fisheries. (2013). *Louisiana Department of Wildlife and Fisheries 2012–2013 Annual Report*, Baton Rouge.

Love, M. S. and Westphal, W. (1990). Comparison of fishes taken by a sportfishing party vessel around oil platforms and adjacent natural reefs near Santa Barbara, California. *Fishery Bulletin US*, 88, 599–605.

Love, M. S., Schroeder, D. M. and Lenarz, W. H. (2005). Distribution of bocaccio (*Sebastes paucispinis*) and cowcod (*Sebastes levis*) around oil platforms and natural outcrops off California with implications for larval production. *Bulletin of Marine Science*, 77(3), 397–408.

Love, M. S., Caselle, J. and Snook, L. (1999). Fish assemblages on mussel mounds surrounding seven oil platforms in the Santa Barbara Channel and Santa Maria Basin. *Bulletin of Marine Science*, 65(2), 497–513.

Lowe, C. G., Anthony, K. M., Jarvis, E. T., Bellquist, L. F. and Love, M. S. (2009). Site fidelity and movement patterns of groundfish associated with offshore petroleum platforms in the Santa Barbara Channel. *Marine and Coastal Fisheries: Dynamics, Management, and Ecosystem Science*, 1(1), 71–89.

Mackinson, S., Curtis, H., Brown, R., McTaggart, K., Taylor, N., Neville, S. and Rogers, S. (2006). A report on the perceptions of the fishing industry into the potential socio-economic impacts of offshore wind energy developments on their work patterns and income. *Science Series Technical Report-Centre for Environment Fisheries and Aquaculture Science*, 133 CEFAS, Lowestoft.

Marine Management Organisation. (2015). Monthly sea fisheries statistics April 2015. www.gov.uk/government/statistical-data-sets/monthly-sea-fisheries-statistics-april-2015. Accessed 8 July 2015.

McDonough, M. and Cowan, J. (2007). Tracking red snapper movements around an oil platform with an automated acoustic telemetry system. *Proceedings of the Gulf and Caribbean Fisheries Institute*, 59, 159–163.

McKindsey, C. W., Archambault, P., Callier, M. D. and Olivier, F. (2011). Influence of suspended and off-bottom mussel culture on the sea bottom and benthic habitats: A review. *Canadian Journal of Zoology*, 89(7), 622–646.

Mee, L. (2006). *Complementary Benefits of Alternative Energy: Suitability of Offshore Wind Farms as Aquaculture Sites*. Inshore Fisheries and Aquaculture Technology Innovation and Development, SEAFISH-Project Ref: 10517.

Michler-Cieluch, T. and Kodeih, S. (2008). Mussel and seaweed cultivation in offshore wind farms: An opinion survey. *Coastal Management*, 36(4), 392–411.

Michler-Cieluch, T. and Krause, G. (2008). Perceived concerns and possible management strategies for governing 'wind farm–mariculture integration'. *Marine Policy*, 32(6), 1013–1022.

Michler-Cieluch, T., Krause, G. and Buck, B. H. (2009). Reflections on integrating operation and maintenance activities of offshore wind farms and mariculture. *Ocean and Coastal Management*, 52(1), 57–68.

Milon, J. W. (1989). Artificial marine habitat characteristics and participation behaviour by sport anglers and divers. *Bulletin of Marine Science*, 44(2), 853–862.

Moura S., Lipsky A. and Morse M. (2015). *Options for Cooperation between Commercial Fishing and Offshore Wind Energy Industries: A Review of Relevant Tools and Best Practices* Seaplan Report. November 2015.

Mueller-Blenkle, C., McGregor, P. K., Gill, A. B. Andersson, M. H., Metcalfe, J., Bendall, V., Sigray, P., Wood, D. T. and Thomsen, F. (2010). *Effects of Pile-driving Noise on the Behaviour of Marine Fish*. Ref: Fish 06–08, Technical Report 31 March 2010. Collaborative Offshore Wind Research into the Environment (COWRIE).

Murray, J. D. and Betz, C. J. (1994). User views of artificial reef management in the southeastern US. *Bulletin of Marine Science*, 55(2–3), 970–981.

National Marine Fisheries Service. (2015). *Final Supplemental Information Report to the 2009 Final Programmatic Environmental Impact Statement (Fishery Management Plan for Offshore Marine Aquaculture in the Gulf of Mexico).* July 2015.

National Oceanic and Atmospheric Administration. (2016). *A Guide to the Application Process for Offshore Aquaculture in U.S. Federal Waters of the Gulf of Mexico.* Updated June 3 2016. sero.nmfs.noaa.gov/sustainable_fisheries/gulf_fisheries/aquaculture. Accessed 16 July 2016.

National Oceanic and Atmospheric Administration. (2013). Oil and Gas Exploration. Ocean Explorer. National Oceanic and Atmospheric Administration. oceanexplorer.noaa.gov/explorations/06mexico/background/oil/oil.html. Accessed 15 June 2015.

Oil and Gas Authority. (2015). Full list of Offshore Fields in Production. www.gov.uk/oil-and-gas-uk-field-data. Accessed 15 June 2015.

Oil and Gas UK. (2013). Decommissioning of Pipelines in the North Sea Region.

Olsen, S. and Valdemarsen, J. W. (1977). *Fish distribution studies around offshore installations*. ICES Report C.M.1977/B:41. 4 pp. International Council for the Exploration of the Sea.

O'Keeffe, A. and Haggett, C. (2012). An investigation into the potential barriers facing the development of offshore wind energy in Scotland: Case study–Firth of Forth offshore wind farm. *Renewable and Sustainable Energy Reviews*, 16(6), 3711–3721.

Osmundsen, P. and Tveterås, R. (2003). Decommissioning of petroleum installations—major policy issues. *Energy Policy*, 31(15), 1579–1588.

Ramos, J., Santos, M. N., Whitmarsh, D. and Monteiro, C. C. (2006). Patterns of use in an artificial reef system: A case study in Portugal. *Bulletin of Marine Science*, 78(1), 203–211.

Reggio, V. C. (1987). *Rigs-to-Reefs: The use of obsolete petroleum structures as artificial reefs*. US Department of the Interior, Minerals Management Service, Gulf of Mexico OCS Regional Office.

Reilly, K., O'Hagan, A. M. and Dalton, G. (2015). Attitudes and perceptions of fishermen on the island of Ireland towards the development of marine renewable energy projects. *Marine Policy*, 58, 88–97.

Reubens, J., Rijcke, M., Degraer, S. and Vincx, M. (2014). Diel variation in feeding and movement patterns of juvenile Atlantic cod at offshore wind farms. *Journal of Sea Research*, 85, 214–221.

Reubens, J., Braeckman, U., Vanaverbeke, J., van Colen, C., Degraer, S. and Vincx, M. (2013a). Aggregation at windmill artificial reefs: CPUE of Atlantic cod (*Gadus morhua*) and pouting (*Trisopterus luscus*) at different habitats in the Belgian part of the North Sea. *Fisheries Research*, 139, 28–34.

Reubens, J., Pasotti, F., Degraer, S. and Vincx, M. (2013b). Residency, site fidelity and habitat use of Atlantic cod (*Gadus morhua*) at an offshore wind farm using acoustic telemetry. *Marine Environmental Research*, 90, 128–135.

Roberts, C. M., Bohnsack, J. A., Gell, F., Hawkins, J. P. and Goodridge, R. (2001). Effects of marine reserves on adjacent fisheries. *Science*, 294(5548), 1920–1923.

Roycroft, D., Kelly, T. C. and Lewis, L. J. (2007). Behavioural interactions of seabirds with suspended mussel longlines. *Aquaculture International*, 15(1), 25–36.

Santos, M. N. and Monteiro, C. C. (2007). A fourteen-year overview of the fish assemblages and yield of the two oldest Algarve artificial reefs (southern Portugal). *Hydrobiologia*, 580(1), 225–231.

Scarcella, G., Grati, F. and Fabi, G. (2011). Temporal and spatial variation of the fish assemblage around a gas platform in the northern Adriatic Sea, Italy. *Turkish Journal of Fisheries and Aquatic Sciences*, 11(3), 433–444.

Scheidat, M., Tougaard, J., Brasseur, S., Carstensen, J., van Polanen Petel, T., Teilmann, J. and Reijnders, P. (2011). Harbour porpoises (Phocoena phocoena) and wind farms: A case study in the Dutch North Sea. *Environmental Research Letters*, 6(2), 025102.

Seaman Jr, W., Lindberg, W. J., Gilbert, C. R. and Frazer, T. K. (1989). Fish habitat provided by obsolete petroleum platforms off southern Florida. *Bulletin of Marine Science*, 44(2), 1014–1022.

Shellfish Association of Great Britain. (2012). *EFF Project - Shellfish Aquaculture in Welsh Offshore Wind farms – Colocation Potential Scoping Meeting.* Park Inn by Radisson, Cardiff, Tuesday 4 December 2012 Report. Shellfish Association of Great Britain. Catrin Ellis Associates.

Simon, T., Pinheiro, H. T. and Joyeux, J. C. (2011). Target fishes on artificial reefs: Evidences of impacts over nearby natural environments. *Science of the Total Environment*, 409(21), 4579–4584.

Soldal, A. V., Svellingen, I., Jørgensen, T. and Løkkeborg, S. (2002). Rigs-to-reefs in the North Sea: Hydroacoustic quantification of fish in the vicinity of a "semi-cold" platform. *ICES Journal of Marine Science*, 59(suppl), S281–S287.

Southgate, T. and Myers, A. A. (1985). Mussel fouling on the Celtic Sea Kinsale field gas platforms. *Estuarine, Coastal and Shelf Science*, 20(6), 651–659.

Spanier, E., Lavalli, K. L. and Edelist, D. (2011). An overview of their application for fisheries enhancement, management, and conservation. In: Bortone, S. A., Brandini, F. P., Fabi, G., and Otake, S. (Eds.). *Artificial reefs in fisheries management.* CRC Press. Boca Raton.

Stanley, D. R. and Wilson, C. A. (1989). Utilization of offshore platforms by recreational fishermen and scuba divers off the Louisiana coast. *Bulletin of Marine Science*, 44(2), 767–776.

Stuiver, M., Soma, K., Koundouri, P., van den Burg, S., Gerritsen, A., Harkamp, T., Dalsgaard, N., Zagonari, F., Guanche, R., Schouten, J. J. and Hommes, S. (2016). The Governance of multi-use platforms at sea for energy production and aquaculture: Challenges for policy makers in European seas. *Sustainability*, 8(4), 333.

Syvret, M., FitzGerald, A., Wilson, J., Ashley, M. and Ellis Jones, C. (2013). *Aquaculture in Welsh Offshore Wind Farms: A Feasibility Study into Potential Cultivation in Offshore Wind Farm Sites.* Report for the Shellfish Association of Great Britain, 250 pp. Aquafish Solutions Ltd.

Thomsen, F., Lüdemann, K., Kafemann, R. and Piper, W. (2006). *Effects of off-shore wind farm noise on marine mammals and fish.* Report prepared on behalf of Collaborative Offshore Wind Research into the Environment (COWRIE) Ltd. Biola, Hamburg, Germany.

Upton H. F. and Buck E. H. 2010. *Open Ocean Aquaculture.* Congressional Research Service Report for Congress 7-5700. RL32694. Congressional Research Service, Washington, DC.

Valdemarsen, J. W. (1979). *Behaviour aspects of fish in relation to oil platforms in the North Sea.* ICES Report C.M. 1979/B:27 International Council for the Exploration of the Sea.

Vandendriessche, S., Derweduwen, J. and Hostens, K. (2014). Equivocal effects of offshore wind farms in Belgium on soft substrate epibenthos and fish assemblages. *Hydrobiologia,.*

Vandendriessche, S., Reubens, J., Derweduwen, J., Degraer, S. and Vincx, M. (2013a). Offshore wind farms as productive sites for fishes? In: Degraer, S., Brabant, R., and Rumes, B. (2013). *Environmental Impacts of Offshore Wind Farms in the Belgian Part of the North Sea: Learning from the Past to Optimise Future Monitoring Programs.* Royal Belgian Institute of Natural Sciences, Brussels, pp. 153–161.

Vandendriessche, S., Hostens, K., Courtens, W. and Stienen, E. (2013b). Fisheries activities change in the vicinity of offshore wind farms. In: Degraer, S., Brabant, R., and Rumes, B. (2013). *Environmental Impacts of Offshore Wind Farms in the Belgian Part of the North Sea: Learning from the Past to Optimise Future Monitoring Programs.* Royal Belgian Institute of Natural Sciences, Brussels, pp. 81–85.

Wever, L., Krause, G. and Buck, B. H. (2015). Lessons from stakeholder dialogues on marine aquaculture in offshore wind farms: Perceived potentials, constraints and research gaps. *Marine Policy,* 51, 251–259.

Whitmarsh, D., Santos, M. N., Ramos, J. and Monteiro, C. C. (2008). Marine habitat modification through artificial reefs off the Algarve (southern Portugal): An economic analysis of the fisheries and the prospects for management. *Ocean and Coastal Management,* 51(6), 463–468.

Wilhelmsson, D. and Malm, T. (2008). Fouling assemblages on offshore wind power plants and adjacent substrata. *Estuarine, Coastal and Shelf Science,* 79(3): 459–466.

Winter, H., Aarts, G. and van Keeken, O. (2010). *Residence Time and Behaviour of Sole and Cod in the Offshore Wind Farm Egmond aan Zee.* Report by IMARES - Wageningen UR and Noordzeewind. p. 50. IMARES, Wageningen.

Wolfson, A., Van Blaricom, G., Davis, N. and Lewbel, G. S. (1979). The marine life of an offshore oil platform. *Marine Ecology Progress Series,* 1, 81–89.

Yates, K. L., Schoeman, D. S. and Klein, C. J. (2015). Ocean zoning for conservation, fisheries and marine renewable energy: Assessing trade-offs and co-location opportunities. *Journal of Environmental Management,* 152, 201–209.

Chapter 11

Compatibility of offshore energy installations with marine protected areas

Ruth H. Thurstan, Katherine L. Yates and Bethan C. O'Leary

Introduction

Central to marine environmental policy is the preservation and maintenance of biodiversity and ecosystem services, which includes the mitigation of negative anthropogenic activities and their effects. Designating marine protected areas is one of the principal tools advocated to achieve these objectives, and as such they have been incorporated into international commitments (e.g., the Convention on Biological Diversity and Sustainable Development Goal 14) and national legislation (e.g., the Marine Strategy Framework Directive for the European Union). The term 'marine protected area' itself refers to an individual site or sites within the marine environment where human activities have been restricted to varying degrees with the aim of protecting living and non-living resources. There is no one internationally accepted definition, although all variants tend to overlap at least partially. The OSPAR Convention, which guides international cooperation on the protection of the marine environment of the North-East Atlantic, defines a marine protected area as:

> An area within the [...] maritime area for which protective, conservation, restorative or precautionary measures, consistent with international law have been instituted for the purpose of protecting and conserving species, habitats, ecosystems or ecological processes of the marine environment.
>
> (OSPAR, 2003)

The United Nations Convention on Biological Diversity target currently commits the 168 signatory governments[1] to protect $\geq$ 10% of marine environments within marine protected areas by 2020 (Convention on Biological Diversity, 2010). In line with this target, the number and coverage of marine protected areas has increased steadily (Pita and Pierce, 2011; Lubchenco and Grorud-Colvert, 2015). However, with only an estimated 5.7% of the sea currently protected in any form of marine protected area[2], progress still needs to be made to meet global targets. Some have argued that coastal and offshore energy installations can provide marine conservation benefits as a result of their forming *de facto* marine protected areas (Inger et al., 2009). Thus, a possible solution to increasing the coverage of protected areas in an ever more-crowded seascape is for marine protected areas and energy generation sites to be co-located, that is, purposefully sited together to achieve dual objectives. However, for future marine spatial planning processes to incorporate opportunities for marine protected area and energy generation co-location effectively, we first need to understand the costs, benefits and limitations.

In this chapter, we define marine protected areas and their common objectives. We explore the realised and potential conservation benefits and costs of energy-generation sites. We then consider how marine energy-generation sites relate to the objectives of protected areas, and we examine whether the co-location of conservation and energy-generation sites can be used to build towards global targets for marine protected areas.

Objectives of marine protected areas

Marine protected areas range from highly protected, no-take areas (i.e., closed to all forms of extractive use) to partially protected areas that only restrict particular activities. They are most commonly established for biodiversity protection and nature conservation; however, more recently they are being recognised as a tool for the management and recovery of commercial fish stocks (DEFRA, 2005; Gaines et al., 2010). The term 'marine protected area' is often used as an umbrella phrase for different types of protected areas, such as Special Areas of Conservation and Special Protection Areas in Europe.

Marine protected areas can be established individually, or as part of a larger network, within which individual sites might restrict different activities and be afforded varying protection. Individual protected areas are assigned a set of specific conservation objectives; however, broader objectives are also likely for the network within which it sits. Conservation objectives for specific sites usually detail nature conservation aspirations, and they consist of statements describing the desired ecological state and quality of a feature or species for which the area is protected, as well as its current state and the pressures to which the feature or species is vulnerable (e.g., Marine Conservation Zone Project, 2010). The objectives establish whether the feature or species in question meets the desired state and should be maintained, or whether it falls below it and should be recovered (Marine Conservation Zone Project, 2010). Objectives of the wider network often include: (1) representativity (representation of all biogeographic areas, habitats or species), (2) connectivity (networks should be designed to enable linkages between protected areas for population persistence), and (3) replication (the duplication of features in separate protected areas to spread risk and ensure that natural variation is covered) (Roberts et al., 2003; Gaines et al., 2010). Conservation objectives therefore determine not only the management regime of a protected area (for example, which activities should be restricted), but also inform the design and designation of new sites to complement broader network objectives. The conservation objectives of a site or network are important for determining whether offshore energy installations can be co-located with marine protected areas.

Principle of co-location

Co-location can be defined as two or more activities designed to run concurrently, either in time or space (Christie et al., 2014). Co-location of offshore energy installations with protected areas (this chapter and Chapter 13), and activities such as fishing (Chapter 10) have been proposed as an option to reduce conflict and ease competing demands on space (Christie et al., 2014; Yates et al., 2015).

Co-location of marine protected areas and energy installations has the potential to offer a win-win situation to stakeholders if these sites can simultaneously meet the objectives of both nature conservation and socio-economics. Furthermore, if offshore energy installations can be considered compatible with protected areas, the overall demand for space for purely conservation purposes could be reduced. Evidence suggests that stakeholders are amenable to the principle of co-location, although concerns remain regarding the loss of grounds for prospecting, displacement of existing activities (particularly fishing), and subsequent socio-economic costs (Christie et al., 2014).

Potential benefits and costs of co-location

Energy generation in the marine environment comes in several forms, all of which affect marine ecosystems and the human activities therein differently (Chapter 8). These effects are likely to vary depending on the type of energy installation, local hydrodynamic regimes, and the habitats, species and human activities that occupy the site (Gill, 2005). As the objectives of marine protected areas focus commonly on the maintenance or improvement of biodiversity and habitat integrity, alterations to the environment have implications for whether energy installations are compatible for co-location with marine protected areas.

The environmental and socio-economic changes exerted by offshore energy installations (e.g., Table 11.1) will also depend on the stage of development considered. For example, biodiversity costs can be greater during the construction and decommissioning stages, but be relatively benign throughout operation. As such, when considering whether such structures can function as marine protected areas, the entire lifespan of an installation needs to be considered. Energy technologies that have been operational for several decades (e.g., oil and gas installations, offshore wind farms and tidal barrages) offer insight into how such structures influence and alter surrounding environments over the long term, and inform discussions on the potential for offshore-energy sites to function as protected areas.

Table 11.1 Potential biodiversity and socio-economic benefits and costs from marine developments of renewable-energy infrastructure

	Benefits	Costs
Biodiversity	Creation of alternative substrata (i.e., hard substrata in an otherwise soft-substratum environment) can increase species richness and density Long lifespan of structures and possibility for sites to remain in place once decommissioned offer long-term protection once construction is complete Excluding activities detrimental to habitat integrity or vulnerable species (e.g., bottom trawling) either on structure or in surrounding exclusion zone	Localised destruction and fragmentation of naturally occurring habitat types and associated communities Reduction in species richness if particular species establish dominance Facilitating the spread of alien, invasive, or otherwise undesirable species Noise, pollution, and intrusive activities associated with construction affect habitats and species across a wide area Areas suitable for energy installations might not be those most in need of protection
Socio-economic	Provide more area for other activities, thereby limiting the immediate socio-economic impacts of implementing energy installations and protected areas Pre-construction surveys and routine monitoring of installation sites during operation could support monitoring and management and reduce overall economic costs of establishment Provide stakeholder engagement opportunities for developers, improving sustainability credentials and realising best environmental practice	Additional protected area designation restricts activities compatible with energy installations but not protected areas (e.g., static gear fishing) Establishing protected area has additional monitoring or management requirements (and therefore, costs) at installations that might not be attractive to developers

In the following sections, we explore three types of energy installations, and discuss their benefits and impacts (both known and potential) within the context of co-location with marine protected areas.

Oil and gas rigs

Offshore oil and gas rigs have formed the basis for petroleum energy extraction for the last four decades (Ferrier and Bamberg, 1982). Oil and gas development and operations are typically broken down into four stages: seismic exploration, exploratory drilling and installation, operation, and de-commissioning (Khan and Islam, 2008). Exploration and installation can take place over a period of 3–5 years, while the operation and production of oil can occur over a period of 25–35 years. One of the most studied impacts of these structures when active has been the effect of pollutants on marine eco-systems. These include impacts from chronic pollution (e.g., waste products such as oil/mud drillings, industrial wastes, and naturally occurring radio-active materials) and acute incidents resulting directly from the installation itself (e.g., blowouts such as the Deepwater Horizon oil spill in the Gulf of Mexico in 2010; Rabalais, 2014), or indirectly through spillage of oil prod-ucts during transport offsite such as the Exxon-Valdez oil spill in Alaska in 1989 (Peterson et al., 2003). Pollutants can take the form of suspensions or solid wastes that can coat organisms, organic substances that can promote eu-trophic or oxygen-deficient effects, and toxic substances that can have toxic, carcinogenic or mutative effects (Patin, 1999). Noise pollution during the ex-ploratory, installation and operational stages can also be problematic for ma-rine life, particularly marine mammals (e.g., Rossi-Santos, 2015, Chapter 8).

Despite the identified problems with chemical and noise pollution, the effects of chronic pollution are generally localised, with most pollutants restricted to within 500 m of the rig (Khan and Islam, 2008). Acute pollu-tion events are, however, by definition unpredictable and can be widespread, likely ruling out operational oil and gas rigs as suitable for co-location with marine protected areas.

The large structure of oil and gas rigs does, however, prevent other poten-tially damaging activities (such as fishing) from taking place in the immediate area (Khan and Islam, 2008). Oil and gas rigs (in common with other offshore energy installations) also introduce hard substrata, presenting opportunities for encrusting organisms to attach. These organisms can form the basis of substan-tially different communities to those present within the surrounding area, po-tentially increasing local biodiversity (Feary et al., 2011). Moreover, the physical complexity of these structures and abundance of interstitial spaces offer refuge to larger animals such as fish and invertebrates (e.g., Friedlander et al., 2014). However, such hard substrata are not natural and might not facilitate the forma-tion of natural species assemblages[3]. Indeed, communities formed on rigs differ from local natural hard substrata communities even several decades following construction (Burt et al., 2009). It is also possible that introducing hard substrata improves the competitiveness of a few species or facilitates the spread of particular species via the provision of 'stepping stones' for larval transport, potentially altering regional community dynamics (Macreadie et al., 2011).

Many oil and gas rigs are now coming to the end of their operational lives (Macreadie et al., 2011), thus presenting a new challenge—what to do with these structures once they are no longer active. Decommissioning procedures are expensive and often subject to stringent international and/or national regulations (Osmundsen and Tveterås, 2003), while the removal of the base structure impacts the surrounding seabed, in addition to destroying communities formed on the rig during its lifetime. In some countries therefore, programmes to convert shallow-water rigs to artificial reefs have been implemented (Kaiser and Pulsipher, 2005). A growing literature is now dissecting the benefits of keeping decommissioned structures—both in shallow and in deep water—in place as artificial habitats (e.g., Claisse et al., 2014; Friedlander et al., 2014; Kolian et al., 2017), although this remains controversial (Fowler et al., 2015). These abandoned structures also continue to present physical obstacles to destructive activities and can act as a physical demarcation of an activity-restricted zone. Co-location of marine protected areas over decommissioned structures would avoid many of the environmental concerns of an active installation, although some environmental impacts (e.g., remains of drill cuttings) will inevitably not align with the objectives of protected areas.

Another option is to move decommissioned rig structures to new areas to function as artificial reefs (Macreadie et al., 2011; Figure 11.1). Decommissioned

Figure 11.1 **A decommissioned, shallow-water rig that has been converted into tourist accommodation and a dive platform, Sabah, Malaysia.**
Source: Sipadan Scuba Diving, Creative Commons Attribution-Share Alike 3.0 Unported license.

rigs can provide additional habitats for the settlement of larvae, potentially improving the survival of habitat-limited species. This in turn can increase connectivity among populations, particularly in the deep sea where reef systems often form stepping stones for larval transport (Macreadie et al., 2011). However, removing rig structures and placing them in a new area of seabed can invoke similar problems in the new area.

Current (tidal) turbines and barrages

Marine currents offer a consistent source of kinetic energy, and so the inherent predictability of tidal power is attractive for developers and reduces the need for backup plants onshore powered by fossil fuels (Delucchi and Jacobson, 2011; Jacobson and Delucchi, 2011). Tidal turbines are installed on the seabed at locations with high tidal current velocities or strong continuous ocean currents. They extract energy from flowing water using submerged rotors, which harness the power of marine currents to drive generators to produce electricity (Sleiti, 2015). Marine currents can be harnessed through different generating methods, including tidal barrages (permeable dams across the full width of an estuary) and tidal stream generators (such as 'SeaGen' in Strangford Loch; Figure 11.2). Estimates of global potential vary; however, tidal energy capacity could exceed 120 terawatt-hours per year globally (Black and Veatch Ltd., 2011), generating enough electricity to provide energy for an estimated 38 million people/year[4]. In the United Kingdom, the total extractable resource from tidal energy is estimated to be approximately 18 terawatt-hours/year (Black and Veatch Ltd., 2011), equivalent to around 6% of the United Kingdom's electricity demand (Digest of UK Energy Statistics, 2016). Tidal barrages have been in operation on a commercial scale since 1967 with the opening of La Rance in France, a facility that produces approximately 480 gigawatt-hours/year (Moan and Smith, 2007), enough to meet the electricity demand for approximately 69,000 people[5]. Tidal power therefore offers substantial opportunity for renewable energy generation in suitable locations.

Despite its potential, tidal power has not yet been widely developed due to high construction costs and concerns regarding their negative environmental impacts. The presence of tidal barrages inevitably results in changes in estuary basins, channels or coastal areas due to the modification of flow, vertical mixing of waters, and resuspension of sediments (Gao et al., 2013). Reduced sediment resuspension can cause localised increases in light penetration, and a reduction in mixing can freshen waters due to reduced saline penetration and decreased water quality (Gao et al., 2013; Hooper and Austen, 2013). In addition, tidal barrages can result in the loss of habitats, especially intertidal mudflats and saltmarshes that are important for some species of nationally and internationally protected birds. Benthic habitats can also change due to alterations to seafloor currents as well as damage during construction

(Frid et al., 2012). A well-publicised impact of tidal barrages is the potential for the structure to impede migratory fish, while marine animals can suffer from direct collision with the barrage and turbines (Hooper and Austen, 2013). Thus, the extensive environmental changes expected from the establishment of tidal barrages make them unlikely candidates for co-location with marine protected areas.

The technology of tidal current turbines is currently limited because of high costs; however, its potential environmental impacts are low compared to other analogous technologies, and increasing investment, government incentives, and decreasing capital costs give it the potential to be commercially competitive (Douglas et al., 2008). Unlike the dam-like structure of tidal barrages, tidal turbines comprise twin rotors attached to the seabed and connected by a cross-beam that is also supported by a seabed structure (Marine Current Turbines, 2011). Potential environmental impacts include altering local hydrodynamics, with corresponding changes to sediment dynamics (Neill et al., 2009, 2012) and associated benthic communities, and increased noise and collision potential for marine mammals and seabirds (Fraenkel, 2006).

Installed in 2008, SeaGen in Strangford Lough (Northern Ireland) is a working example of a tidal current turbine (Figure 11.2). Strangford Lough itself is designated as a national Marine Conservation Zone and an international European Special Area of Conservation and Special Protection Area, primarily its important benthic and coastal habitats as well as the bird and seal populations it supports (Department of the Environment, 2015). Prior to SeaGen's construction, the environmental impact assessment considered the potential for environmental effects on designated sites and populations as

Figure 11.2 SeaGen tidal power plant, Strangford, County Down, Northern Ireland, June 2011 (showing blades raised for maintenance).
Source: Ardfern, Creative Commons Attribution-Share Alike 3.0 Unported license.

well as socio-economic effects through baseline surveys and assessment. Development was ultimately allowed subject to mitigation and continued monitoring through construction and operation to ensure sustained ecological integrity of the site and its features (Marine Current Turbines, 2011; Savidge et al., 2014). Marine mammals appear to avoid passing too near the device, although the high variability of the data means that undetected impacts cannot yet be ruled out. In addition, the benthic community lost through the construction of the turbine has been replaced with the colonisation of its foundations (Marine Current Turbines, 2011). Strangford Lough thus provides an example of co-location of a offshore energy installation within an already highly protected site. Consequently, the construction and operation of the installation had to occur in line with existing environmental designations, providing the potential for advancing good environmental practice in the development of tidal turbines.

Wind power

For offshore wind installations to convert enough energy to be commercially viable, large expanses of sea are required for the foundations and associated infrastructure of many individual turbines, although the actual footprint on the seabed (in terms of area modified) is low (Miller et al., 2013). Other infrastructure includes scour protection (e.g., large boulders, gravel or cobble) and subsurface cabling (both connecting turbines to offshore substation(s) and to the shore) (Gill, 2005). The construction phase necessitates the destruction of habitats in the immediate infrastructure area, and turbidity can increase during construction from resuspended sediments, potentially smothering organisms in the immediate vicinity and down current from the site. Noise is also a concern during the pile-driving process, where foundations are hammered into the seabed (Gill, 2005), although alternative seabed fixings are available (Miller et al., 2013). However, once in place, the foundations and scour-protection structure provide hard-attachment substrata for epifauna (Wilson and Elliott, 2009). Additionally, scour protection surrounding the structure can provide a more heterogeneous habitat than was present prior to construction (Petersen and Malm, 2006).

Once in operation, wind-farm installations emit noise and electromagnetic fields (Miller et al., 2013). While noise production throughout operation will be less acute than during construction, there are concerns that many animals using acoustics (from marine mammals, to finfish and crustaceans) will exhibit behavioural modifications or suffer auditory injury from operational noise (e.g., Bailey et al., 2010). Sharks and rays, in addition to some marine mammals, are sensitive to electromagnetic fields (Gill et al., 2009). There are also concerns about animals, particularly birds, colliding with the structures (Brabant et al., 2015). In addition, infrastructure foundations can alter local hydrodynamics, and thus the transportation and deposition of sediments

(Gill, 2005). In common with other hard-structure developments, introducing new habitats might also facilitate the spread of non-indigenous species (De Mesel et al., 2015).

Unlike oil and gas developments, the recent construction of offshore wind installations means that decommissioning has not yet occurred, but discussions are happening about what to do with defunct structures (Smyth et al., 2015). Wind turbine structures are expected to function for a period of 20–30 years, at which point mechanical fatigue is likely to set in, or the installation will no longer function at the capacity required (Ortegon et al., 2013, in Smyth et al., 2015). This does not necessarily mean the end of operation because, unlike finite oil and gas reserves, repowering (replacement of mechanical parts) is an option. However, due to improvements in turbine technology and the increasing size of newer turbine towers and blades, new foundations will be required even if a wind farm site is repowered to meet updated spacing requirements (e.g., the Nørrekær Enge Wind Farm in Denmark was repowered in 2009; 77 turbines were replaced with 13 larger machines increasing the installed capacity from 1.7 to 30 megawatts[6]).

The challenges of leaving defunct structures in place as artificial habitat include reduced navigational safety and the spread of non-indigenous species (Smyth et al., 2015). Leaving such structures in place also raises questions as to who is liable or responsible for the structure should an incident occur? Potential socio-economic benefits include the aggregation of fish, which could potentially be exploited by recreational or commercial fishers, or as a recreational resource for non-extractive stakeholders such as divers (Shani et al., 2012).

Summary

Due to the physical similarity of many marine-energy installations, environmental and socio-economic effects are comparable, and particularly those related to the construction of foundations and introduction of hard substrata. Consequently, the same considerations arise regarding co-location of any of these structures with marine protected areas, namely, the destructive construction process, the introduction of unnatural habitat, and the potential for pollution. However, noise pollution rather than chemical pollution is likely the biggest problems in the case of wind farms. Many countries require that environmental impact assessments are done prior to constructing an energy installation (Wilson and Elliott, 2009), with the goal of identifying adverse impacts and avoiding, mitigating or offsetting these. In some cases, the localised destruction of habitats will not be able to be avoided or mitigated, thus challenging their co-location with marine protected areas. However, some (e.g., Wilson and Elliott, 2009) have advocated that the benefits of creating new habitats and the subsequent protection of soft-sediment habitats between structures outweigh the costs. For example, Wilson and Elliott (2009) calculated that adding a turbine foundation provides up to 2.5 times the area lost by the placement of the structure. Thus, while site-level objectives of marine protected area might not necessarily be met, increases in regional biological diversity are possible.

Discussion

Whether offshore energy-generation sites can legally be co-located with marine protected areas depends on the environmental objectives of the marine protected area in question. For example, the European Habitats Directive does not prohibit the development of energy installations within its network of marine protected areas (*Natura* 2000 sites). In principle therefore, energy installations can be fitted retrospectively within protected sites under European Union legislation. The question then becomes whether development is allowed under domestic law. Christie et al. (2014) concluded that development within United Kingdom's *Natura* 2000 sites is legally feasible, as long as the adverse effects on environmental integrity of a site are assessed appropriately, and adequate mitigation measures are put in place (Chapter 13). Legal mechanisms for co-locating energy installations also exist for other forms of protected areas (e.g., Marine Conservation Zones in England); however, protected area designations providing higher protection and excluding all 'damaging' activities (i.e., reference areas) will not be compatible with renewable-energy infrastructure (Christie et al., 2014). This could also apply in other parts of the world where 'sanctuary' or 'highly protected' sites exist within a broader protected-area network.

From an ecological perspective, the feasibility and desirability of co-location depends on many system aspects, including the conservation costs and operational benefits, and whether the installation predates the protected-area designation or is being applied retrospectively. Arguably, the greatest inconsistencies with marine protected area objectives occur during the construction and decommissioning phases, which have the greatest potential to compromise the ecological integrity of the site in the short term. Whether this is viewed as being incompatible with marine protected area designation will be informed by the specific conservation objectives. For example, if the goal of the designation is to increase overall biodiversity, the addition of hard substrata could facilitate this over the long term. If, however, the goal is to maintain or recover a vulnerable species, an energy installation probably would be incompatible at any stage of operation, unless the resulting offshore energy infrastructure provides suitable habitat for that species.

There are several other, broader considerations for co-location opportunities, such as the potential for pollution or the alteration of the substratum beyond the local scale. For example, oil and gas rigs might align with marine protected area objectives, but the likelihood of pollution could outweigh any benefits of co-location. Another hurdle is that energy installations modifying existing habitats and supporting novel communities could end up protecting something that was not originally there.

In situations where energy infrastructure has been in place for several decades, marine protected-area designations will have to be fitted retrospectively. This negates the environmental issues of constructing and operationalising the marine installation, which are perhaps the most costly in terms of ecological integrity. However, it does question whether an installation can be considered a 'natural' protected area. In such situations, the potential for incompatibility with protected-area objectives might also arise should the installation (if still operational) threaten existing populations via pollution or collision. Alternatively, site integrity can be comprised if the structure is removed at the end of its working life. A structure can also affect the potential for a site to be designated as a protected area because the surrounding seabed will not always return to its original condition after removing infrastructure (Mazik and Smyth, 2013).

Evaluating the impacts of co-location is highly complex because of the many potential socio-economic, technological and environmental interactions. Variation in the intention of different protected-area designations means that opportunities for co-location are not guaranteed in each case. Finally, while not all areas are suitable for co-location, many could in effect function as temporary *de facto* protected areas because they halt other environmentally damaging activities (Christie et al., 2014). Even though such sites cannot be considered within networks of protected areas, they can still potentially have net positive environmental benefits.

Additional challenges of co-location

Much of the data required to inform evaluations of co-location potential are unavailable, but stakeholder consultation could offset these lacunae (e.g., Yates and Schoeman, 2013). However, this approach relies on full disclosure and participation. In addition, co-locating conservation areas with industrial infrastructure could potentially restrict future development, repowering opportunities, or decommissioning options. Without incentives such as relevant changes in policy or offsetting opportunities, developers may be unwilling to commit to the additional requirements and restrictions that come with co-location.

Future developments and areas for research

Determining best-practice options for decommissioning is a growing research area (e.g., Kaiser and Pulsipher, 2005). Future issues not yet fully considered are that infrastructure designs might require alteration to maximise co-location potential. For instance, designs could be modified to ensure longer-term benefits to local ecosystems when it is possible to leave decommissioned structures in place, such as engineering micro-scale surfaces to enhance the resilience of encrusting organisms. Future technologies could also be more compatible with the goals of marine protected areas. For example, floating platforms for offshore wind turbines suitable for deep-water locations are now being built and trialled (Pitcher, 2014; Carrington, 2014). These structures have fewer, direct impacts on the benthos, and so could be more compatible with marine protected areas.

Conclusions

Networks of marine protected areas are generally established to ensure ecological coherence, ecological integrity, and representation of habitats and/or biogeographic regions. All energy installations offer some temporary *de facto* protection to their surrounding marine area simply by excluding certain activities. Planned co-location intended to achieve conservation targets and reduce conflict in crowded seascapes is possible if the various phases of development are compatible with the conservation objectives of the protected areas in question. While co-location is an option in some instances, the direct and indirect environmental impacts of energy installations and issues of protecting 'natural' habitats will limit the extent to which co-located sites can contribute to network targets for protection. Thus, incorporating co-location opportunities into marine spatial-planning processes will require a better understanding of the impacts of these infrastructures at both local and network scales. As improved technologies emerge that have a reduced environmental impact, the opportunities for co-location within a marine planning context are likely to grow.

Highlights

- Energy installations have both environmental costs and benefits, but all modify their environments at the local scale.
- Co-location of energy installations is possible if development is compatible with the conservation objectives of the protected area.
- All energy installations offer some temporary *de facto* protection to the surrounding area by excluding some activities.
- Maximising co-location opportunities within marine planning requires better understanding of environmental impacts at local and network scales.
- As offshore energy technologies continue to develop and pose fewer environmental impacts, co-location opportunities will grow.

Notes

1 Full list of Parties and Signatories available at www.cbd.int/information/parties. shtml Accessed 28 June 2017.
2 From www.cbd.int/doc/press/2017/pr-2017-06-05-mpa-pub-en.pdf. Accessed 28 June 2017.
3 Some may argue that most marine ecosystems can no longer be considered fully natural, having been modified over the decades and centuries by human activities such as fishing and coastal development.
4 Average world annual electricity demand was 3,144.4 kilowatt hours (0.0031444 gigawatt-hours) per capita in 2014. Data from The World Bank: http://data. worldbank.org/indicator/EG.USE.ELEC.KH.PC. Accessed 28 June 2017.
5 Average annual electricity demand in France was 6,944 kilowatt hours (0.006944 gigawatt-hours) per capita in 2014. Data from The World Bank: http://data. worldbank.org/indicator/EG.USE.ELEC.KH.PC. Accessed 28 June 2017.
6 Nørrekær Enge Wind Farm, operated by Vattenfall, overview available at http://corporate.vattenfall.dk/om-os/vores-virksomhed/vattenfall-i-danmark/ vores-vindmoller/norrekar-enge-vindmollepark/. Accessed 28 June 2017.

References

Bailey, H., Senior, B., Simmons, D., Rusin, J., Picken, G. and Thompson, P.M. (2010). Assessing underwater noise levels during pile-driving at an offshore windfarm and its potential effects on marine mammals. *Marine Pollution Bulletin*, 60(6), pp. 888–897.

Black and Veatch Ltd. (2011). UK tidal current resource and economics. *Report CTC799. The Carbon Trust, London.* www.carbontrust.com/media/174041/phaseiitidalstreamresourcereport2005.pdf.

Brabant, R., Vanermen, N., Stienen, E.W.M. and Degraer, S. (2015). Towards a cumulative collision risk assessment of local and migrating birds in North Sea offshore wind farms. *Hydrobiologia*, 756(1), pp. 63–74.

Burt, J., Bartholomew, A., Usseglio, P., Bauman, A. and Sale, P.F. (2009). Are artificial reefs surrogates of natural habitats for corals and fish in Dubai, United Arab Emirates? *Coral Reefs*, 28(3), pp. 663–675.

Carrington, D, (2014). Drifting off the coast of Portugal, the frontrunner in the global race for floating windfarms. *The Guardian, 23 Jun 2014.* www.theguardian.com/environment/2014/jun/23/drifting-off-the-coast-of-portugal the-frontrunner-in-the-global-race-for-floating-windfarms.

Christie, N., Smyth, K., Barnes, R. and Elliott, M. (2014). Co-location of activities and designations: a means of solving or creating problems in marine spatial planning? *Marine Policy*, 43, pp. 254–261.

Claisse, J.T., Pondella, D.J., Love, M., Zahn, L.A., Williams, C.M., Williams, J.P. and Bull, A.S. (2014). Oil platforms off California are among the most productive marine fish habitats globally. *Proceedings of the National Academy of Sciences*, 111(43), pp. 15462–15467.

Convention on Biological Diversity. (2010). *COP Decision X/2. Strategic plan for biodiversity 2011–2020.* www.cbd.int/decision/cop/?id=12268.

DEFRA. (2005). *Securing the Benefits – the Joint UK Response to the Prime Minister's Strategy Units Net Benefits Report on the Future of the Fishing Industry in the UK.* Department of the Environment, Farming and Rural Affairs, UK.

Delucchi, M.A. and Jacobson, M.Z. (2011). Providing all global energy with wind, water, and solar power, Part II: Reliability, system and transmission costs, and policies. *Energy Policy*, 39(3), pp. 1170–1190.

De Mesel, I., Kerckhof, F., Norro, A., Rumes, B. and Degraer, S. (2015). Succession and seasonal dynamics of the epifauna community on offshore wind farm foundations and their role as stepping stones for non-indigenous species. *Hydrobiologia*, 756(1), pp. 37–50.

Department of the Environment. (2015). Strangford Lough Marine Conservation Zone. *Northern Ireland*. www.doeni.gov.uk.

Digest of UK Energy Statistics. (2016). Chapter 5: Electricity. *Department for Business, Energy & Industrial Strategy*. www.gov.uk/government/statistics/electricity-chapter-5-digest-of-united-kingdom-energy-statistics-dukes.

Douglas, C.A., Harrison, G.P. and Chick, J.P. (2008). Life cycle assessment of the SeaGen marine current turbine. *Proceedings of the Institution of Mechanical Engineers, Part M: Journal of Engineering for the Maritime Environment*, 222(1), pp. 1–12.

Feary, D.A., Burt, J.A. and Bartholomew, A. (2011). Artificial marine habitats in the Arabian Gulf: review of current use, benefits and management implications. *Ocean and Coastal Management*, 54(10), pp. 742–749.

Ferrier, R.W. and Bamberg, J.H. (1982). *The History of the British Petroleum Company*. Cambridge: Cambridge University Press, pp. 201–203.

Fowler, A.M., Macreadie, P.I. and Booth, D.J. (2015). Should we "reef" obsolete oil platforms? *Proceedings of the National Academy of Sciences*, 112(2), pp. E102.

Fraenkel, P.L. (2006). Tidal current energy technologies. *Ibis*, 148(s1), pp. 145–151.

Frid, C., Andonegi, E., Depestele, J., Judd, A., Rihan, D., Rogers, S.I. and Kenchington, E. (2012). The environmental interactions of tidal and wave energy generation devices. *Environmental Impact Assessment Review*, 32(1), pp. 133–139.

Friedlander, A.M., Ballesteros, E., Fay, M. and Sala, E. (2014). Marine communities on oil platforms in Gabon, West Africa: high biodiversity oases in a low biodiversity environment. *PLoS ONE*, 9(8), pp. e103709.

Gaines, S.D., White, C., Carr, M.H. and Palumbi, S.R. (2010). Designing marine reserve networks for both conservation and fisheries management. *Proceedings of the National Academy of Sciences*, 107(23), pp. 18286–18293.

Gao, G., Falconer, R.A. and Lin, B. (2013). Modeling effects of a tidal barrage on water quality indicator distribution in the Severn Estuary. *Frontiers of Environmental Science & Engineering*, 7(2), pp. 211–218.

Gill, A.B. (2005). Offshore renewable energy: ecological implications of generating electricity in the coastal zone. *Journal of Applied Ecology*, 42(4), pp. 605–615.

Gill, A.B., Huang, Y., Gloyne-Philips, I., Metcalfe, J., Quayle, V., Spencer, J. and Wearmouth, V. (2009). COWRIE 2.0 Electromagnetic Fields (EMF) Phase 2: EMF-sensitive fish response to EM emissions from subsea electricity cables of the type used by the offshore renewable energy industry. COWRIE Ltd Final Report (COWRIE-EMF-1-06), March 2009.

Hooper, T. and Austen, M. (2013). Tidal barrages in the UK: ecological and social impacts, potential mitigation, and tools to support barrage planning. *Renewable and Sustainable Energy Reviews*, 23, pp. 289–298.

Inger, R., Attrill, M.J., Bearhop, S., Broderick, A.C., Grecian, W.J., Hodgson, D.J., Mills, C., Sheehan, E., Votier, S.C., Witt, M.J. and Godley, B.J. (2009). Marine renewable energy: potential benefits to biodiversity? An urgent call for research. *Journal of Applied Ecology*, 46(6), pp. 1145–1153.

Jacobson, M.Z. and Delucchi, M.A. (2011). Providing all global energy with wind, water, and solar power, Part I: Technologies, energy resources, quantities and areas of infrastructure, and materials. *Energy Policy*, 39(3), pp. 1154–1169.

Kaiser, M.J. and Pulsipher, A.G. (2005). Rigs to Reef programs in the Gulf of Mexico. *Ocean Development and International Law*, 36(2), pp. 119–134.

Khan, M.I. and Islam, M.R. (2008). Sustainable management techniques for offshore oil and gas operations. *Energy Sources, Part B: Economics, Planning, and Policy 3*, 2(4), pp. 121–132.

Kolian, S.R., Sammarco, P.W. and Porter, S.A. (2017). Abundance of corals on offshore oil and gas platforms in the Gulf of Mexico. *Environmental Management*, 60(2), pp. 357–366. doi:10.1007/s00267-017-0862-z.

Lubchenco, J. and Grorud-Colvert, K. (2015). Making waves: the science and politics of ocean protection. *Science*, 350(6259), pp. 382–383.

Macreadie, P.I., Fowler, A.M. and Booth, D.J. (2011). Rigs-to-reefs: will the deep sea benefit from artificial habitat? *Frontiers in Ecology and the Environment*, 9(8), pp. 455–461.

Marine Conservation Zone Project. (2010). Ecological network guidance. *Natural England and Joint Nature Conservation Committee, June 2010*. http://jncc.defra.gov.uk/pdf/100705_ENG_v10.pdf.

Marine Current Turbines. (2011). *SeaGen Environmental Monitoring Programme Final Report, 16 Jan 2011*. Royal Haskoning Enhancing Society.

Mazik, K. and Smyth, K. (2013). Is 'minimising the footprint' an effective intervention to maximise the recovery of intertidal sediments from disturbance? Phase 1: Literature review. *Natural England Commissioned Reports, Number 110*. http://publications.naturalengland.org.uk/publication/5091106.

Miller, R.G., Hutchison, Z.L., Macleod, A.K., Burrows, M.T., Cook, E.J., Last, K.S. and Wilson, B. (2013). Marine renewable energy development: assessing the Benthic Footprint at multiple scales. *Frontiers in Ecology and the Environment*, 11(8), pp. 433–440.

Moan, J.L. and Smith, Z.A. (2007). *Energy Use Worldwide: A Reference Handbook*. Santa Barbara, CA: ABC-CLIO, Inc.

Neill, S.P., Litt, E.J., Couch, S.J. and Davies, A.G. (2009). The impact of tidal stream turbines on large-scale sediment dynamics. *Renewable Energy*, 34(12), pp. 2803–2812.

Neill, S.P., Jordan, J.R. and Couch, S.J. (2012). Impact of tidal energy converter (TEC) arrays on the dynamics of headland sand banks. *Renewable Energy*, 37(1), pp. 387–397.

Osmundsen, P. and Tveterås, R. (2003). Decommissioning of petroleum installations – major policy issues. *Energy Policy*, 31(15), pp. 1579–1588.

OSPAR. (2003). *OSPAR Convention for the Protection of the Marine Environment of the North-East Atlantic, Annex 9 A-4.44a*. Meeting of the OSPAR Commission, Bremen, 23–27 June 2003.

Petersen, J.K. and Malm, T. (2006). Offshore windmill farms: threats to or possibilities for the marine environment. *Ambio: A Journal of the Human Environment*, 35(2), pp. 75–80.

Peterson, C.H., Rice, S.D., Short, J.W., Esler, D., Bodkin, J.L., Ballachey, B.E. and Irons, D.B. (2003). Long-term ecosystem response to the Exxon Valdez oil spill. *Science*, 302(5653), pp. 2082–2086.

Pita, C., Pierce, G.J., Theodossiou, I. and Macpherson, K. (2011). An overview of commercial fishers' attitudes towards protected areas. *Hydrobiolgia*, 670(1), pp. 289–306.

Pitcher, G. (2014). Floating wind turbines planned. *New Civil Engineer, 12 Sep 2014. Emap Ltd, London.* www.nce.co.uk/floating-wind-turbines-planned/8669600. article.

Rabalais, N. (2014). Assessing early looks at biological responses to the Macondo event. *BioScience*, 64(9), pp. 757–759.

Roberts, C.M., Branch, G., Bustamante, R.H., Castilla, J.C., Dugan, J., Halpern, B.S., Lafferty, K.D., Leslie, H., Lubchenco, J., McArdle, D., Ruckelshaus, M. and Warner, R.R. (2003). Application of ecological criteria in selecting marine reserves and developing reserve networks. *Ecological Applications*, 13, pp. S215–S228.

Rossi-Santos, M.R. (2015). Oil industry and noise pollution in the humpback whale (*Megaptera novaeangliae*) soundscape ecology of the Southwestern Atlantic breeding ground. *Journal of Coastal Research*, 31(1), pp. 184–195.

Savidge, G., Ainsworth, D., Bearhop, S., Christen, N., Elsaesser, B., Fortune, F., Inger, R., Kennedy, R., McRobert, A., Plummer, K.E., Pritchard, D.W., Sparling, C.E. and Whittaker, T.J.T. (2014). Strangford Lough and the SeaGen tidal turbine. In: M.A. Shields and A.I.L Payne, eds., *Marine Renewable Energy Technology and Environmental Interactions*. Dordrecht: Springer, pp. 153–172.

Shani, A., Polak, O. and Shashar, N. (2012). Artificial reefs and mass marine ecotourism. *Tourism Geographies: An International Journal of Tourism Space, Place and Environment*, 14(3), pp. 361–382.

Sleiti, A.K. (2015). Overview of tidal power technology. *Energy Sources, Part B: Economics, Planning and Policy*, 10(1), pp. 8–13.

Smyth, K., Christie, N., Burdon, D., Atkins, J.P., Barnes, R. and Elliott, M. (2015). Renewables-to-reefs? – Decommissioning options for the offshore wind power industry. *Marine Pollution Bulletin*, 90(1), pp. 247–258.

Wilson, J.C. and Elliott, M. (2009). The habitat-creation potential of offshore wind farms. *Wind Energy*, 12(2), pp. 203–212.

Yates, K.L. and Schoeman, D.S. (2013). Spatial access priority mapping (SAPM) with fishers: a quantitative GIS method for participatory planning. *PLoS ONE*, 8(7), pp. e68424.

Yates, K.L., Schoeman, D.S. and Klein, C.J. (2015). Ocean zoning for conservation, fisheries and marine renewable energy: assessing trade-offs and co-location opportunities. *Journal of Environmental Management*, 152, pp. 201–209.

Chapter 12

Marine spatial planning and stakeholder collaboration

Advancing offshore wind energy and ocean ecosystem protection in New England

Priscilla M. Brooks and Tricia K. Jedele

Introduction

Marine spatial planning has gained prominence around the world over the past decade, as resource managers look to tackle the challenges of increasing commercial uses and protection of the ocean (Crowder, 2006; Gopnik, 2015). Marine spatial planning in the northeastern United States began with state-led initiatives in Massachusetts and Rhode Island. These initiatives were prompted in part by the states' respective interests in enabling the development of offshore wind energy to satisfy renewable energy targets. Both Massachusetts and Rhode Island engaged in multi-year marine spatial planning driven by stakeholders and science that resulted in the first comprehensive marine spatial plans in the country (Massachusetts Executive Office of Energy and Environmental Affairs, 2009; Rhode Island Coastal Resources Management Council, 2010). These state planning processes then catalysed a federal mandate for a comprehensive, regional marine spatial plan to cover all federal and state ocean waters in the New England region[1] (Northeast Regional Planning Body, 2016).

We describe a case study of the marine spatial-planning efforts of Massachusetts, Rhode Island, and the New England region as a whole. We explore how the region's experience and expectations regarding marine spatial planning facilitated the development of offshore wind, protected important ecological areas and marine resources, and fostered a lasting relationship between environmental advocates and an offshore wind developer to protect the endangered North Atlantic right whale (*Eubalaena glacialis*).

Massachusetts ocean plan

In the early 2000s, coastal regulators were addressing many proposals for new industrial uses in Massachusetts' marine waters. Proposals included several offshore wind-energy facilities, wave-and tidal-energy facilities, a natural gas pipeline, sand and gravel extraction, and an offshore liquid natural gas port. The then Secretary of Environmental Affairs, Ellen Roy Herzfelder, likened Massachusetts' ocean waters to "the Wild West" saying that "... everyone is trying to put their stake in the ground. We need to assert control" (Associated

Press, 2003). Projects like sand and gravel mining, natural gas pipelines, and the proposal to build the 130-turbine, offshore Cape Wind project in 2001 were perceived as new industrial activities to Massachusetts (and New England), and received a mixed greeting from coastal communities, fishers, and other stakeholders (Ziner, 2002; Daley, 2006, 2013).

In response to concerns raised about the State's ability to manage these new ocean uses, in 2003 Massachusetts Governor Mitt Romney created the Massachusetts Ocean Management Task Force and charged the group to "... examine current trends and issues, identify data and information gaps, review existing ocean governance mechanisms and draft recommendations for administrative, regulatory and statutory changes, if deemed necessary" (Massachusetts Ocean Management Task Force, 2004, p. 3). The Task Force, comprising state regulators, local officials, environmental nongovernment organisations, regional planning agencies, businessmen, and scientists, issued its final recommendations in 2004 (Massachusetts Ocean Management Task Force, 2004). The recommendations called for new legislation to authorise development of a comprehensive ocean management plan to "... streamline governance of the public trust ocean resources" and "... establish basic standards for allowable uses, impact control and resource protection—including which different uses and impacts [are] allowed and/or controlled in particular areas of the state's oceans ..." (Massachusetts Ocean Management Task Force, 2004, p. 29).

In 2008, Governor Deval Patrick signed into law the Massachusetts Ocean Act of 2008, mandating that the State develop a comprehensive marine spatial plan, referred to locally as an 'ocean management plan' (Massachusetts Ocean Act, 2008). The Act required among other provisions that the ocean management plan "... identify and protect special, sensitive and unique estuarine and marine life and habitats," "... foster sustainable uses that capitalize on economic opportunity without significant detriment to the ecology and natural beauty of the area," and "... identify appropriate locations and performance standards for activities, uses, and facilities ..." allowed under existing law. These uses and facilities included, among other uses, "... appropriate scale renewable energy facilities," which had been prohibited throughout most of State ocean waters until the passage of this Act. The Act required the formation of and consultation with two advisory committees—the Ocean Advisory Commission, comprising several state officials and stakeholders, and the Ocean Science Advisory Council, comprising academic, government, and non-government organisation scientists. These legally mandated advisory committees gave stakeholders a formal role in the ocean planning process and were effective in enabling stakeholders to project a strong and influential voice in the development of the ocean plan. The two advisory councils were convened many times during the development of the Massachusetts Ocean Plan to provide comments on draft elements of the plan, which were then formally considered and addressed by state ocean planners.

Upon passage of the Massachusetts Ocean Act, a 16-month planning process began. The process included extensive public and stakeholder outreach including public listening sessions, public workshops, regular meetings of the advisory committees, formal public hearings following the release of the draft Massachusetts Ocean Management Plan, and hundreds of smaller meetings with various stakeholders. A baseline assessment was done that entailed the collection and analysis of extensive biological, physical, geological, oceanographic, and human use data—all of which is accessible on a public data portal known as the 'Massachusetts Ocean Resources Information System' (Massachusetts Office of Coastal Zone Management, 2017). The final Massachusetts Ocean Management Plan was released on 31 December 2009, identifying wind-energy areas and special, sensitive, and unique areas of marine life and habitat along with a regulatory framework for implementing the plan (Massachusetts Executive Office of Energy and Environmental Affairs, 2009). Since then, the Ocean Advisory Commission and Science Advisory Council have continued to advise the State on updates to the scientific data underlying the plan and plan revisions.

Rhode Island ocean plan

A desire to build offshore wind-energy facilities was also a primary catalyst for developing the Rhode Island marine spatial ocean plan. In 2004, the Rhode Island legislature passed the Renewable Energy Standard (Rhode Island Renewable Energy Standard, 2004) requiring that renewable energy provide 16% of the State's electrical power need by 2019, and in 2007, Rhode Island's Office of Energy Resources determined that the State would have to invest in offshore wind-energy facilities to achieve the then Rhode Island Governor Donald Carcieri's mandate that offshore wind resources supply 15% of the State's electrical power by 2020 (Rhode Island Coastal Resources Management Council, 2010, p. 25).

Rhode Island used its authority under the federal Coastal Zone Management Act to initiate a marine spatial planning process (United States Coastal Zone Management Act of 1972, 1972a) and the subsequent development of Rhode Island's marine spatial plan, known as the 'Ocean Special Area Management Plan' (Rhode Island Coastal Resources Management Council, 2010). A Special Area Management Plan is a regulatory tool available to coastal states that allows for comprehensive planning for natural resource protection and reasonable, coastal-dependent economic growth. The Special Area Management Plan includes standards and criteria to guide public and private uses of lands and waters. Once approved by the federal government, a Special Area Management Plan becomes federally enforceable (United States Coastal Zone Management Act of 1972, 1972b). Rhode Island's plan was also driven by a recognition that "… the uses of marine resources in Rhode Island were intensifying [and] that optimizing the potential of this intensification would require intentional action driven by design rather than accident" (Rhode Island Coastal Resources Management Council, 2010, p. 25). The primary goals of the Ocean Special Area Management Plan were to promote and enhance existing and future ocean uses, foster and manage adaptively a properly functioning ecosystem, encourage economic development consistent with and complementary to the State's overall economic development, social, and environmental needs and goals, and build a framework for coordinated decision-making between state and federal management agencies (Rhode Island Coastal Resources Management Council, 2010, pp. 20–21).

The Ocean Special Area Management Plan was developed with intensive stakeholder engagement. It also relied on scientific data, and approximately 30 research projects were commissioned throughout the planning process. Finalised in October 2010, the Ocean Special Area Management Plan identified wind-energy zones as well as protected areas to protect marine life and habitat along with a regulatory framework for its implementation. Recognising the value of stakeholder engagement and input into ocean management decision-making, the marine spatial plan also created two standing stakeholder advisory panels. The Habitat Advisory Board was created to provide advice to the Council on the ecological function, restoration, and protection of the marine resources and habitats (Rhode Island Coastal Resources Management Council 2010, p. 961). The Fishery Advisory Board was created to provide advice to the Council on the potential adverse impacts of other uses on commercial and recreational fishers and fisheries activities (Rhode Island Coastal Resources Management Council, 2010, pp. 966–967). Both advisory councils were designed to provide advice on specific proposed projects, including the siting of individual wind turbines within a wind farm to identify the best site for each individual structure. Since the completion of the Rhode Island ocean plan, the two advisory councils have been convened regularly to consider and provide feedback on various offshore wind-energy proposals.

President calls for a National Ocean Policy

While Massachusetts and Rhode Island were engaged in developing their marine spatial plans, the federal government was examining the management of federal marine resources. On 19 July 2010, the federal Interagency Ocean Policy Task Force—a body established by President Barack Obama in 2009 "… to better meet our Nation's stewardship responsibilities for the oceans, coasts, and Great Lakes" (United States Office of the President, 2009)—issued its final recommendations for the protection, maintenance, and restoration of the health of the ocean, coastal, and Great Lakes ecosystems and resources (United States White House Council on Environmental Quality, 2010). On that same day, President Obama established the nation's first National Ocean Policy through Executive Order 13547, entitled 'Stewardship of the Ocean, Our Coasts, and the Great Lakes' (United States Executive Order 13547, 2010). The Executive Order mandated the implementation of the Interagency Ocean Policy Task Force recommendations and established a National Ocean Policy. It also established the National Ocean Council (comprising the heads of multiple federal agencies) and charged it with implementing the new ocean policy. Importantly, the Executive Order specifically called for the development of regional coastal and marine spatial plans

"… to enable a more integrated, comprehensive, ecosystem-based, flexible, and proactive approach to planning and managing sustainable multiple uses across sectors and improve the conservation of the ocean, our coasts, and the Great Lakes" (United States Executive Order 13547, 2010).

Similar to the marine spatial planning in Massachusetts and Rhode Island, the regional marine spatial planning process in the Northeast advanced within a framework of extensive stakeholder engagement and data collection and analysis of the marine environment and related human uses. The Northeast Regional Planning Body was formed in 2012 to guide the development of the regional ocean plan and included representatives from the six New England states, six federally recognised indigenous tribes, nine federal agencies, and the New England Fishery Management Council (Northeast Regional Planning Body, 2017). Stakeholder outreach was extensive with stakeholders engaged regularly over four years through various meetings, stakeholder forums and workshops, state-based advisory committees, formal calls for public comment on various documents, and extensive electronic outreach through a dedicated website. The Northeast Ocean Plan was finalised in December 2016 (Northeast Regional Planning Body, 2016).

Setting the stage for offshore wind

In 2009, the federal government promulgated final regulations for the Atlantic Outer Continental Shelf Renewable Energy Program (United States Minerals Management Service, 2009), which was authorised by the Energy Policy Act of 2005 (United States Energy Policy Act of 2005, 2005). The regulations provided a framework for issuing leases for Atlantic Outer Continental Shelf activities that support production and transmission of energy from sources other than oil and natural gas. But the regulatory framework alone would not be enough to advance offshore wind energy efficiently in the United States.

On 23 November 2010, four months after issuing the Executive Order calling for the creation of a National Ocean Policy, the United States Department of Interior, the federal agency that houses the Bureau of Ocean Energy Management and the lead agency permitting offshore wind-energy, launched its so-called 'Smart from the Start' wind-energy initiative for the Atlantic Outer Continental Shelf (United States Department of the Interior, 2010). Recognising that offshore wind was a new industry in which developers were faced with unique timing, engineering, and financing obstacles, Smart from the Start was created to facilitate siting, leasing, and construction of new projects. By proactively identifying areas on the Atlantic Outer Continental Shelf most suitable for commercial wind-energy activities (Wind Energy Areas) and completing an environmental impact assessment for those areas, the Bureau of Ocean Energy Management's goal was effectively to guide the issuance of leasing instruments for all assessment and site-characterisation activities for wind-energy facilities proposed within the designated Wind Energy Areas, while avoiding negative environmental consequences.

The Bureau of Ocean Energy Management's renewable energy program grew out of Smart from the Start and included four phases: (1) leasing, (2) site assessment, (3) construction, and (4) operation. The leasing phase was sub-divided into another seven sub-phases: call for information and nomination, area identification, environmental assessment of identified area, proposed lease-sale notice, final lease-sale notice, lease sale (via live auction), and awarding of a lease (United States Department of the Interior, 2017a).

On 18 August 2011, the Bureau of Ocean Energy Management began its offshore leasing and environmental review process in Rhode Island and Massachusetts when it identified the Rhode Island/Massachusetts Wind Energy Area (Figure 12.1), and began taking public comments about the area.

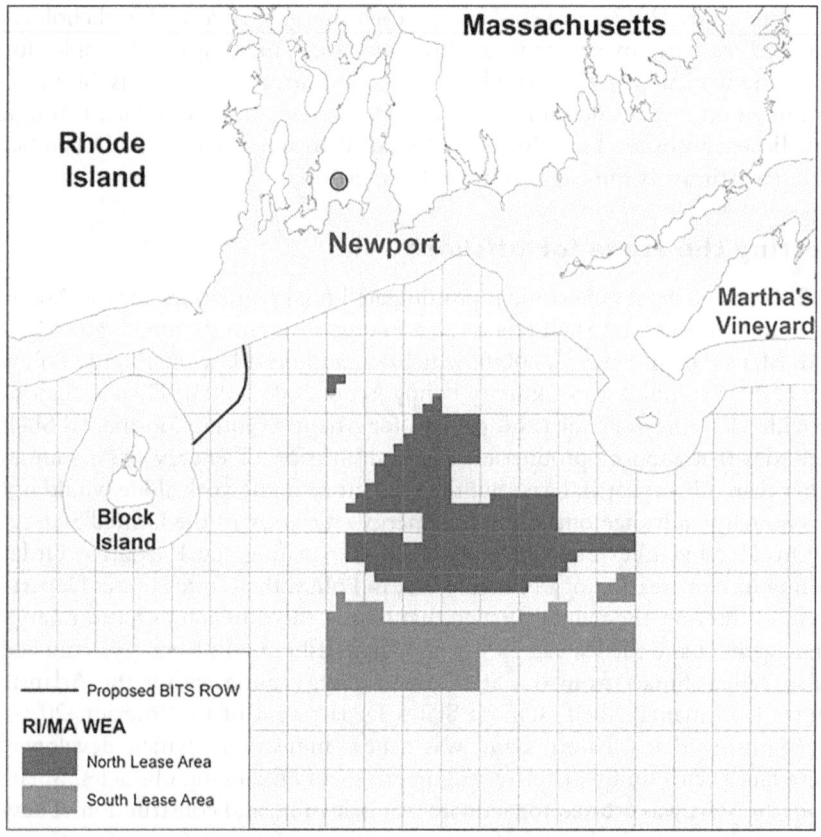

Figure 12.1 Map of the Rhode Island and Massachusetts Wind Energy Area. The square blocks are individual lease blocks. The Block Island Transmission System Right of Way (BITS ROW) indicates the approximate location of a transmission cable servicing the Block Island Wind Farm, located adjacent to Block Island. United States Department of Interior, Bureau of Ocean Energy Management (2017b).

Separately and several months later, an area offshore Massachusetts and adjacent to the Rhode Island/Massachusetts Wind Energy Area was also identified by the Bureau of Ocean Energy Management to be considered as a potential future wind-energy leasing (United States Department of the Interior, 2011). This area, known as the Massachusetts Wind Energy Area, was ultimately leased in 2015.

Accustomed to working together as stakeholders, making science-based decisions, and compromising following robust dialogue, there was an expectation among New England stakeholders that as the federal government moved forward with planning and siting of offshore wind facilities, the public and stakeholder engagement would at least equal that of Massachusetts and Rhode Island marine spatial planning. Because this expectation was created before the federal government began its offshore-wind initiative, a reconciliation of sorts between the Smart from the Start offshore wind-energy initiative and marine spatial planning, as New Englanders had to come to understand it, had to happen to secure success.

Ultimately, the renewable energy program of the Bureau of Ocean Energy Management was able to adapt in a manner that allowed advocates and stakeholders to infuse some of the core tenets of marine spatial planning into the leasing and permitting process, especially in the identification phase of the leasing process. These core tenets included extensive stakeholder outreach and engagement and scientific data compilation and analysis of the ecological, cultural, and human uses of the wind energy area, and this information was incorporated into the Final Environmental Assessment for the Rhode Island/Massachusetts Wind Energy Area (United States Department of the Interior, 2013). In the early phases of the leasing, the Bureau of Ocean Energy Management analysed the ecological, cultural, and human-use characteristics of the Wind Energy Area, invited public comments, and held public meetings to inform stakeholder groups about progress and next steps. In addition, Massachusetts formed Fisheries and Habitat Advisory Committees to provide stakeholder input on Wind Energy Area identification and leasing, while Rhode Island consulted with its Fishery Advisory Board and Habitat Advisory Board already established under Rhode Island's state ocean plan.

The Conservation Law Foundation and several other environmental organisations filed comments with the Bureau of Ocean Energy Management during the environmental assessment phase for the Rhode Island/Massachusetts Wind Energy Area. Their primary recommendation was that the section of the Final Environmental Assessment document called 'Standard Operating Conditions' be amended to include more protective measures for endangered species like the North Atlantic right whale (Figure 12.2) that would govern the site assessment and characterisation taking place within the leased area. The Final Environmental Assessment and Standard Operating Conditions did not ultimately incorporate all the recommendations of the Conservation Law Foundation's and other advocacy organisations. Recognising that the

Figure 12.2 North Atlantic right whale (*Eubalaena glacialis*). Photo by Brian Skerry.

co-occurrence of right whales and offshore wind could present a stumbling block for offshore-wind development, advocacy organisations like the Conservation Law Foundation, in the spirit of stakeholder-engaged marine spatial planning, began to work with the Bureau of Ocean Energy Management and developers to identify additional measures that developers could employ to provide more protection for the North Atlantic right whale.

North Atlantic right whale

Concerns about potential impacts to the North Atlantic right whale population during site assessment and ultimately construction and operation ranged from the following: displacement of these whales into shipping lanes, interruption of foraging, interference with right whale communication, and confusion and harassment resulting from high-resolution geophysical surveying, and pile driving to install meteorological measuring devices by prospective lessees in designated Wind Energy Areas. Sub-bottom profiling types of high-resolution surveying are used for several purposes, including to characterise ocean-bottom topography and subsurface geology, to investigate potential benthic biological communities, to identify archaeological resources, to characterise the potential location for the installation of a meteorological tower, and to gather information necessary to submit the required Site Assessment Plan.

There was particularly great concern about potential impacts to right whales from sub-bottom profiling types of high-resolution geophysical surveying to characterise seafloor sediment. This technology has the potential to affect right whales negatively due to its pulse rate and sound intensity, and the potential for simultaneous, multiple surveys across a single Wind Energy Area during the time when right whales are present (Rhode Island Coastal Resources Management Council, 2010, p. 122 and pp. 739–758). According to the Bureau of Ocean Energy Management's analysis, sub-bottom profilers generate sound within the hearing thresholds of most marine mammals that could be present in the Wind Energy Area and surrounding waters (U.S. Department of the Interior 2013, p. 4(64–67).

North Atlantic right whales are one of world's rarest and most endangered species of baleen whales (Clapham et al., 1999). They have been listed as Endangered across their range since 1973 under the United States Endangered Species Act (NOAA Fisheries, 2017a; United States Endangered Species Act of 1973, 1973). According to the North Atlantic Right Whale Western Atlantic Stock assessment report by the National Oceanographic and Atmospheric Administration in February 2017, the minimum size of the western population was 440 based on a photo-identification census from 2012 (NOAA Fisheries, 2017b). The stock assessment also underscored that any mortality or serious injury for this species could result in population decline because "… the small population size and low annual reproductive rate of right whales suggest that human sources of mortality may have a greater effect relative to population growth rates than for other whales" (NOAA Fisheries, 2017b, p. 6). According to the report, "… the principal factors believed to be retarding growth and recovery of the population are ship strikes and entanglement with fishing gear. Between 1970 and 1999, a total of 45 right whale mortalities was recorded" (NOAA Fisheries, 2017b, p. 6).

For this reason, the identification and leasing for the Wind Energy Area had to consider the potential for adverse impacts of artificial sound to right whales. To the extent that site characterisation and assessment could displace right whales farther offshore, these animals could be exposed to additional sources of mortality, including increased potential for ship strikes, and higher energy expenditure in response to avoiding noise sources (Rhode Island Coastal Resources Management Council, 2010, pp. 739–758; Nowack et al., 2004). These concerns were heightened by the proximity (within two nautical miles) of the Wind Energy Area to the adjacent shipping lanes (Rhode Island Coastal Resources Management Council, 2010, p. 566). At the time of the environmental assessment, there was little scientific evidence regarding the behavioural response of right whales to sound. The assessment had to rely on a 2010 report concerning whale behaviour in response to seismic activity, a 1988 report about behavioural responses to geophysical surveying in the Alaska Beaufort Sea, a 2000 report about whale avoidance of seismic air guns in Western Australia, and a 2001 report identifying 'swimming away' behaviour in response to seismic surveys in the Alaska

Beaufort Sea (United States Department of the Interior, Bureau of Ocean Energy Management, 2013, p. 4(64)–4(67)). As a result, the Bureau of Ocean Energy Management's Final Environmental Assessment concluded that "… it is expected that marine mammals would avoid the area around the [high resolution geophysical] survey activities, thereby limiting potential effects," and that given the high mobility of the species, "… they have the ability to move away from the sound if disturbance occurs" (United States Department of the Interior, Bureau of Ocean Energy Management, 2013, p. 4(66)).

A study in December 2003 by the Woods Hole Oceanographic Institution in the Bay of Fundy and specific to North Atlantic right whales offered evidence of another response that raised concerns about both ship strikes and the potential for depletion of energy stores. The study concluded that because five of the six whales observed after a sound stimulus

> … [swum] to and remain[ed] near the surface, instead of staying at depth, the whales most probably increased their risk of being struck. Under ideal conditions (e.g. favorable sighting weather and skilled lookouts), forcing the whales to the surface might assist collision mitigation, but by staying just below the surface, the whales were vulnerable but seldom visible.
>
> (Nowack et al., 2004, pp. 228–229)

The study also concluded that the high-powered ascent to the surface "… could cost these whales significant energy, especially if repeated often." (Nowack et al., 2004, p. 230). This energetic expenditure is compounded by the time spent responding to noise stimuli that can limit foraging time (Nowak et al., 2004).

The Conservation Law Foundation argued that given the species' endangered status and the paucity of relevant data regarding its specific responses to high-resolution geophysical surveys, a seasonal restriction (discussed below) prohibiting or otherwise limiting potentially harmful sound generation (including sub-bottom profiling and pile driving) during the times of year that North Atlantic right whales were in the Wind Energy Area, and other appropriate mitigation measures, would be most effective at minimising potential impacts. They also advocated that these mitigation measures should be reflected in the Standard Operating Conditions for site characterisation and assessment, and that they be required in any specific lease issued to a developer. The Conservation Law Foundation's advice was not heeded in the release of the Final Environmental Assessment, leading it and partner organisations to negotiate directly with the prospective developer.

A voluntary mitigation agreement

New England's extensive experience with science and stakeholder-driven marine spatial planning, and the mutual interest of developers and environmental advocates in advancing offshore wind with as few delays and impacts as possible gave life to the possibility of forging stand-alone agreements among wind-energy developers and environmental organisations, emphasising accords that contained conditions beyond federal government requirements under the Standard Operating Conditions specified in the Final Environmental Assessment. Advocacy organisations immediately recognised that a window of opportunity had been opened to begin discussions with wind developers directly about what more could be done to protect marine mammals, especially right whales. On 7 May 2014, the Conservation Law Foundation, the Natural Resources Defense Council, and National Wildlife Federation were able to secure a private and voluntary agreement with Deepwater Wind (the company that had expressed the most interest in the Rhode Island/Massachusetts Wind Energy Area and that ultimately won the lease to this area) to go beyond lease requirements (Conservation Law Foundation et al., 2014). Additional signatories to the agreement included Environment America, the International Fund for Animal Welfare, Oceana, and Sierra Club.

To minimise the risk of ship strikes, Deepwater Wind agreed that all vessels of any length associated with site assessment and characterisation be limited to 10 knots during the seasonal restriction. Deepwater Wind agreed to maintain a minimum 500-m exclusion to protect all marine mammals and sea turtles, with an exception for bow-riding dolphins, and agreed to provide a minimum of two, trained observers at each sub-bottom profiling site and four, trained visual observers at each pile-driving site. Furthermore, Deepwater Wind agreed to use the best commercially available technology such as bubble curtains (a system that produces bubbles in a deliberate arrangement in the water column to dampen sound), cushion blocks (disks of wood or other material placed on top of the piling to minimise noise), temporary noise-attenuation pile design, vibratory pile drivers, and/or press-in pile drivers to reduce sound and horizontal propagation (CSA Ocean Sciences Inc., 2014).

Deepwater Wind also agreed to several seasonal restrictions and precautionary periods. For example, the agreement includes the seasonal prohibition on the installation of meteorological towers set forth in the final Environmental Assessment. The agreement goes further, however, and includes a precautionary period from 1 to 14 May, with additional mitigation requirements for pile driving, and imposes seasonal restrictions prohibiting sub-bottom profiling from 1 February to 30 April. There are also other protective measures for right whales from sub-bottom profiling from 1 to 31 January and 1 to 14 May (Conservation Law Foundation et al., 2014). The Conservation Law Foundation, the Natural Resources Defense Council, and the National Wildlife Federation are now working to incorporate similar mitigation measures into site assessment survey plans for the neighbouring Massachusetts Wind Energy Area, and expect also to promote including these measures into the construction and operation plans for any wind-energy facilities.

Conclusion

New England's collective experience with marine spatial planning, and the associated stakeholder engagement and science-based decision-making, helped to form productive and lasting relationships between stakeholders. As a result of the trust that had been built through years of marine spatial planning in the region, environmental organisations and an offshore wind developer were able to come together and review the data about how North Atlantic right whales might use the lease area and learn about the technological, information, and financial needs and obstacles for the developer. From that they were able to craft an agreement that worked for the developer and was more protective of the North Atlantic right whale. The first offshore wind project in the United States has now been completed; the Block Island Wind Farm is a five-turbine, 30-megawatt facility located in Rhode Island waters adjacent to the Rhode Island/Massachusetts Wind Energy Area (Figure 12.1) in the renewable energy zone. This five-turbine project began generating electricity in 2016. Deepwater Wind is now preparing to develop a larger wind-energy facility in the Rhode Island/ Massachusetts Wind Energy for which it holds the lease and continues to collaborate actively with the Conservation Law Foundation and its partners to mitigate the negative impacts of offshore wind energy on the North Atlantic right whale.

Highlights

- Scientific data collection and analysis and stakeholder engagement were hallmarks of the Massachusetts and Rhode Island state processes of marine spatial planning.
- A desire to build offshore wind-energy facilities was a primary catalyst for developing the Massachusetts and Rhode Island Ocean plans.
- A common understanding developed in Rhode Island and Massachusetts about the role of science and stakeholder participation in ocean management decision-making.
- Knowing that right whales in wind energy areas could be an obstacle for wind development, environmental advocates used marine spatial planning to provide heightened protection for right whales while advancing wind development.
- Trust built over years of marine spatial planning led environmental advocates and an offshore wind developer to craft an agreement that worked for the developer and that was more protective of right whales.

Note

1 New England is a region of the United States located in the northeast of the country and includes the states of Maine, Vermont, New Hampshire, Massachusetts, Rhode Island and Connecticut.

References

Associated Press. (2003). State task force to consider zoning ocean use. *Portsmouth Herald*. 15 June. Available at www.seacoastonline.com/article/20030615/NEWS/306159969

Clapham, P.J., Young, S.B. and Brownell, R.L. (1999). Baleen whales: conservation issues and the status of the most endangered populations: *Mammal Review*, 291(1). pp. 35–60.

Conservation Law Foundation, Natural Resources Defense Council and National Wildlife Federation. (2014). Proposed mitigation measures to protect North Atlantic right whales from site assessment and characterization activities of offshore wind energy development in the Rhode Island and Massachusetts wind energy area. *Letter to Maureen Bornholdt, U.S. Bureau of Ocean Energy Management. May 7, 2014*. www.clf.org/wp-content/uploads/2014/05/050714-NARW-Letter-to-BOEM-re-RI-MA-WEA-850.pdf

Crowder, L.B., Osherenko, G., Young, O.R., Airamé, S., Norse, E.A., Baron, N., Day, J.C., Douvere, F., Ehler, C.N., Halpern, B.S. and Langdon, S.J. (2006). Resolving mismatches in U.S. ocean governance. *Science*, 313(5787). pp. 617–618.

CSA Ocean Sciences Inc. (2014). *Quieting technologies for reducing noise during seismic surveying and pile driving workshop*. Summary Report for the U.S. Department of the Interior, Bureau of Ocean Energy Management BOEM 2014-061 (Contract Number M12PC00008). Springfield, VA: National Technical Information Service.

Daley, B. (2006). LNG plan for site off Gloucester gets 1st OK. *Boston Globe, 11 July*. http://archive.boston.com/news/local/massachusetts/articles/2006/07/11/lng_plan_for_site_off_gloucester_gets_1st_ok/

Daley, B. (2013). The costly sand wars in Massachusetts. *Worcester Telegram*. 15 December. www.telegram.com/article/20131215/NEWS/312159935

Gopnik, M. (2015). *From the Forest to the Sea – Public Lands Management and Marine Spatial Planning*. New York, NY: Routledge.

Massachusetts Executive Office of Energy and Environmental Affairs. (2009). *Massachusetts Ocean Management Plan, 2009*. www.mass.gov/eea/waste-mgnt-recycling/coasts-and-oceans/mass-ocean-plan/2009-final-ocean-plan.html

Massachusetts Ocean Act of 2008 (Massachusetts State Government). (2008). Chapter 114 of the Acts of 2008, State of Massachusetts and Massachusetts General Law Section Chapter 21A, Section 4C. https://malegislature.gov/Laws/SessionLaws/Acts/2008/Chapter114

Massachusetts Ocean Management Task Force. (2004). *Waves of Change: The Massachusetts Ocean Management Task Force Report and Recommendations*. Boston: Massachusetts Executive Office of Energy and Environmental Affairs. www.mass.gov/eea/docs/czm/oceans/waves-of-change/waves-of-change.pdf

Massachusetts Office of Coastal Zone Management. (2017). *Massachusetts Ocean Resource Information System*. www.mass.gov/eea/agencies/czm/program-areas/mapping-and-data-management/moris/

NOAA Fisheries. (2017a). *North Atlantic right whales (Eubalaena glacialis)*. www.nmfs. noaa.gov/pr/species/mammals/whales/north-atlantic-right-whale.html

NOAA Fisheries. (2017b). *Stock assessment for the North Atlantic right whale (Eubalaena glacialis) Western Atlantic Stock*. www.nefsc.noaa.gov/publications/tm/tm241/8_ F2016_rightwhale.pdf

Northeast Regional Planning Body. (2016). *Northeast Ocean Plan*. neoceanplanning. org/wp-content/uploads/2016/10/Northeast-Ocean-Plan_Full.pdf

Northeast Regional Planning Body. (2017). *Ocean Planning in the Northeast: Northeast Regional Planning Body*. http://neoceanplanning.org/about/northeast-rpb/

Nowack, D.P., Johnson, M.P. and Tyack, P.L. (2004). North Atlantic right whales *(Eubalaena glacialis)* ignore ships but respond to altering stimuli. *Proceedings of the Royal Society B: Biological Sciences*, 271(1536), pp. 227–231.

Rhode Island Coastal Resources Management Council. (2010). *Rhode Island Ocean Special Area Management Plan*. www.crmc.ri.gov/samp_ocean.html

Rhode Island Renewable Energy Standard. (2004). *State of Rhode Island General Laws, Title 39, Chapters 39–26*. http://webserver.rilin.state.ri.us/Statutes/title39/39-26/ INDEX.HTM

United States Coastal Zone Management Act of 1972. (1972a). United States Code, Title 16, Section 1453-304(17). https://coast.noaa.gov/czm/act/sections/#304

United States Coastal Zone Management Act of 1972. (1972b). United States Code, Title 16, Section 1453-304(6a). https://coast.noaa.gov/czm/act/sections/#304

United States Department of the Interior. (2010). *Salazar launches 'smart from the start' initiative to speed offshore wind energy development off the Atlantic Coast*. Press release issued on November 23, 2010. www.doi.gov/news/pressreleases/Salazar-Launches-Smart-from-the-Start-Initiative-to-Speed-Offshore-Wind-Energy-Development-off-the-Atlantic-Coast

United States Department of the Interior. (2011). *Interior launches leasing process for commercial wind energy offshore Rhode Island and Massachusetts*. Press release issued on August 17, 2011. www.doi.gov/news/pressreleases/Interior-Launches-Leasing-Process-for-Commercial-Wind-Energy-Offshore-Rhode-Island-and-Massachusetts.cfm

United States Department of the Interior, Bureau of Ocean Energy Management. (2017a). *Regulatory framework*. www.boem.gov/Regulatory-Development-Policy-and-Guidelines/

United States Department of the Interior, Bureau of Ocean Energy Management. (2017b). *Map of the Rhode Island/Massachusetts wind energy area*. www.boem.gov/ Rhode-Island/

United States Department of the Interior, Bureau of Ocean Energy Management. (2013). *Commercial Wind Lease Issuance and Site Assessment Activities on the Atlantic Outer Continental Shelf Offshore Rhode Island and Massachusetts Revised Environmental Assessment*. Washington, DC: Bureau of Ocean Energy Management, Office of Renewable Energy Resources. www.boem.gov/uploadedFiles/BOEM/ Renewable_Energy_Program/State_Activities/BOEM%20RI_MA_Re-vised%20EA_22May2013.pdf

United States Department of the Interior Minerals Management Service. (2009). Renewable energy and alternate uses of existing facilities on the outer continental shelf. *Federal Register*, 74(81). pp. 19638–19871. www.gpo.gov/fdsys/pkg/FR-2009-04-29/pdf/E9-9462.pdf

United States Endangered Species Act of 1973. (1973). United States Code, Title 16, Chapter 35, Sections 1531–1544. www.nmfs.noaa.gov/pr/laws/esa/text.htm

United States Energy Policy Act of 2005. (2005). United States Code, Title 42, Chapter 149, Sections 15801–16636. www.congress.gov/109/plaws/publ58/PLAW-109publ58.pdf

United States Executive Order 13547. (2010). Stewardship of the ocean, our coasts, and the Great Lakes. *Federal Register*, 75(140), pp. 43023–43027. www.gpo.gov/fdsys/pkg/FR-2010-07-22/pdf/2010-18169.pdf

United States Office of the President. (2009). *Memorandum for the Heads of Executive Departments and Agencies, Subject: National Policy for the Oceans, Our Coasts, and the Great Lakes*. Washington, DC: The White House, Office of the Press Secretary. https://obamawhitehouse.archives.gov/the-press-office/memorandum-heads-executive-departments-and-agencies-subject-government-contracting

United States White House Council on Environmental Quality. (2010). *Final Recommendations of the Interagency Ocean Policy Task Force*. Washington, DC: White House Council on Environmental Quality. www.nsf.gov/geo/opp/opp_advisory/briefings/nov2010/optf_finalrecs.pdf

Ziner, K.L. (2002). Offshore harvest of wind is proposed for Cape Cod. *New York Times*, 16 April 2002. www.nytimes.com/2002/04/16/science/offshore-harvest-of-wind-is-proposed-for-cape-cod.html

Chapter 13

Co-locating offshore wind farms and marine protected areas

A United Kingdom perspective

Matthew Ashley, Melanie Austen,
Lynda Rodwell and Stephen C. Mangi

Introduction

In many cases, a clash occurs between developing an economy and the desire to protect marine biodiversity through a more ecologically focused management of the sea (Qiu and Jones, 2012). Co-locating compatible activities to meet both these goals simultaneously provides multiple potential benefits. For example, co-locating offshore wind farms with marine protected areas could help contribute to the energy needs of society and the economic development of the renewable energy industry, while concurrently offering biodiversity protection. By co-locating these activities, there might also be benefits from reducing the overall area restricted to other economic uses of the sea. However, realising these potential benefits depends on many factors, including the specifics of the marine protected area (Chapter 11). In this chapter we explore the co-location potential of two types of marine protected area in the United Kingdom, and consider the lessons that can be learnt for future planning processes.

Policy, planning and licensing approaches to co-locate offshore wind farms and marine protected areas

Co-locating wind farms within marine protected areas is, in principle, feasible within the United Kingdom, as is designating marine protected areas at sites already containing wind farms. However, any overlapping activity or development must be compatible with the protected site's conservation objectives and should not negatively affect site 'integrity' (HM Government, 2011, 2014). The emphasis is ultimately on prior assessment to establish if the ecological effects of construction, maintenance and presence of offshore wind farms are compatible with the specific objectives of the relevant marine protected areas. Increased monitoring before and after co-location, and assessment of mitigation options are also required to maximise opportunities to reduce risk of adverse effects on protected features.

The United Kingdom Marine Policy Statement, published in 2011 as the guiding document for marine planning in United Kingdom regions, encourages co-location of activities where they are compatible (HM Government, 2011). Planning decisions are required to consider a need for developments to "... avoid harm to marine ecology, biodiversity and geological conservation interests" (HM Government, 2011). The options suggested to avoid harm to features of conservation interest include "... location, mitigation and consideration of reasonable alternatives. For cases where significant harm cannot be avoided the appropriate compensatory measures should be sought" (HM Government, 2011).

The Offshore Energy Strategic Environmental Assessment done in 2008–2009 by the United Kingdom Government assessed the implications of more offshore wind-farm leasing. The report recognised the potential for co-locating wind farms with marine protected areas to reduce potential spatial conflict: "Where the objectives of the conservation sites and renewable energy development are coincident, preference should be given to locating wind farms in such areas to reduce the potential spatial conflict with other users" (DECC, 2009). It also stated that marine protected areas are not intended to be "... strict no-go areas for other activities" (DECC, 2009); although wind farm developers are made aware that "... designation may necessitate, subject to the conclusions of any appropriate assessment, suitable mitigation measures so as to avoid adverse effects on a designated site or species" (DECC, 2009).

Marine protected area designations

In the United Kingdom, the main types of marine protected area that are likely to occupy the same sea space as offshore wind farms are European Marine Sites, designated under European Law, and Marine Conservation Zones under United Kingdom law. European Marine Sites include Special Areas of Conservation for habitats and species of European importance designated under the Habitats Directive (EEC, 1992), Special Protection Areas for bird species of European importance designated under the Birds Directive (EC, 2009), and Sites of Community Importance. Sites of Community Importance are areas that have been proposed to the Commission for protection, and they are given transitional protection while that proposal is being reviewed. Once approved, Sites of Community Importance can be designated as Special Areas of Conservation by the Member State. At a European scale, these sites form a network of protected areas that supports the aim of the Marine Strategy Framework Directive to achieve Good Environmental Status of European Union waters by 2020.

On the other hand, Marine Conservation Zones are designated under the United Kingdom Marine and Coastal Access Act 2009 to protect nationally important habitats and species. As a network of sites, these zones contribute to fulfilling the United Kingdom's obligations under the Convention on Biological Diversity as well as non-binding instruments such as the recommended coherent network of marine protected areas under the OSPAR (Oslo and Paris Conventions) Recommendation 2003/3 (Christie et al., 2014).

All sites require the designated habitat or species features they contain to be recovered or maintained to favourable condition, which is termed "... the condition that would be expected in the absence of significant anthropogenic pressures which have an adverse effect" (JNCC, 2010b; Carr et al., 2016). For a habitat to be in favourable condition, "... the extent is required to be stable or increasing and its structures and functions, its quality and the composition of its characteristic biological communities are such as to ensure that it remains in a condition which is healthy and not deteriorating" (JNCC, 2010b; Carr et al., 2016). Favourable condition in Special Areas of Conservation and thereby, Sites of Community Importance is assessed as whether "... the natural range and area of a habitat feature is stable or increasing and which are necessary for its long-term maintenance are present and are likely to continue to exist for the foreseeable future" (JNCC, 2016).

Environmental Impact Assessments provide the main tool to assess, on a case-by-case basis, if a wind farm will have an adverse effect on the designated features (habitat or species) within a site (DECC, 2012; PINS, 2012). Environmental impact assessments also require developers to consider alternative designs to avoid adverse effects, justify decisions where avoidance is not possible, and consider mitigation to limit adverse effects (DECC, 2012; PINS, 2012). Where a development is proposed in a European Marine Site, the Habitats Regulations require the competent authority, before authorising a project likely to have an effect on a European site, to " ... make an appropriate assessment of the implications for that site in view of that site's conservation objectives" (HM Government, 2010). The developer is required to provide the competent authority with information required for this assessment or to determine whether an appropriate assessment is required. Where marine protected areas are proposed in areas that already have existing activities such as offshore wind farms, the effect of these activities on the conservation objectives for the protected features will need to be assessed on a case-by-case basis, and if required, the relevant regulatory agencies will introduce management or mitigation to prevent adverse effects on site features (Natural England, 2014, 2017).

Case studies

In the United Kingdom, co-location has of offshore energy and marine conservation has already occurred at two locations; we discuss these examples in the following case studies.

Case study 1: Walney

In the eastern Irish Sea west of Walney, there is a Marine Conservation Zone that was designated in 2016 containing three existing wind farms: Walney (93 turbines, operational in 2012), West of Dudden (108 turbines, operational 2014), and Ormonde (30 turbines, operational 2011) (Figure 13.1). Another wind farm is also being constructed within the Marine Conservation Zone

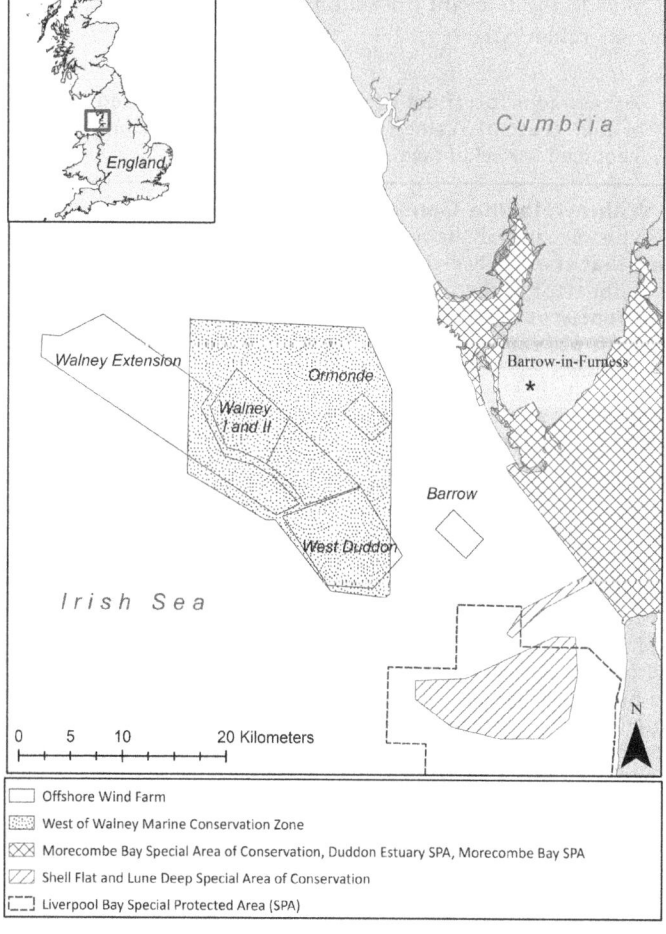

Figure 13.1 Marine conservation designations and offshore wind farms in the Walney region (Eastern Irish Sea, United Kingdom).

(as of 2017)—Walney Extension (207 turbines). The West of Walney Marine Conservation Zone was designated to protect two, broad-scale habitat
features (subtidal sand and subtidal mud), and one feature of conservation
importance (sea pen and burrowing megafauna communities) (Table 13.1).
This area also has three other marine protected areas nearby (all European
conservation designations): Shell Flat and Lune Deep Special Area of Conservation (designated, 2011), Liverpool Bay Special Protected Area (designated,
2010), and Morecombe Bay Special Area of Conservation (designated, 2005)
(Figure 13.1). This case study focuses on why co-location was considered
appropriate within this marine protected area and the mitigation that was
identified in the environmental impact assessment for the Walney Extension
offshore wind farm within the West of Walney Marine Conservation Zone.
As the cable route and construction also affected other conservation designations, we include the relevant mitigation identified for these designations in
the summary (Table 13.1).

Table 13.1 Assessment for effect of co-location on protected features
within West of Walney Marine Conservation Zone and mitigation
required for wind farms under construction

West of Walney Marine Conservation Zone
Feature type: Broadscale habitat
Protected features: Subtidal sand and subtidal mud
Feature type: Habitat feature of conservation importance
Protected features: Sea pen and burrowing fauna communities
Conservation objectives: Recover to favourable condition

Expected effect of co-location on protected feature	Marine protected area management response	Planning and license conditions (including mitigation required)
An area of the features was predicted to be lost following burial beneath the turbines and associated cables, but would be an unimportant proportion of the features' extents across the entire site (DEFRA, 2015)	Co-location offers a potential solution when considering conservation alongside other marine resources. By selecting the site, the total area otherwise closed to fishing would be reduced, because fishing by bottom trawling is limited within wind farms (DEFRA, 2015)	The route of the offshore-cable corridor will be adjusted to avoid sensitive habitats Restrictions to the timing and location of piling during construction will be implemented to reduce potential effects on spawning aggregations of herring and cod Intertidal and onshore construction to be avoided from October to March to reduce impacts on sensitive bird species in winter within Liverpool Bay Special Protected Area (Dong Energy, 2013)

Opportunities from co-location

While site selection of European Marine Sites is based only on the presence of habitat and species of European Importance, site selection of Marine Conservation Zones also includes social and economic impact assessment of the designation effects on existing activities. As such, co-location at Walney was considered in light of opportunities to reduce social and economic impacts on other industries (Christie et al., 2014). During site selection, West of Walney Marine Conservation Zone was identified as having the potential to be a good example of how marine conservation can be co-located with an area of the sea used by offshore wind farms (DEFRA, 2015).

There were several reasons for this. The area of the protected features that would be buried beneath the turbine foundations and associated cables was assessed as an insignificant proportion of the total extent of the features within the site (<1%). The presence of wind-farm infrastructure deters fishing, and so was considered to reduce impacts on protected features from bottom-towed fishing (trawling and dredging). Co-location increased the area of subtidal mud and communities of sea pens and burrowing fauna within the regional network of marine protected areas. Therefore, the presence of the infrastructure was not considered to be detrimental to conservation of the features at the site and within the regional network (DEFRA, 2015). Wind farms within the site are required to be developed and run considering the needs of the features on the surrounding seabed. During site assessment, it was anticipated that the industry would not need to change its operations substantially (DEFRA, 2015). At the Walney site, co-location offered a solution for protecting nationally important habitat features, while reducing area lost to other activities in the region. Fishing by bottom trawling is generally limited within both wind-farm developments and marine protected areas, and so co-locating these activities provided an opportunity to limit the total area that might otherwise be closed to fishing (Natural England, 2014).

Challenges for co-location

The protected features within West of Walney Marine Conservation Zone were assessed to have a conservation objective of 'recover' rather than 'maintain' (Natural England, 2014). Successful co-location would therefore require an improvement in the status of the declining sea pens in the region (DEFRA, 2015). Slow-growing species such as sea pens probably have longer recovery times than many other benthic animal species (Mazik and Smyth, 2013); thus, the loss of even small areas of these features could negatively impact the recovery to a 'favourable' condition (Mazik and Smyth, 2013). Detailed assessments of the effect of wind farms at this site on subtidal mud habitats and communities of sea pens and burrowing fauna have been ongoing since 2016 (Natural England, Kendal, Cumbria, personal communication). The assessment will also investigate the implications of reducing fishing pressure on recovery.

Case study 2: Race Bank

Race Bank is part of Inner Dowsing, Race Bank, and the North Ridge Site of Community Importance and candidate Special Area of Conservation (accepted, 2011) in the North Sea. As such, Race Bank falls under the assessment, planning and licensing requirements of a Special Area of Conservation in relation to co-location with offshore wind farms (Figure 13.2, Table 13.2). The Inner Dowsing, Race Bank, and the North Ridge Site of Community Importance contains two main features of importance, both Annex 1 habitats: Sandbanks and *Sabellaria spinulosa* reefs. The Wash and North Norfolk Special Area of Conservation (designated, 2005) is also situated farther inshore. There are two offshore wind farms already operating within this Site of Community Importance: Lincs (75 turbines, operational, 2013) and Lynn and Inner Dowsing (54 turbines, operational, 2009). Race Bank offshore wind farm is under construction (as of 2017) (91 turbines). A larger wind farm has also received consent to be built offshore of Race Bank wind farm: Triton Knoll (up to 288 turbines, consented 2013) (Figure 13.2). This case study focuses on the rationale for co-locating the activities, and the mitigation identified in the Environmental Statement regarding the co-location of Race Bank and Lincs offshore wind farms and Inner Dowsing, Race Bank, and North Ridge Site of Community Importance because these provide the most recent environmental assessments.

Opportunities from co-location

An area of designated features within the Site of Community Importance is lost to wind-farm infrastructure, specifically the foundations, cable routes, and associated concrete mattresses (Natural England, 2017). The area lost was assessed to be unlikely to cause an adverse effect to site features because it was <1% of the total area of the designated feature. Mitigation included altering the cable routes and adjusting construction times to avoid other negative impacts to protected habitats and species (Table 13.2). Other activities that can be potentially damaging to protected features occur in the region, including bottom-towed fishing and dredging (Natural England, 2017). Co-location was suggested to potentially to reduce localised impacts from these

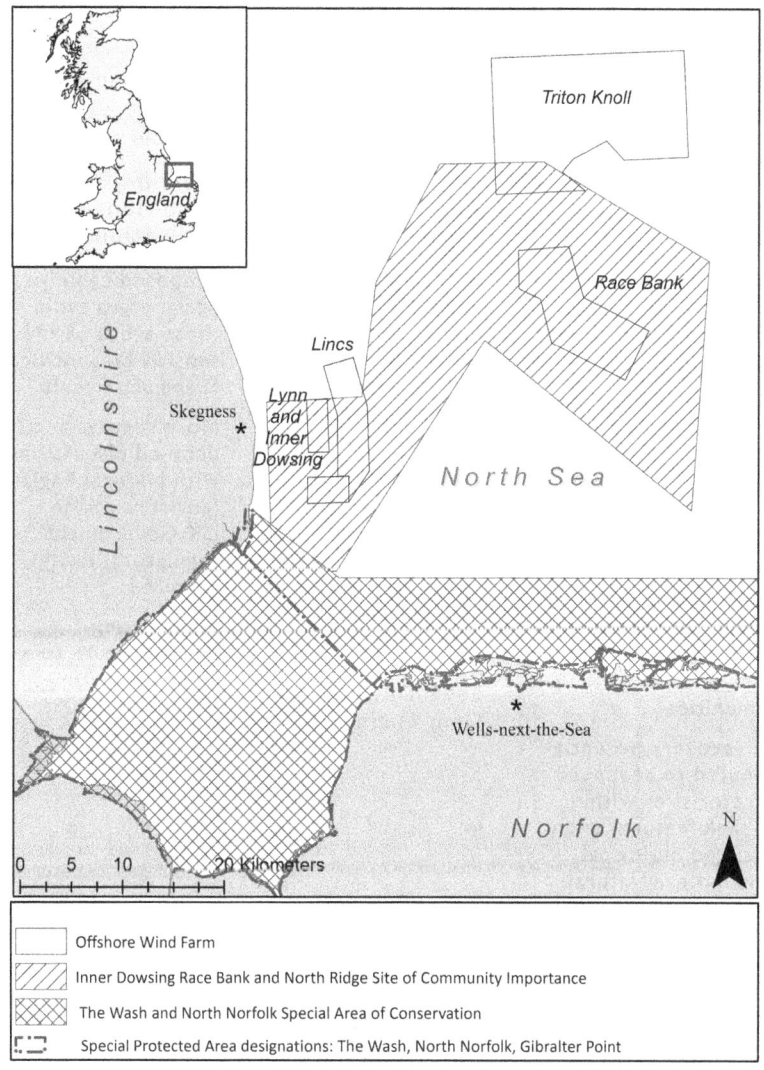

Figure 13.2 Marine conservation designations and offshore wind farms in the Race Bank region (North Sea, United Kingdom).

Table 13.2 Assessment for effect of co-location on protected features within Inner Dowsing, Race Bank and North Ridge Site of Community Importance and candidate Special Area of Conservation and mitigation required for wind farms under construction

Inner Dowsing, Race Bank and North Ridge Site of Community Importance and candidate Special Area of Conservation
Feature type: Annex I habitat
Protected features: Sandbanks (slightly covered by seawater all the time) and reefs
Conservation objectives: Maintain or restore to favourable condition

Expected effect of co-location on protected features	Marine protected area management response	Planning and license conditions (including mitigation required)
Physical obstruction from wind farms (moderate—sandbank; high-level reef)	Habitat features lost at sites of turbine foundations and scour protection assessed as <1% of total area of designated features (sandbanks and reefs)	Proposed cable route designed to avoid those areas of known sensitivity, specifically *S. spinulosa* reefs
Physical loss to features, both removing habitats and obstructing some areas		More surveys and detailed discussions with Natural England (advisors to the UK Government regulators) likely required
Reef feature with higher exposure to obstruction than sandbank because the affected proportion of this habitat is higher; however, total habitat loss is small. Recovery of *Sabellaria spinulosa* reefs possibly slower than for sandbank communities	Losses of sandbank sub-features and biotopes assessed as small in extent and temporary, owing to high recovery potential within highly dynamic environment (Centrica, 2009; DECC, 2012; Dong Energy, 2014)	Micro-siting infrastructure to avoid sensitive features (Centrica, 2009; DECC, 2012; Dong Energy, 2014)
High recovery potential attributed to sediment characteristics within sandbank features and associated biological communities (Natural England and JNCC, 2013)		

other activities. Assessments suggested that restoration of habitats to meet conservation objectives could be achieved through appropriate management of all activities in the site, such as fishing and aggregate dredging, as well as wind-farm construction (JNCC, 2010a).

Challenges from co-location

The environmental impact assessments for existing wind farms present in the site suggested minor adverse impacts (i.e., the impact is undesirable, but of limited concern) on organisms that live within and on the seabed (Centrica, 2007, 2009, 2010). Increased sediment loads and sedimentation across the site were predicted to result in changes in sediment characteristics and fauna communities, but these were also thought to be minor and within the variation observed in the wider region because it is a highly dynamic environment (Centrica, 2007, 2009, 2010).

Hard substrata installed at Race Bank offshore wind farm and other developments for cabling and scour protection were considered to be a temporary impact for the lifetime of the wind farm (Natural England, 2017). During the period that these hard substrata are present, the area of designated features is reduced. Although the lifespan of offshore wind-farm infrastructure is a minimum of 20 years, this does not take into account the potential for replacing or repairing infrastructure, or leaving decommissioned structures in place (Smyth et al., 2015). As a result, assessment of a temporary loss of habitat might in fact be longer than anticipated.

Discussion

There can be benefits to biodiversity from the presence of wind farm infrastructure, including increases in colonising organisms, and the fish and crustacean species that prey on them and seek shelter at the site of foundations (Chapter 11). However, these changes in habitat and species' distributions do not necessarily present benefits for the protection of the pre-existing habitats or species features designated for protection in the case studies of marine protected areas we present here (Tables 13.1 and 13.2). For instance, Lincs, Lynn and Inner Dowsing offshore wind farms occur where protected *Sabellaria spinulosa* (polychaete worm) reefs (Centrica, 2007, 2009) could be enhanced. Each turbine base is surrounded by concrete mattresses, which in addition to the turbine foundation, reduce the extent where *S. spinulosa* reefs occur (Natural England, 2017). Similarly, Race Bank offshore wind farm is being constructed within an area of protected sandbank features. A trade-off has therefore occurred between the loss of extent of a protected feature and the potential for maximising the overall area in regional seas available to other economic activities.

The main challenge to assessing co-location compatibility between marine protected areas and offshore wind farms is identifying the actual area of features of conservation importance lost, or in the unfavourable conditions resulting from the construction and operation of offshore wind farms. Assessment within marine protected areas has suggested the area lost to physical infrastructure is acceptable, because it tends to represent a small (< 1%) area of the total feature protected, and the loss is probably temporary (Tables 13.1 and 13.2). These calculations are based on the area of the features outside of the immediate footprint of infrastructure remaining in favourable condition, or not being lost over time. Also, the temporary nature of the loss relies on the wind-farm infrastructure being removed during the decommissioning phase, and following removal, a return to pre-construction conditions (MMO, 2014; Smyth et al., 2015).

Environmental impact assessments for wind farms have concluded that construction and presence will have only minor adverse impacts on the physical substrata or biological communities (Tables 13.1 and 13.2). However, post-construction monitoring related to the licensing conditions often does not extend beyond three years following construction. This has been criticised as insufficient to assess long-term changes in sediment characteristics and biological communities (Walker and Judd, 2010; MMO, 2014). Pre- and post-construction monitoring has also been criticised for not investigating changes over time. Neither has United Kingdom monitoring investigated fine-scale changes resulting from altered hydrodynamic forces surrounding foundations, nor have any changes to interconnected trophic webs or fouling on infrastructure been assessed (Walker and Judd, 2010; MMO, 2014).

If changes persist at greater footprints than just the area occupied by infrastructure, the total loss of extent of a protected feature, or decline in favourable conditions, might be greater than originally assessed. In our case studies, reporting on the extent and condition of features across the protected area is required at least every six years (Williams, 2006). This monitoring is unrelated to impact assessments, so it will be essential to examine the extent of protected features during the life cycle of co-located wind farms and the condition of the features. Ultimately, long-term monitoring could assess the success of co-location in the United Kingdom.

Compared to other anthropogenic activities that could damage designated features (e.g., substratum removal in large-scale, aggregate dredging), changes in substratum or biological communities around foundations might appear to be minor (DECC, 2012; Natural England, 2013). However, effects from wind farms could be present for the lifespan of the development, rather than being single, one-off events. This has led to considerable uncertainty over the long-term effects of developments on pre-existing physical environments and related fauna communities (MMO, 2014). Studies at a comparable wind farm in Belgium demonstrated changes in sediment characteristics and associated fauna communities had occurred up to 50 m from gravity-base foundations (Coates et al., 2014); however, changes such as these might not necessarily indicate unfavourable conditions. Although fauna communities can be altered, ecological functioning might recover from the effects of disturbance, and the new community could potentially function in a similar way to the original (Bolam et al., 2006; Elliot et al., 2007; Mazik and Smith, 2013). However, the effects of changes to ecological interrelationships between sediment characteristics, predators and prey within fine-scale footprints of infrastructure and over the footprint of an offshore wind farm remain uncertain (MMO, 2014). Such changes, and the effect on extent and ecological functioning of protected features, will be important to consider in future monitoring.

Mitigation, such as micro-siting infrastructure to avoid habitat features, and avoiding construction during sensitive life-cycle stages of protected species provide opportunities to support conservation objectives, alongside management of ongoing activities. For instance, the extent of *S.spinulosa* reef was shown to increase following construction of Thanet offshore wind farm, approximately 160 miles south of the Race Bank sites (Pearce et al. 2014). Wind farm pilings were micro-sited to avoid biogenic reef features at Thanet, suggesting the approach was successful in reducing impact on this sensitive feature. The increase in extent suggests co-location is possible if it were micro-sited to avoid biogenic reef features.

Highlights

- For offshore energy to be co-located with marine protected areas, activities must be assessed as compatible with conservation objectives.
- Marine protected areas have been co-located with wind farms, although some designated habitat features are often lost under infrastructure.
- In the United Kingdom's planning and licensing approaches, the extent of the protected feature lost is assessed compared to overall extent at the site. If the proportion lost is considered negligible, mitigation options can be applied.
- Uncertainty remains regarding the long-term effects of the presence of wind-farm infrastructure on sediment characteristics and associated biological communities.
- Long-term monitoring to assess changes in the extent and health of protected features in marine protected areas is required to ensure the compatibility aims of marine spatial planning aims are met.

References

Bolam, S.G., Schratzberger, M. and Whomersley, P. (2006). Macro and meiofaunal recolonisation of dredged material used for habitat enhancement: temporal patterns in community development. *Marine Pollution Bulletin*, 52, pp. 1746–1755.

Carr, H., Cornthwaite, A., Wright, H. and Davies, J. (2016). *Assessing Progress Towards an Ecologically Coherent MPA Network in Secretary of State Waters in 2016: Methodology*. Joint Nature Conservation Committee, Peterborough.

Centrica. (2007). *Lincs Offshore Wind Farm Environmental Statement Non-Technical Summary*. Centrica Energy Ltd., Windsor.

Centrica. (2009). *Docking Shoal offshore wind farm Race Bank offshore wind farm supplementary environmental information*. Non-Technical Summary, September 2009. http://assets.dongenergy.com.

Centrica. (2010). *LID6 Environmental Statement on behalf of Lincs Wind Farm Limited Non-Technical Summary.* Centrica Energy Ltd., Windsor.

Christie, N., Smyth, K., Barnes, R. and Elliott, M. (2014). Co-location of activities and designations: a means of solving or creating problems in marine spatial planning? *Marine Policy,* 43, pp. 254–261.

Coates, D.A., Deschutter, Y., Vincx, M. and Vanaverbeke, J. (2014). Enrichment and shifts in macrobenthic assemblages in an offshore wind farm area in the Belgian part of the North Sea. *Marine Environmental Research,* 95, pp. 1–12.

DECC. (2009). *Offshore energy strategic environmental assessment,* Department of Energy and Climate Change. Environnemental report, Recommandation 14, pp. 215–216. www.gov.uk/government/publications/uk-offshore-energy-strategic-environmental-assessment-oesea-environmental-report.

DECC. (2012). *Record of the Appropriate Assessment Undertaken for Applications Under Section 36 of the Electricity Act 1989. Projects, Docking Shoal Offshore Wind Farm (as Amended) Race Bank Offshore Wind Farm (as Amended) Dudgeon Offshore Wind Farm.* Department of Energy and Climate Change, London.

DEFRA. (2015). *West of Walney Recommended Marine Conservation Zone January 2015 Consultation on Sites Proposed for Designation in the Second Tranche of Marine Conservation Zones.* Department for Environment, Food and Rural Affairs, London.

Dong Energy. (2013). *Walney Extension Offshore Wind Farm Environmental Statement.* Dong Energy, London.

Dong Energy. (2014). *Race Bank Offshore Wind Farm Information for Habitats Regulations Assessment.* Prepared by Royal Haskoning DHV for DONG Energy, London, 8 December 2014.

EC. (2009). Directive 2009/147/EC of the European Parliament and of the Council of 30 November 2009 on the conservation of wild birds. European Community

EEC. (1992). Council Directive 92/43/EEC on the Conservation of Natural Habitats and of Wild Fauna and Flora. Official Journal of the European Communities No L206 of 22 July 1992. European Economic Community.

HM Government. (2010). *Regulation 61 of the Habitats Regulations and Regulation 25 of the Offshore Marine Regulations. The Conservation of Habitats and Species Regulations 2010.* Her Majesty's Government The Stationary Office, London.

HM Government. (2011). *UK Marine Policy Statement.* Her Majesty's Government The Stationary Office, London.

HM Government. (2014). *East inshore and east offshore marine plans.* Her Majesty's Government Retrieved July 2014 from www.gov.uk/government/publications/east-inshore-and-east-offshore-marine-plans.

JNCC. (2010a). Inner Dowsing, Race Bank and North Ridge Selection Assessment Document, Version 5.0, p. 486. Joint Nature Conservation Committee.

JNCC. (2010b). *Identifying marine conservation zones a quick reference guide.* A report by the Joint Nature Conservation Committee and Natural England 2010. http://jncc.defra.gov.uk/PDF/IdentifyingMarineConservationZones.pdf. Joint Nature Conservation Committee. (2016). *Special areas of conservation (SAC).* Information page accessed on December 2016 from: http://jncc.defra.gov.uk/page-23.

Marine Management Organisation. (2014). *Review of post-consent offshore wind farm monitoring data associated with licence conditions.* A report produced for the Marine Management Organisation, p. 194. Project No: 1031. Marine Management Organisation, Newcastle-Upon-Tyne.

Mazik, M. and Smyth, K. (2013). *Is 'minimising the footprint' an effective intervention to maximise the recovery of intertidal sediments from disturbance? Phase 1: Literature review.* Natural England Commissioned Reports, Number 110, NECR110. Natural England, London.

Natural England. (2014). *Marine Conservation Zones Natural England's Advice to Defra on Recommended Marine Conservation Zones to Be Considered for Consultation in 2015.* Natural England, London.

Natural England. (2017). Natural England Conservation Advice for Marine Protected Areas Inner Dowsing, Race Bank and North Ridge SCI – UK0030370 Supplementary Advice on Conservation Objectives. https://designatedsites. naturalengland.org.uk/. Last updated 20th March 2017.

Natural England and Joint Nature Conservation Committee. (2013). *Inner Dowsing, Race Bank and North Ridge Candidate Special Area of Conservation Formal Advice.* Natural England, London.

Pearce, B., Farinas-Franco, J.M., de Burgh, A. and Somerfield, P. (2014). Repeated mapping of reefs constructed by *Sabellaria spinulosa* at an offshore wind farm site. *Continental Shelf Research*, 83, pp. 3–13.

PINS (2012). *Habitat Regulations Assessment for Nationally Significant Infrastructure Projects, Version 3.* Planning Inspectorate. HM Government Planning Inspectorate, October 2012.

Qiu, W. and Jones, P.J.S. (2013). The emerging policy landscape for marine spatial planning in Europe. *Marine Policy*, 39, pp. 182–190.

Smyth, K., Christie, N., Burdon, D., Atkins, J.P., Barnes, R. and Elliott, M. (2015). Renewables-to-Reefs? – Decommissioning options for the offshore wind power industry. *Marine Pollution Bulletin*, 90, pp. 247–258.

Walker, R. and Judd, A. (2010). *Strategic Review of Offshore Wind Farm Monitoring Data Associated with FEPA Licence Conditions.* Department Environment Food and Rural Affairs. Report ME1117, London.

Williams, J.M. (Ed.). (2006). *Common standards monitoring for designated sites: first six year report.* Summary. Joint Nature Conservation Committee, Peterborough. Available online from: http://jncc.defra.gov.uk/pdf/CSM_06summary.pdf.

Chapter 14

Conservation challenges in the face of new hydrocarbon discoveries in the Mediterranean Sea

Tessa Mazor, Noam Levin, Eran Brokovich and Salit Kark

The Mediterranean Sea marine biodiversity hotspot

The Mediterranean Sea is the largest and deepest semi-enclosed basin in the world (Boudouresque 2004). It constitutes less than 1% of global ocean surface space, but it contains immense biodiversity relative to its size (Bianchi & Morri 2000). This area supports ~ 18% of the world's macroscopic marine species, of which 25–30% are endemic (Cuttelod et al. 2008). Emblematic species of conservation concern include the endemic Mediterranean monk seal *Monachus monachus*, 11 cetacean species (Franzosini et al. 2013), Atlantic bluefin tuna *Thunnus thynnus*, and sea turtles *Caretta caretta* and *Chelonia mydas*. The Mediterranean Sea also features unique ecosystems and habitats such as the endemic seagrass *Posidonia oceanica* that forms large underwater meadows, vermetid reefs built by sea snails *Dendropoma petraeum*, and Mediterranean coralligenous assemblages (Bianchi & Morri 2000; Boudouresque 2004). The rich and unique marine species and habitats of the region distinguish it as a global Biodiversity Hotspot (Cuttelod et al. 2008).

At the same time, the biodiversity of the Mediterranean Sea is threatened by a wide range of sea-based and land-based anthropogenic activities (Coll et al. 2010; Micheli et al. 2013a). The coastal areas are highly populated, with ~ 600 cities and a population of ~ 250 million inhabitants, along with ~ 250 million tourists that visit annually (Cuttelod et al. 2008). The Mediterranean Sea is surrounded by over 20 countries from Asia, Africa and Europe. Anthropogenic perturbations such as habitat degradation, pollution, by-catch, invasive species, climate change and exploitation of marine species and resources threaten the unique ecosystems (Bianchi & Morri 2000). In addition, social (e.g., cultural, religious and political) and economic divisions, such as the large contrast between Europe and North Africa, greatly challenge both the ability to minimise and control threats to the surrounding environment and to coordinate conservation efforts (Fraschetti et al. 2009).

To protect the biodiversity of this global Biodiversity Hotspot, marine spatial plans that meet specific goals and fit within realistic socio-economic constraints must be developed. Conservation initiatives and spatial planning processes are increasing in the region, with much focus on the promulgation of marine reserves and networks (Portman et al. 2012; Micheli et al. 2013b). Surprisingly, most initiatives ignore how marine spatial planning can address the rapidly expanding, and potentially vastly damaging, threat for the region's biodiversity: hydrocarbon exploration and extraction.

Recent hydrocarbon discoveries in the Mediterranean Sea

Some of the largest natural gas fields globally have recently been discovered in the Mediterranean Sea and they are being developed at a rapid pace. There are currently 152 existing offshore natural gas and oil fields in the region, and 172 planned for development over the next four years (Pruett 2013; Infield 2013). Two-thirds of both present and potential natural gas and oil fields lie beyond the 12-nautical mile territorial waters (Figure 14.1a). The U.S. Geological Survey estimates that the eastern Mediterranean holds 9.8 trillion cubic metres of recoverable natural gas, 417 billion litres of recoverable oil and 715 billion litres of natural gas liquids (Schenk et al. 2010; Kirschbaum et al. 2010). Recent gas discoveries by Israel and Cyprus, including the large Leviathan gas field, are estimated to hold a combined 980 billion cubic metres of natural gas (see Figure 14.2). These discoveries have spurred a flurry of exploration, especially in the Levantine Basin, which is situated in the eastern and south-eastern Mediterranean (Darbouche et al. 2012; Figure 14.1).

The present distribution of drilling wells is mostly limited to shallow waters (< 500 m) (Figure 14.1a). However, technological advances are enabling offshore operations in the Mediterranean to expand into deeper waters, as seen by the location of oil and gas concessions (Figure 14.1a). The spatial distribution of past oil spills appears closely related to shipping routes and ports, rather than to the location of drilling wells (Abdulla & Linden, 2008). However, some concessions are located in seismically sensitive areas (e.g., in the Aegean and Ionian Seas), where higher risks of earthquakes must be considered (Figure 14.1c). These active areas also provide important habitats for many marine species, such as the globally Endangered Mediterranean monk seal and green sea turtles (International Union for Conservation of Nature 2012).

Hydrocarbon operations and their conservation challenges

Research on the ecological impacts of offshore oil and gas exploration is often focused on the negative consequences of oil and gas spills on marine ecosystems and species (e.g., Davies et al. 2007; Fisher et al. 2014). While these potential impacts (Chapter 8) are crucial to account for in marine spatial planning processes (Cordes et al., 2016), oil and gas exploitation might also provide some benefits and opportunities for conservation that should also be considered (Kark et al. 2015; Chapters 3, 10, and 11). Below we describe four

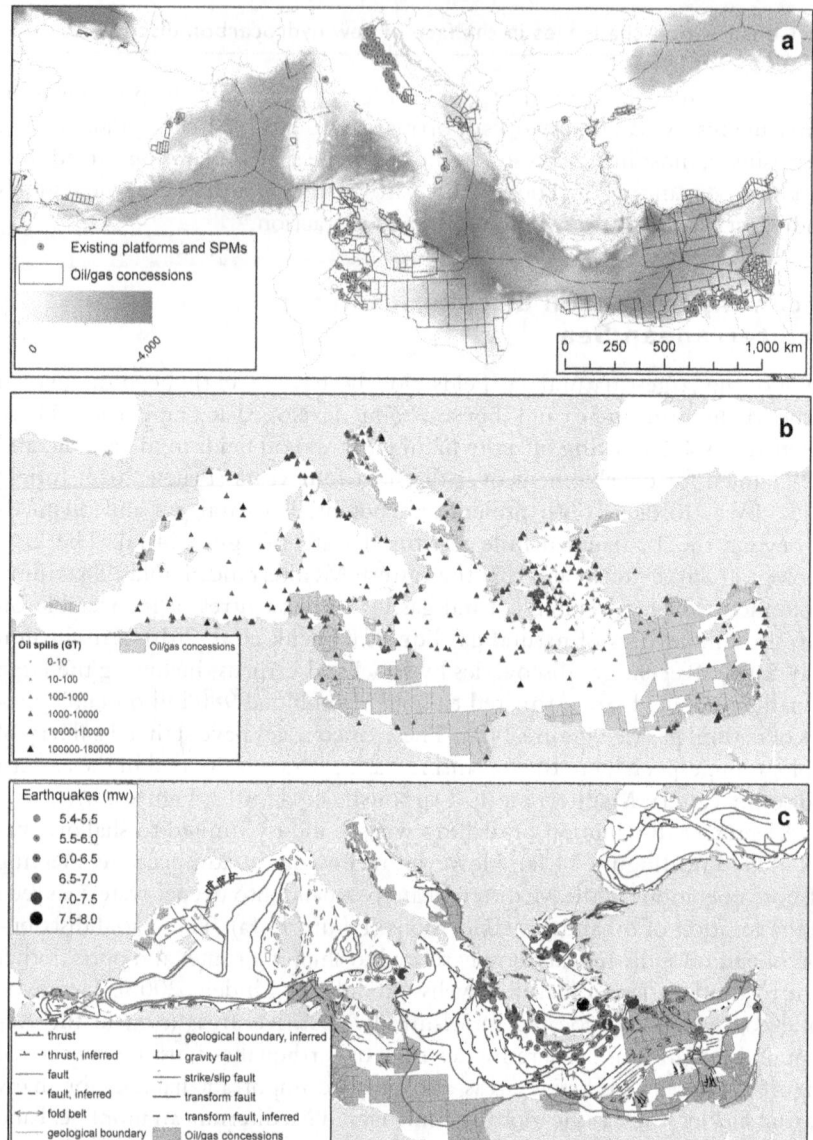

Figure 14.1 Oil and natural gas concession areas, existing platforms and offshore single-point moorings (SPM) in the Mediterranean Sea, based on data from Infield Offshore Energy Database (Infield 2013): (a) location of concessions areas and existing drilling with respect to the bathymetry and marine boundaries of Exclusive Economic Zones (thick black lines) (The Flanders Marine Institute 2012), (b) recorded oil spills in Mediterranean concessions in the 35 years between January 1977 and July 2013 (*n* = 778; Supporting Information; REMPEC 2013) and (c) seismic activity (major earthquakes in the 20th Century) and major faults in the Mediterranean relative to oil and gas concessions (fault data, Asch 2003; earthquake data, ISC 2013).

Source: Adapted from Emerging conservation challenges and prospects in an era of offshore hydrocarbon exploration and exploitation, Kark *et al.*, Conservation Biology, 2015, doi:10.1111/cobi.12562, Copyright © [2015], Published by Wiley on behalf of Society for Conservation Biology, Wiley Periodicals, Inc.

Figure 14.2 Tamar gas field production platform in the Mediterranean Sea. Photo by Ilan Nisim.

challenges that hydrocarbon operations present for the Mediterranean Sea's biodiversity and explore possible collaborations and opportunities that could be gained for conservation and marine planning in the region.

Challenge 1: Oil spills and gas leaks

The Mediterranean Sea is one of the regions with the highest risk of oil pollution (Martini & Patruno 2005) and has been identified as an area of global concern for oil spills from tankers (O'Rourke & Connolly 2003). A total of 778 oil spills was recorded in the Mediterranean Sea between January 1977 and July 2013 (International Maritime Organisation 2013). Using remote-sensing methods, 897 to 2,297 oil slicks (covering 1382 to 3885 square degrees) were identified annually between 1999 and 2004 (Ferraro et al. 2009). Surprisingly, oil-spill preparedness in this region is limited (Moller et al. 2003). While the probability of a large oil spill is low, such a spill could have extensive negative impacts on biodiversity. Impacts of oil spills could be even greater in the Mediterranean Sea than elsewhere, due to its enclosed geography and its deep isolated basins.

Many preparedness plans exist for oil spills of Mediterranean Basin governments; however, cooperative international plans rarely exist due to conflicts between nations, particularly in the eastern Mediterranean (Khadduri 2012). Cooperative agreements, such as bilateral and trilateral partnerships between countries, could improve the readiness of the eastern Mediterranean for oil spills (Martini & Patruno 2005). The shared risk of the region might encourage collaboration to facilitate the safe exploration and exploitation of hydrocarbon resources as evident in other regions around the globe (e.g., US–Mexico Transboundary Hydrocarbon Agreement; The Bureau of Safety and Environmental Enforcement 2015). Such agreements and treaties that help forge ties between countries could be a catalyst for strengthening collaborative potential of improved disaster prevention as well as future marine conservation plans and actions (Levin et al. 2013).

Challenge 2: Deepwater exploration

International oil companies are now drilling in the Mediterranean to depths of > 6,000 m (e.g., Hodoa field, Egypt,Khadduri 2012). The scarcity of data on the ecology and oceanography of the deep waters of the Mediterranean (Costello et al. 2010; Levin et al. 2014) and the unknown impacts to biodiversity resulting from developing marine oil and natural gas operations challenges conservation efforts and environmental decisions. Previously, the deep Mediterranean Sea was once thought to hold few species; however, it has now been estimated that approximately 2,805 species inhabit the Mediterranean's deep-sea, with ~ 66% of these species still yet to be discovered (Danovaro et al. 2010).

In general, there is poor understanding of how deep-water ecosystems recover from disturbances (Gates & Jones 2012). Research has suggested that deep-sea ecosystems provide critical habitat for global ocean functioning, so any large disturbances could alter biological processes of the whole marine ecosystem (Danovaro et al. 2008; Loreau 2008). Deep-sea drilling could provide an opportunity to gain research funding and access to operate remote underwater vehicles. This in turn could enable us to enhance our limited knowledge of the deep-sea regions and to study the interaction of drilling within deep ecosystems (e.g., Before-After Control-Impact studies). Furthermore, deep-sea operations could encourage countries to devise multi-national marine spatial plans, which extend beyond territorial waters and aim to protect deeper habitats.

Challenge 3: Marine borders and cross-country collaboration

Cross-country collaboration of conservation efforts in the Mediterranean Sea is highly valuable for efficient planning and protection of marine biodiversity (Mazor et al. 2013). Hydrocarbon operations can either hinder or facilitate such collaboration potential. A momentum of Exclusive Economic Zone declaration can be observed in recent years in the eastern Mediterranean as increased offshore exploitation and hydrocarbon discoveries occur in the region. Cyprus and Israel, for example, have recently reached an agreement on their maritime boundaries following the discovery of large offshore gas in Israel (Leviathan: the largest gas field found in the Mediterranean; Wählisch 2011; Energy Information Administration 2013). The prospect of joint exploration of oil and gas reserves and collaboration in a proposed natural gas pipeline to export gas to Europe (via Greece and Italy) between Cyprus and Israel enhances socio-economic relations (Shaffer 2011). Cooperation arrangements between national companies regarding joint exploration, drilling and production, as well as preparedness for the mitigation of possible oil spills, even when maritime claims are not settled, could strengthen socio-political ties that could in turn enhance conservation and spatial planning collaborations (Buszynski & Sazlan 2007; Levin et al. 2013).

On the other hand, several countries have not officially declared their Exclusive Economic Zones (The Flanders Marine Institute 2012; Katsanevakis et al. 2015), and not all marine boundaries have been agreed. Disagreements over marine borders are one of the causes for political tension between countries, for example, between Cyprus and Turkey (Naylor 2011). Oil and gas discoveries also provide countries with great economic and political independence (Shaffer 2011), but given the complicated political history of the region, they also have the potentail to stimulate conflicts and impede collaborative planning.

Challenge 4: Hydrocarbon infrastructure

Hydrocarbon infrastructure in the sea could facilitate indirect protection for marine biodiversity. The International Maritime Organization (2013) enforces a 500-m safety zone around drilling platforms, which excludes all fishing and recreational activities. This safety zone is followed by all Mediterranean countries and is extended in some regions with security issues (e.g., 5-km radius is enforced in Israel). If correctly managed in collaboration with conservation managers, this safety zone has the potential to act as a *de facto* marine protected area (Chapter 11). Another possible benefit is that hydrocarbon infrastructure could act as artificial reefs. In many areas of the Mediterranean, rocky substrata are scarce, especially in the deep-sea where clay and silt are more common. Hydrocarbon pipes and rigs could enhance the recruitment and reproduction of rare and endangered fish, algae or coral species (e.g., Gass & Roberts 2006; Davies et al. 2007, Chapters 10 and 11). Platforms might also have other functions for conservation, such as providing stopping sites for migratory bird species, or as potential stations for monitoring biodiversity and gaining new biological information. Nevertheless, challenges remain. Hydrocarbon infrastructure can have many negative impacts on the marine environment, including pollution (Henderson et al. 1999; Ko & Day 2004), habitat damage (Grant & Briggs 2002; Davies et al. 2007), and increased prevalence of invasive species (Rivas et al. 2010), all of which can lead to biodiversity loss and reduced ecosystem resilience (see Chapter 8 for more details). The extent to which potential biodiversity benefits or negative impacts are realised will depend to some extent on where instillations are situated in the seascape (Chapter 11).

Conservation plans and initiatives

Marine conservation initiatives in the Mediterranean are increasing, mostly in the form of networks of marine protected areas at local, national and regional scales (Levin et al. 2013). Yet, broad-scale conservation initiatives have rarely considered the negative impacts of oil and gas exploration

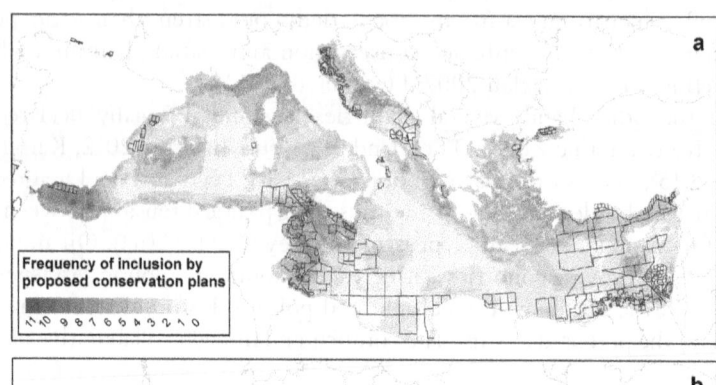

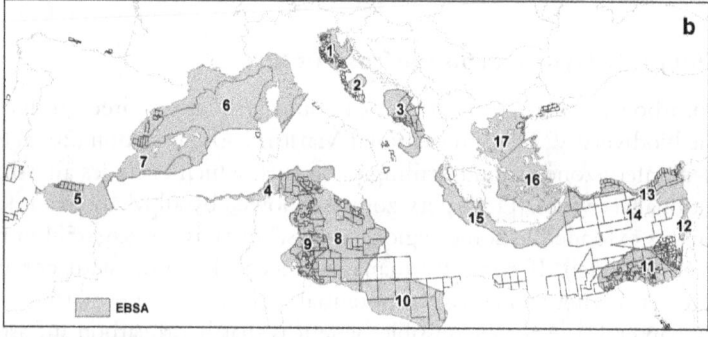

Ecologically or Biologically Significant Marine Areas (EBSA):
1: Northern Adriatic
2: Jabuka / Pomo Pit
3: South Adriatic Ionian Strait
4: Algero Tunisian Margin
5: Alboran Sea and Connected Areas
6: North West Mediterranean Pelagic Ecosystem
7: North Western Mediterranean Benthic Ecosystem
8: Sicilian Channel
9: Le Golfe de Gabos
10: Gulf of Sirte
11: Nile Delta Fan
12: East Levantine Canyons (ELCA)
13: North East Levantine Sea
14: Akamas and Chrysochou Bay
15: Hellenic Trench
16: Central Aegean Sea
17: North Aegean

Figure 14.3   Oil and natural gas concession areas (black outline; Infield 2013) in the Mediterranean Sea and their overlap with (a) conservation priority areas shared between five or more proposed schemes (adapted from Micheli et al. 2013b), and (b) Ecologically or Biologically Significant Marine Areas (EBSA; www.cbd.int/ebsa).

explicitly, and how they overlap or compete with conservation objectives (Micheli et al. 2013b). For instance, a synthesis of priority marine conservation areas in the Mediterranean (areas selected by five or more initiatives; Micheli et al. 2013b) has a 27% overlap (130,000 km^2) with gas and oil concessions, equal to 17% of the total size of the concession areas (Figure 14.3a).

Furthermore, 46% (566,200 km^2) of Ecologically or Biologically Significant Marine Areas (EBSA) in the Mediterranean overlap with gas and oil concessions (Figure 14.3b). Future work in the Mediterranean Sea would benefit from incorporating emerging oil and gas plans into spatial planning (e.g., Mazor et al. 2014). One of the major challenges for marine spatial planning will be to optimise the siting of many new installations to minimise negative impacts and maximise potential conservation benefits. Table 14.1 outlines some possible tools and techniques that could be used to do so.

Table 14.1 Proposed tools and their approaches for incorporating offshore energy information into marine spatial plans, with examples from literature. Given the limited inclusion of offshore energy operations in marine conservation planning, examples from the terrestrial literature were included.

Tools	Approach	Literature
Cost or threat layer	Including oil and gas exploration and extraction as a cost layer or as a threat layer within systematic conservation planning tools such as *Marxan*, *Marxan with Zones*, *Marxan with Probability* or *Zonation*. The cost layer could include areas that are licenced or leased to oil and gas companies, actual hydrocarbon fields or drilling platforms.	Schneider et al. (2011, 2012); Mazor et al. (2014)
Discovery scenarios	Accounting for uncertainty in the location of future oil and gas developments. Modelling scenarios predict future discoveries and then are incorporated into conservation plans.	Wilson et al. (2013)
Avoid strategy	Avoiding potential hydrocarbon exploration areas from the marine plan. For example, the 'lock out' term used in *Marxan* where planning units are left out of the analysis. Thus, conservation priorities can be selected around hydrocarbon features to avoid conflicting priorities.	Kiesecker et al. (2009)
Distance and buffers	Creating a buffer zone around drill sites or calculating a 'safe' distance from hydrocarbon features could minimise some local impacts (e.g., pile cuttings released during drilling operation restricted to a zone within 100 m of the discharge; Neff 2010).	Rogers et al. (2005)

(Continued)

Tools	Approach	Literature
Trade-off analysis	Quantifying the value of hydrocarbon reserves and other economic activities of the sea.	Mazor et al. (2014); Chapter 2
	Examining the trade-offs for reaching conservation goals (e.g., protecting a percentage of a species distribution) and maintaining energy-generation operations.	
Simulation analysis	Incorporating a simulation model or various scenarios of oil spill/disaster events into a marine plan (e.g., *Marxan with Probability*) to avoid potential spill areas.	Goldman et al. (2015); Gopal et al. (2015)
Food web modelling	Hydrocarbon threats can be examined in the context of a food web model to estimate and predict cumulative impacts on biodiversity. Spatial plans can then prioritise actions that minimise specific components of hydrocarbon operations.	Giakoumi et al. (2015)
Including linear features	Avoiding planning marine reserves near pipelines and transportation routes from drill sites. These features can be difficult to plan around given their linear structure.	Mazor et al. (2014)
Co-location options	Examining the impacts of co-locating offshore energy operations and other activities in marine spatial plans (e.g., aquaculture farms, marine reserves, and fishing).	Yates et al. (2015); Chapters 10 and 13.

Ways forwards for marine spatial plans

Hydrocarbon operations are rapidly moving ahead in the Mediterranean Sea, while multi-sector marine spatial plans are only slowly emerging. To protect the region's marine biodiversity, transparent marine spatial plans incorporating hydrocarbon operations need to be proposed quickly. While the incorporation of hydrocarbon activities and impacts into marine plans demands some estimates and approximations, including such developments is important for sufficiently protecting marine biodiversity from the threats they pose. To move forward, planning initiatives for the region should aim to leverage opportunities (e.g., funding for deep-sea biological surveys; Kark et al. 2015) from current and future hydrocarbon operations (Chapter 3). Marine spatial plans need to incorporate objectives of different sectors (e.g., conservation, commercial fisheries, offshore energy) and quantify trade-offs between sectors for efficient decision-making and successful uptake of plans (Yates et al 2015).

Highlights

- The Mediterranean Sea Biodiversity Hotspot is facing rapid exploration and exploitation of hydrocarbon resources.
- Hydrocarbon activities pose many challenges for biodiversity of the Mediterranean region due to its enclosed basin shared by >20 countries.
- Opportunities can be gained by collaborating with the hydrocarbon industry: funding to explore the deep-sea, *de facto* marine protected areas, and strengthened ties between countries.
- Future marine spatial plans should include hydrocarbon activities to improve biodiversity protection.
- There are a range of tools and approaches available for explicitly accounting for hydrocarbon activities in marine spatial plans.

References

Abdulla, A. and Linden, O. (2008). *Maritime traffic effects on biodiversity in the Mediterranean Sea: Review of impacts, priority areas and mitigation measures.* IUCN Centre for Mediterranean Cooperation, Malaga, Spain.

Bianchi, C.N. and Morri, C. (2000). Marine biodiversity of the Mediterranean Sea: Situation, problems and prospects for future research. *Marine Pollution Bulletin*, 40(5), pp. 367–376.

Boudouresque, C. F. (2004). Marine biodiversity in the Mediterranean; status of species, populations and communities. *Travaux scientifiques du Parc national de Port-Cros*, 20, pp. 97–146.

Buszynski, L. and Sazlan, I. (2007). Maritime claims and energy cooperation in the South China Sea. *Contemporary Southeast Asia: A Journal of International and Strategic Affairs*, 29(1), pp. 143–171.

Coll, M., Piroddi, C., Steenbeek, J., Kaschner, K., Lasram, F.B.R., Aguzzi, J., Ballesteros, E., Bianchi, C.N., Corbera, J., Dailianis, T. and Danovaro, R. (2010). The biodiversity of the Mediterranean Sea: Estimates, patterns, and threats. *PloS One*, 5(8), pp. e11842.

Cordes, E.E., Jones, D.O., Schlacher, T.A., Amon, D.J., Bernardino, A.F., Brooke, S., Carney, R., Deleo, D.M., Dunlop, K.M., Escobar-Briones, E.G. and Gates, A.R. (2016). Environmental impacts of the deep-water oil and gas industry: A review to guide management strategies. *Frontiers in Environmental Science*, 4, pp. 1–26.

Costello, M.J., Coll, M., Danovaro, R., Halpin, P., Ojaveer, H. and Miloslavich, P. (2010). A census of marine biodiversity knowledge, resources, and future challenges. *PloS One*, 5(8), pp. e12110.

Cuttelod, A., García, N., Malak, D.A., Temple, H.J. and Katariya, V. (2009). *The Mediterranean: A biodiversity hotspot under threat*. Wildlife in a changing world – An analysis of the 2008 IUCN Red List of Threatened Species, 89.

Danovaro, R., Corinaldesi, C., D'Onghia, G., Galil, B., Gambi, C., Gooday, A.J., Lampadariou, N., Luna, G.M., Morigi, C., Olu, K. and Polymenakou, P. (2010).

Deep-sea biodiversity in the Mediterranean Sea: the known, the unknown, and the unknowable. *PLoS One*, 5(8), pp. e11832.

Danovaro, R., Gambi, C., Dell'Anno, A., Corinaldesi, C., Fraschetti, S., Vanreusel, A., Vincx, M. and Gooday, A.J. (2008). Exponential decline of deep-sea ecosystem functioning linked to benthic biodiversity loss. *Current Biology*, 18(1), pp. 1–8.

Darbouche, H., El-Katiri, L. and Fattouh, B. (2012). *East Mediterranean Gas: What kind of a game-changer?* Oxford Institute for Energy Studies, Oxford, pp. 1–33.

Davies, A.J, Roberts, J.M, and Hall-Spencer, J. (2007). Preserving deep-sea natural heritage: emerging issues in offshore conservation and management. *Biological Conservation*, 138(3), pp. 299–312.

Energy Information Administration. (2013). *Overview of oil and natural gas in the Eastern Mediterranean region.* U.S. Department of Energy Information Administration, Washington, DC.

Ferraro, G., Meyer-Roux, S., Muellenhoff, O., Pavliha, M., Svetak, J., Tarchi, D. and Topouzelis, K. (2009). Long term monitoring of oil spills in European seas. *International Journal of Remote Sensing*, 30(3), pp. 627–645.

Fisher, C.R., Demopoulos, A.W., Cordes, E.E., Baums, I.B., White, H.K. and Bourque, J.R. (2014). Coral communities as indicators of ecosystem-level impacts of the Deepwater Horizon spill. *BioScience*, 64(9), pp. 796–807.

Franzosini, C., Genov, T. and Tempesta, M. (2013). *Cetacean manual for MPA managers.* ACCOBAMS, MedPAN and UNEP/MAP-RAC/SPA. Ed. RAC/SPA., Tunis, pp. 1–77.

Fraschetti, S., D'Ambrosio, P., Micheli, F. and Pizzolante, F. (2009). Design of marine protected areas in a human dominated seascape. *Marine Ecology Progress Series*, 375, pp. 13–24.

Gass, S.E. and Roberts, J.M. (2006). The occurrence of the cold-water coral *Lophelia pertusa* (Scleractinia) on oil and gas platforms in the North Sea: Colony growth, recruitment and environmental controls on distribution. *Marine Pollution Bulletin*, 52(5), pp. 549–559.

Gates, A.R. and Jones, D.O.B. (2012). Recovery of Benthic Megafauna from Anthropogenic Disturbance at a Hydrocarbon Drilling Well (380 m Depth in the Norwegian Sea). *PloS One* 7(10), pp. e44114.

Giakoumi, S., Halpern, B.S., Michel, L.N., Gobert, S., Sini, M., Boudouresque, C.F., Gambi, M.C., Katsanevakis, S., Lejeune, P., Montefalcone, M. and Pergent, G. (2015). Towards a framework for assessment and management of cumulative human impacts on marine food webs. *Conservation Biology*, 29(4), pp. 1228–1234.

Goldman, R., Biton, E., Brokovich, E., Kark, S. and Levin, N. (2015). Oil spill contamination probability in the southeastern Levantine basin. *Marine Pollution Bulletin*, 91(1), pp. 347–356.

Gopal, S., Kaufman, L., Pasquarella, V., Ribera, M., Holden, C., Shank, B. and Joshua, P. (2015). Modeling coastal and marine environmental risks in Belize: The marine integrated decision analysis system (MIDAS). *Coastal Management*, 43(3), pp. 217–237.

Grant, A. and Briggs, A.D. (2002). Toxicity of sediments from around a North Sea oil platform: Are metals or hydrocarbons responsible for ecological impacts. *Marine Environmental Research,* 53(1), pp. 95–116.

Henderson, S.B., Grigson, S.J.W., Johnson, P. and Roddie, B.D. (1999). Potential impact of production chemicals on the toxicity of produced water discharges from North Sea oil platforms. *Marine Pollution Bulletin*, 38(12), pp. 1141–1151.

Infield. (2013). Infield Energy Gateway GIS mapping. [online] Available at: www.infield.com/.

International Maritime Organisation. (2013). *International Maritime Organisation.* United Nations, London, UK.

International Union for Conservation of Nature. (2012). *Marine mammals and sea turtles of the Mediterranean and Black Seas.* International Union for Conservation of Nature, Gland, Switzerland and Malaga, Spain.

Kark, S., Brokovich, E., Mazor, T. and Levin, N. (2015). Emerging conservation challenges and prospects in an era of offshore oil and natural gas discoveries. *Conservation Biology,* 29(6), pp. 1573–1585.

Katsanevakis, S., Levin, N., Coll, M., Giakoumi, S., Shkedi, D., Mackelworth, P., Levy, R., Velegrakis, A., Koutsoubas, D., Caric, H. and Brokovich, E. (2015). Marine conservation challenges in an era of economic crisis and geopolitical instability: The case of the Mediterranean Sea. *Marine Policy,* 51, pp. 31–39.

Khadduri, W. (2012). East Mediterranean Gas: Opportunities and challenges. *Mediterranean Politics,* 17(1), pp. 111–117.

Kiesecker, J.M., Copeland, H., Pocewicz, A., Nibbelink, N., McKenney, B., Dahlke, J., Holloran, M. and Stroud, D. (2009). A framework for implementing biodiversity offsets: Selecting sites and determining scale. *BioScience,* 59(1), pp. 77–84.

Kirschbaum, M.A., Schenk, C.J., Charpentier, R.R., Klett, T.R., Brownfield, M.E., Pitman, J.K., Cook, T.A. and Tennyson, M.E. (2010). *Assessment of undiscovered oil and gas resources of the Nile Delta Basin Province, Eastern Mediterranean.* US Geological Survey, No. 2010–3027.

Ko, J.Y. and Day, J.W. (2004). A review of ecological impacts of oil and gas development on coastal ecosystems in the Mississippi Delta. *Ocean & Coastal Management,* 47(11), pp. 597–623.

Levin, N., Coll, M., Fraschetti, S., Gal, G., Giakoumi, S., Göke, C., Heymans, J.J., Katsanevakis, S., Mazor, T., Öztürk, B. and Rilov, G. (2014). Biodiversity data requirements for systematic conservation planning in the Mediterranean Sea. *Marine Ecology Progress Series,* 508, pp. 261–281.

Levin, N., Tulloch, A.I., Gordon, A., Mazor, T., Bunnefeld, N. and Kark, S. (2013). Incorporating socioeconomic and political drivers of international collaboration into marine conservation planning. *BioScience,* 63(7), pp. 547–563.

Loreau, M. (2008). Biodiversity and ecosystem functioning: The mystery of the deep-sea. *Current Biology,* 18(3), pp. R126–R128.

Martini, N. and Patruno, R. (2005). Oil pollution risk assessment and preparedness in the east Mediterranean. *International Oil Spill Conference Proceedings,* pp. 259–264.

Mazor, T., Possingham, H.P., Edelist, D., Brokovich, E. and Kark, S. (2014). The crowded sea: Incorporating multiple marine activities in conservation plans can significantly alter spatial priorities. *PloS One,* 9(8), pp. e104489.

Mazor, T., Possingham, H.P. and Kark, S. (2013). Collaboration among countries in marine conservation can achieve substantial efficiencies. *Diversity Distributions,* 19(11), pp. 1380–1393.

Micheli, F., Halpern, B.S., Walbridge, S., Ciriaco, S., Ferretti, F., Fraschetti, S., Lewison, R., Nykjaer, L. and Rosenberg, A.A. (2013a). Cumulative human impacts on Mediterranean and Black Sea marine ecosystems: Assessing current pressures and opportunities. *PloS One,* 8(12), pp. e79889.

Micheli, F., Levin, N., Giakoumi, S., Katsanevakis, S., Abdulla, A., Coll, M., Fraschetti, S., Kark, S., Koutsoubas, D., Mackelworth, P. and Maiorano, L. (2013b). Setting priorities for regional conservation planning in the Mediterranean Sea. *PLoS One*, 8(4), pp. e59038.

Moller, T.H., Molloy, F.C. and Thomas, H.M. (2003). Oil spill risks and the state of preparedness in the regional seas. *International Oil Spill Conference Proceedings*, 2003, pp. 919–922.

Naylor, H. (2011). *Vast gas fields found off israel's shores cause trouble at home and abroad.* The National. [online] Avaialble at: www.thenational.ae/news/world/middle-east/vast-gas-fields-found-off-israels-shores-cause-trouble-at-home-and-abroad#full.

Neff, J.M. (2010). *Fate and effects of water based drilling muds and cuttings in cold water environments.* A Scientific Review Prepared for Shell Exploration and Production Company. Shell Exploration and Production Company, Houston.

O'rourke, D. and Connolly, S. (2003). Just oil? The distribution of environmental and social impacts of oil production and consumption. *Annual Review of Environment and Resources*, 28(1), pp. 587–617.

Portman, M.E., Nathan, D., and Levin, N. (2012). From the Levant to Gibraltar: A regional perspective for marine conservation in the Mediterranean Sea. *Ambio*, 41(7), pp. 670–681.

Pruett, L. (2013). *Global Maritime Boundaries Database. Global GIS Data Services LLC, Herndon, Virginia.* [online] Available at: www.globalgisdata.com/.

Rivas, G., Moore, A., Dholoo, E. and Mitchell, P. (2010). Alien invasive species: Risk and management perspectives for the oil and gas industry. In: *SPE international conference on health, safety and environment in oil and gas exploration and production.* Society of Petroleum Engineers.

Rogers, S.I., Eastwood, P.D., Houghton, C.A. and Mills, C. (2005). *Developing the ecosystem approach: key tasks and practical progress towards integrated planning in the UK.* International Council for the Exploration of the Sea.

Schenk, C.J., Kirschbaum, M.A., Charpentier, R.R., Klett, T.R., Brownfield, M.E., Pitman, J.K., Cook, T.A. and Tennyson, M.E. (2010). *Assessment of undiscovered oil and gas resources of the Levant Basin Province, Eastern Mediterranean.* U.S. Geological Survey Fact Sheet 2010–2014, pp. 1–4.

Schneider, R.R., Hauer, G., Dawe, K., Adamowicz, W. and Boutin, S. (2012). Selection of reserves for woodland caribou using an optimization approach. *PloS One*, 7(2), pp. e31672.

Schneider, R.R., Hauer, G., Farr, D., Adamowicz, W.L. and Boutin, S. (2011). Achieving conservation when opportunity costs are high: Optimizing reserve design in Alberta's oil sands region. *PLoS One*, 6(8), pp. e23254.

Shaffer, B. (2011). Israel – New natural gas producer in the Mediterranean. *Energy Policy*, 39(9), pp. 5379–5387.

The Bureau of Safety and Environmental Enforcement. (2015). *International & interagency collaboration, the Bureau of Safety and Environmental Enforcement, USA.* [online] Available at: www.bsee.gov/.

The Flanders Marine Institute (VLIZ). (2012). *Maritime boundaries geodatabase, version 6.1.* [online] Available at: www.vliz.be/vmdcdata/marbound.

Wählisch, M. (2011). Israel-Lebanon offshore oil & gas dispute – Rules of international maritime law. *ASIL Insights*, 15.

Wilson, R.R., Liebezeit, J.R. and Loya, W.M. (2013). Accounting for uncertainty in oil and gas development impacts to wildlife in Alaska. *Conservation Letters*, 6(5), pp. 350–358.

Yates, K.L., Schoeman, D.S. and Klein, C.J. (2015). Ocean zoning for conservation, fisheries and marine renewable energy: Assessing trade-offs and co-location opportunities. *Journal of Environmental Management,* 152, pp. 201–209.

Chapter 15

Siting offshore energy arrays

A case study using interactive marine planning

Karen A. Alexander, Ron Janssen and Timothy G. O'Higgins

Introduction

In 2010, Jack Dangermond, president of Environmental Systems Research Institute (Esri) spoke to the TED.com conference about 'GeoDesign', a concept based on a book written in the 1960s by landscape architect Ian McHarg. In this book, 'Design with Nature', McHarg pioneered the concept of ecological planning, showing how scientific information, maps, and more could be combined and integrated with the design process to make harmonious and responsible plans. The book inspired Jack Dangermond and it was this idea that led to the creation of the organisation Esri—the world's most successful spatial analysis company that builds, and enables software users to build, geographic information systems (GIS).

Yet, the president of Esri did not believe this was enough. He suggested that we need to be able to harness everything we can measure—from ecology and biology to geology, hydrology, meteorology and other scientific data—and use them to design a better world, a world that considers the character of the environment and integrates it into what we design. While the name is new, the concept is ancient. The role of cartography and planning in human development has been immense (Harley and Woodward, 1987), the overall aim of which is to infuse design with science-based and value-based information to enable planners and stakeholders to facilitate holistic planning decisions.

Geodesign is 'design in geographic space' that provides the framework for exploring issues from an interdisciplinary point of view by combining science- and value-based designs. It is a set of technological ideas that combine geography with design. It does this by providing tools, such as simulation models, multi-criteria analysis, visualisation, spatial optimisation and real-time feedback. Using a case study, we describe the use of geodesign tools to facilitate collaborative marine planning based around a prospective tidal energy extraction site on the west coast of Scotland.

Case study background

The £10 million Saltire Prize Challenge was launched in Scotland in 2011 to accelerate the commercial development of wave and tidal energy technology.

As part of this, Marine Scotland (managers of Scotland's seas) and the Crown Estate (owners of the United Kingdom's seabed) made available seven search areas to be included in a leasing round to support competitors. One such potential lease site was located at the south end of the Kintyre Peninsula, on the south-west coast of Scotland (Figure 15.1), a site that is now being leased by two Scottish companies (Nautricity Ltd. and Argyll Tidal Ltd.) to investigate the possibility of locating up to six tidal turbines in the area. Diverse industries and activities operate around and within the site. This includes fishing (mostly fixed gear, includes gillnets, long lines, pots, traps, and any other gear that is anchored at least at one end), shipping, tourism (wildlife boat tours and paddle steamer tours), and recreational sailing and diving, all of which might be affected negatively by tidal energy development (Chapter 6). In addition, three cetacean species live in the area, including harbour porpoise (*Phocoena phocoena*), bottlenose dolphin (*Tursiops truncatus*) and minke whale (*Balaenoptera acutorostrata*).

This particular site was also of interest because conflict was already simmering within the local community due to a previously proposed offshore energy development. In 2010, the announcement of the potential development of an offshore wind farm array at Machrihanish, on the south-west Kintyre coast, led to conflict within the community and between the

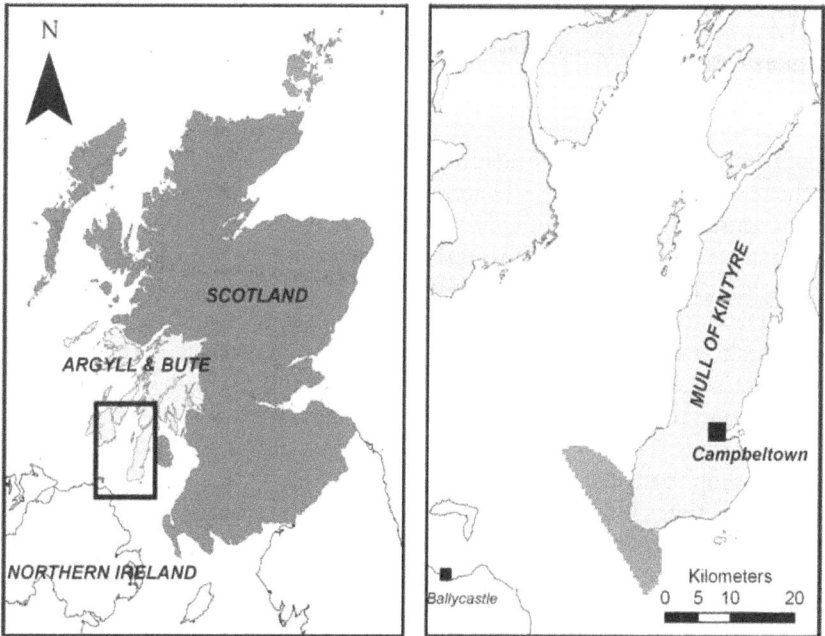

Figure 15.1 Case study site: southern tip of the Mull of Kintyre peninsula, Scotland. Sourced from Alexander et al., 2012.

community and the developers. Following public consultation on offshore wind farms, Marine Scotland received 297 responses regarding the Kintyre Inshore Windfarm proposal. Concerns raised by the local community included the impact on tourism and on visitors to a local golf club and its views of the western seaboard. This, combined with other factors such as the wind resource and proximity to the local airport (8.4 km), led to eventual abandonment of the project. The pre-existing conflict, in addition to the myriad ongoing uses of the site, provided the perfect opportunity for a study into the use of geodesign tools to identify conflicting stakeholder positions and to facilitate collaborative marine spatial planning.

Our aims were to assess the potential for geodesign tools to gather information at the fine spatial scales (e.g., ≤ 1 km^2) relevant to local marine spatial planning, to identify potential conflicts arising from proposed offshore renewable energy development and to support negotiation between marine stakeholders (for more information on stakeholders in the marine spatial planning process, see Chapter 9). Should geodesign tools achieve these aims, they would potentially benefit the offshore industry in particular and marine spatial planners and managers in general.

Geodesign tools and workshop format

The geodesign tools we used were based on the method of Arciniegas et al. (2011) to combine GIS, spatial multi-criteria analysis, and a touch-table to facilitate stakeholder dialogue in two workshops. A touch-table is an interactive touch screen, more recently developed as a large-scale (e.g., 46", or 117 cm screen) tablet computer built in a table format that allows simultaneous input from up to 60 users while identifying which person is touching where. In this case study, we used a DiamondTouch table™, the physical setup of which consisted of the DiamondTouch device connected to a personal computer via USB cable, a video projector suspended above the table and aimed down onto the touch surface, and receiver mats (placed on chairs) connected to the DiamondTouch unit by cables. Only four participants could use this touch-table at any one time. We ran the device concurrently with ArcGIS® with the CommunityViz (www.communityviz.com) extensions for interactive planning to enable the presentation of map data and the spatial multi-criteria analysis. The touch-table provided the interface between the data and the workshop participants (Figure 15.2).

The workshops followed the steps of the geodesign framework as defined by Steinitz (2012). Steinitz distinguishes three iterations through the framework, phrasing the questions as 'Why?' in the first iteration, as 'How?' in the second iteration and as 'What? Where? and When?' in the third and final iteration. The workshops go through these iterations in one afternoon. To be able to do this so quickly, the workshops relied heavily on stakeholder input.

Figure 15.2 Participants in the local-knowledge workshop using the touch-table to detail the areas most important for recreational use.

We did no extensive fieldwork prior to the workshops. We believed that the information about the region was not complete and might even be incorrect, but that stakeholders would correct the information.

The first workshop (a 'local knowledge' workshop) used GIS to present existing data in map format and to collect new spatial data using a 'map valu-ation tool' that allowed users to 'draw' onto the GIS display to input features of importance not identified on the original maps, and to change the val-ues of these identified areas according to relative importance. We generated six initial stakeholder value maps (*i.* tidal energy, *ii.* commercial shipping, *iii.* commercial fishing, *iv.* recreational shipping, *v.* tourism and *vi.* environ-ment) using existing data from organisations such as the British Ordnance Survey, the Royal Yachting Association and Historic Scotland, among oth-ers. We aggregated data by stakeholder group to generate stakeholder value maps using weighted summation, a commonly used method for spatial multi-criteria analysis. For example, we calculated the value for tidal devices as the weighted sum of tidal flow, depth in metres, type of seabed and distance to port. We set and specified weightings using expert judgement. We assigned values from 1 to 10 to a grid of 500 × 500 m cells based on the size of the study area and the likely size of tidal devices to be installed.

A second workshop (a 'negotiation' workshop) used GIS to present the updated data (from the first workshop) in combination with a 'multi-criteria analysis trade-off' tool that used spatial multi-criteria analysis for compari-son and ranking of three offshore energy-development scenarios, making it possible to structure and aggregate the information to facilitate negotiation.

The tool displayed the 'best' and 'worst' areas (i.e., those with the highest and lowest values) for particular stakeholder groups (tidal energy, commercial and social—grouped to allow information presented to be more manageable for the workshop participants), allowing sea uses to be compared and therefore facilitating spatial trade-offs. We asked stakeholders to trade negotiable cells as follows: two cells (0.25 km^2: ~ 40 megawatts [MW]), five cells (0.25 km^2: ~ 100 MW) and 10 cells (0.25 km^2: ~ 200 MW). The multi-criteria analysis trade-off tool provided feedback regarding how the overall value of the area for each stakeholder group changed, depending on how participants allocated cells as tidal energy sites. We displayed this information to participants using a bar chart on a separate screen.

Both workshops followed the same sequence of sessions. The first session described the research problem being addressed in the study (identifying an optimal location for tidal energy deployment within an established lease site), what would be involved in addressing the problem, and the stakeholders involved. The second allowed participants to familiarise themselves with the tools. The third session involved participant use of the tools to fulfil the objectives of the research.

'Local knowledge' workshop

Five stakeholders participated in this workshop: a fisher, a yacht owner, a local wildlife tour operator and two recreational divers, all of whom were invited based on their considerable knowledge and personal/professional interest of the area. Unfortunately, we were unable to access stakeholders with local knowledge of tidal energy, shipping or the environment, and so the maps relevant to these stakeholders remained in the initial format developed in advance of the workshop.

The participants updated three stakeholder-value maps during the 'local knowledge' workshop based on the knowledge of the local fisheries, recreational shipping and tourism uses (Figure 15.3). The fisher identified that fixed-gear fishing for crab (*Cancer pagurus*) and lobster (*Homarus gammarus*), scallop (*Pectinidae* spp.) dredging and scallop diving occurred within the area and suggested that some areas within the study site had tides too strong for fishing. The fisher also pointed out two routes for fishing vessel transit around the Mull of Kintyre. The fisher allocated the highest possible value to all fishing areas. Participants identified that the area around the Mull of Kintyre was mainly a passage route for recreational vessels and marked a route hugging the coast of the peninsula as the most important. The local recreational sailor found it difficult to give values to cells within the study site because sailing around the Mull involved much tacking, and poor weather means that routes can change; depending on the experience of the yachtsman, routes are not always followed closely. Finally, discussion of the tourism map revealed that no regular marine tourism occurred around the Mull of Kintyre, leaving much

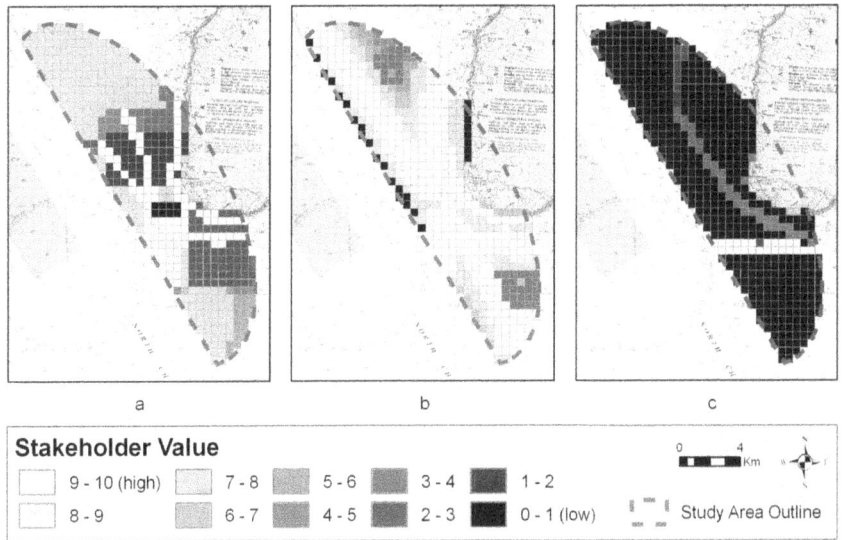

Stakeholder Value

☐ 9 - 10 (high)	7 - 8	5 - 6	3 - 4	1 - 2
☐ 8 - 9	6 - 7	4 - 5	2 - 3	0 - 1 (low)

0   4 Km

⌐ ¬ Study Area Outline

Figure 15.3 The results of the 'local-knowledge' workshop: (a) fisheries output, (b) recreational sailing output and (c) tourism output.

of the study site marked as 'low value'. Furthermore, all local stakeholders agreed that no kayaking occurs within the area due to the currents, and that the dive sites are unpopular.

In addition to collecting local data during this workshop, we identified some conflicts between marine users. In particular, the fishers who had a long history and tradition of working in the area indicated a strong sense of ownership and general unwillingness to concede space. Participants also suggested that the area around the Mull of Kintyre was a passage route for recreational vessels and that boats were 'funnelled' through the area from the Clyde marinas, causing the area to become a bottleneck. The local tour operator pointed out that although the area cannot be used regularly, when placed in a situation of potential loss of access, it is likely that all stakeholders would exaggerate the importance of the area and suddenly claim that most of their income arises from that area.

'Negotiation' workshop

Six stakeholders participated in the negotiation workshop: a representative of the local sailing club, a wildlife tour operator, a local fisherman, the representative of the local fishing association, a consultant from Scottish Renewables (representing the tidal energy industry) and a member of the Argyll & Bute Council. We invited participants personally based on their role as a

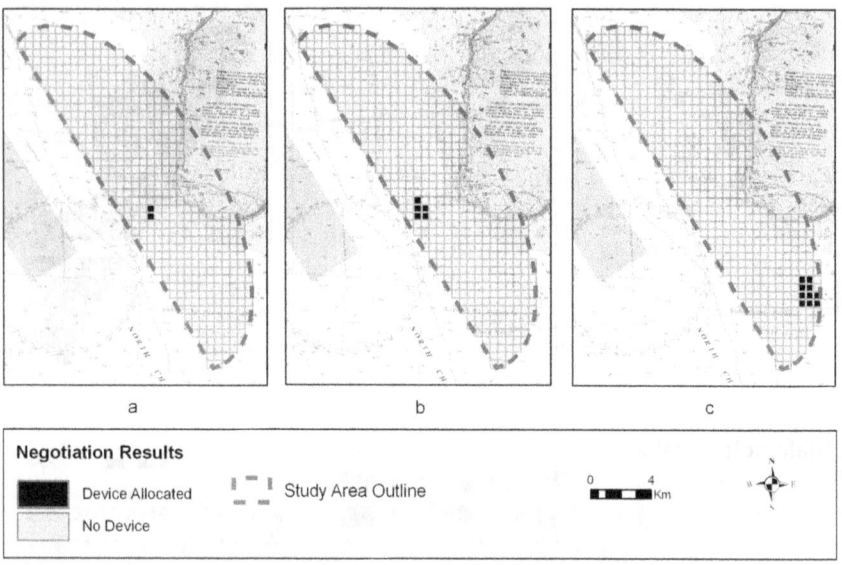

Figure 15.4 The results of the 'negotiation' workshop: (a) device allocation for a ~ 40 MW alternative, (b) device allocation for a ~ 100 MW alternative, and (c) device allocation for a ~ 200 MW alternative.

high-level stakeholder (with the exception of the local fisher, who was interested in the use of the information from the previous workshop).

We created three more maps during this workshop, presenting the agreed (negotiated) locations of tidal energy developments under the three scenarios described earlier (Figure 15.4). Participants found it easiest to allocate two cells for the 40 MW scenario, choosing cells that were offshore to avoid conflicts with other stakeholders, but that still had good tidal flows and were not too deep. Participants appeared to find it more difficult to allocate the five-cell (100 MW development) and 10-cell (200 MW development) allocations. The 100 MW allocation was similar to the results of the 40 MW scenario, although the selected cells were shifted to the west to allow for tacking of recreational vessels between land and the tidal devices to avoid sailing around the devices when in transit. All participants with the exception of the fishing representative (who insisted that the whole area was important to fisheries) concurred with this allocation of devices. As the tidal energy-development scenario increased, negotiation became more difficult. At this point, the participants raised the issue of exclusion zones particularly as they related to their impact on the fishing industry both in terms of loss of access and safety implications for vessels (i.e., turbines acting as winches and pulling fishing boats under). The fisher noted that one large block of cells would lose less space for fisheries than several smaller ones (due to the resulting exclusion zone, which in terms of this study would have a device/exclusion zone ratio of 1:8 cells, reducing to 1:5 then 1:4, etc., as the number of cells in a block increases). Additionally, the recreational shipping representative suggested that 10 cells would interfere with yachting around the Mull of Kintyre if the cells were kept in the same area as previous alternatives. We suggested that a site farther to the south was a more appropriate location. All stakeholders agreed that the third and largest tidal energy development scenario should be allocated in this part of the study site.

As negotiations progressed, participants could view the effects of each of the proposal allocations in a bar chart displayed on a separate screen next to the touch-table. Each chart showed the overall stakeholder value for all six stakeholders of the Mull of Kintyre site. As a starting position, the current situation (no devices) was considered optimal for all current stakeholders with a value of 10 and no value for 'tidal energy'. Locating tidal devices increased the value of tidal energy at the expense of all other stakeholders. The 40 MW scenario increased the value of 'tidal energy' at the expense of mainly 'tourism'. The 100 MW scenario had a considerable effect on the 'commercial shipping' and 'fishing' stakeholders. Finally, the scores for the 200 MW alternative showed that the most affected stakeholder was 'commercial shipping'.

Perhaps most interesting was some of the discussion held during the negotiations. One area of focus was the quality of the data upon which negotiations were to be based. First, participants suggested that the published tidal data were inaccurate. The fisheries association representative advised that fixed-gear fisheries used all areas of the proposed site, thus contradicting the 'local knowledge' map from the previous workshop. The fishing-association representative also suggested that dealing with commercial fishing as a single sector was simplistic and that different types of fishing should be considered. During this workshop, the fishing association representative pointed out that "… the negotiations are based upon an assumption that tidal energy would take precedence over other stakeholders". Some participants suggested that areas farther offshore would be preferable for tidal devices, to avoid conflict with other stakeholders, yet they also raised concerns that these areas might be too deep for cost-effective development. We also noted that the fishing-association representative stressed that fishers do not want to share information on fishing practices with each other, which suggests the potential for inter-industry conflict.

Conclusions

We identified four main issues during the workshops that we consider relevant in the development of marine renewable energy. First, existing data regarding human uses of the marine environment do not currently exist at a scale that is useful for siting installations—the negotiations made clear that data collection at a relevant scale for development is imperative. Second, offshore renewable energy development can effectively place particular areas of the common sea space under private control (e.g., through exclusion zones) and place spatial restrictions on many ongoing activities—the fishing industry is most likely to be affected in this way (Chapter 6). Third, the proposed need for safety exclusion zones raises the question of liability: fishers may 'push' the boundaries of exclusion zones, particularly under economic stress. There are currently no known guidelines concerning the identification of liability and insurance costs in the case of accidents involving renewable devices and this should be resolved prior to installing devices. Finally, the suggestion that the tidal energy industry takes precedence over any other stakeholder could be a reason for potential conflict between developers and existing users—this standpoint should be borne in mind when consulting with stakeholders.

We conclude that the most important result of our study was the successful transfer of an approach, originally developed for land-use planning, to a marine-planning problem. In the United Kingdom, Marine Scotland has begun mapping spatial uses of Scottish Territorial Waters including fishing activity using an interview-based method (known as Scotmap). However, this is a time-consuming method of data collection and one that does not take into account user values. Also of importance, the companies interested in developing tidal energy at this site contacted us to request our data and results; we do not yet know if the data have been or will be taken into consideration in any development decisions for this location.

In the United Kingdom and throughout Europe, spatial data infrastructure is developing rapidly; with the emergence of Open Geospatial Consortium standards, spatial data are becoming ubiquitous on the web and data constraints are beginning to evaporate, perhaps heralding a new age in geodesign. Paradoxically, the movement from proprietary software toward open-source data formats and software is driving this revolution in geodesign. The main benefits of the geodesign tools we have described are as follows: they are quick and easy to use, they include 'added value' by means of identifying areas used as well as the value that stakeholders place on these uses, and they provide feedback on the overall value of the plan, thereby helping to find an optimal rather than just an acceptable solution. These geodesign tools engaged all stakeholders during both workshops. Participants had little or no experience with this type of tool, and yet engaged easily with the process. Furthermore, the approach enabled participants to view the 'bigger picture' and enhanced interactivity and communication across sectors. Participants shared ideas, asked each other questions and brought up points that were then developed by others. The interactivity stimulated by the use of these tools is central to this approach and would not occur when using online data gathering or face-to-face interviewing methods.

Highlights

- Offshore industrial developments might lead to *de facto* 'privatisation of the sea', but this is a complex issue that should be addressed by decision-makers.
- Precedence of one marine industry over another can lead to conflict and so trade-offs should be explicitly identified in the marine spatial planning process.
- Marine and coastal use data are often not available at appropriate scales for local marine planning.
- Geodesign tools are useful for identifying sea-user conflict and viewing the 'bigger picture'.

References

Alexander, Karen A., et al. "Interactive marine spatial planning: siting tidal energy arrays around the Mull of Kintyre." PLoS One 7.1 (2012): e30031.

Arciniegas, G., Janssen, R. and Omtzigt, N. (2011). Map-based multicriteria analysis to support interactive land use allocation. *International Journal of Geographical Information Science*, 25(12), pp. 1931–1947.

Harley, J.B. and Woodward, D. (1987). *The History of Cartography. Volume 1: Cartography in Prehistoric, Ancient, and Medieval Europe and the Mediterranean.* Chicago, IL: The University of Chicago Press, p. 622.

Steinitz, C. (2012). *A Framework for Geodesign: Changing Geography by Design.* Redlands, CA: ESRI Press.

Chapter 16

The future of marine spatial planning

Corey J. A. Bradshaw, Lucy Greenhill and Katherine L. Yates

Introduction

Space in the marine realm is, it seems, still the final frontier. While we have yet to explore most of the ocean's depths (Copley 2014) and discover up to two-thirds of its extant species (Appeltans et al. 2012), we are rapidly increasing our exploitation of the marine environment. Space in the inshore realm faces rising competition from not just 'traditional' uses like shipping and fishing in the quest for food security, but also from a rising demand for offshore-energy generation, both fossil-fuel and increasingly, renewable-energy production. Like the challenge of realising the United Nations' Sustainable Development Goals (sustainabledevelopment.un.org), long-term, sustainable use of the marine environment requires reconciling competing-sector goals and conserving biodiversity as the consumer demands of the 7.5 billion-strong and rising human population accelerate (Bradshaw and Brook 2014). In many ways, this increasing competition for dwindling space has thrust upon society the need for a more integrated and structured approach to the planning and management of marine resources (Chapter 1).

Offshore energy is an important and growing sector that needs to be included within marine spatial planning, and this new volume comprehensively outlines how the expansion of this sector in particular can best be incorporated into the many facets of multi-sector marine spatial planning. Indeed, one can even argue that offshore-energy development and prospecting have, in many regions, driven the development of marine spatial planning (Chapter 1, Jay 2010; Jones et al. 2013). As such, the experts in this volume have sketched out how effective marine spatial planning can minimise the negative impacts of offshore-energy developments and realise new opportunities for multi-sector benefits. Collectively, the chapters herein have also highlighted some of many challenges that remain for marine spatial planning, and suggest how increased collaboration, more meaningful participation, and more aligned governance structures combined with transparent planning tools can contribute to future successes.

Instead of a random, case-by-case process of *post hoc* conflict resolution, marine spatial planning is primarily a foresight-focussed approach that explicitly aims to develop optimal planning for using the marine environment, and to mitigate both *user–user* and *user–nature* conflicts. Some of the most complex real and potential conflicts in this regard are those that pit traditional and developing industries—fisheries, oil and gas exploration and exploitation, wind turbines, and other existing and emerging energy-generation technologies—against the integrity of the marine environment (*user–nature* conflict). Pessimists might be convinced that viable trade-offs aiming to safeguard the marine environment from industrially sourced degradation are impossible, and historical evidence would, in many respects, support their arguments (e.g., oil spills, overfishing, and pollution). However, irrespective of the ethical and philosophical arguments against development of the marine environment, industry and governments will continue to push the technological boundaries to provide the resources our growing economies demand. In that sense, morally based opposition alone is untenable, and so participatory and balanced trade-offs are preferable to provide the best possible outcomes for the environment and society simultaneously (Chapter 2). However, there is nothing in this sentiment that assures achieving single-minded, sector-specific goals—trade-offs imply compromise (Chapter 6), whether they embody lower profits, suboptimal locations, some *acceptable* environmental damage, or lower harvest yields. Offshore-energy development, and especially the renewables sector, has arguably driven the development of marine spatial planning in many places, especially Europe (Chapter 1, Jay 2010). As a tool for rationalising management of the sea, balancing different objectives, and encouraging ecosystem-based management, marine spatial planning has the potential to lead to more positive outcomes for environmental integrity. Thus, it could be argued that the expansion of offshore renewable energy has in fact been one of the main motivations for improving marine management that could lead to long-term environmental benefits.

From the perspective of maintaining environmental integrity in particular, two important observations emerge from the various chapters presented in this volume. Perhaps the most recent and fundamental shift in environmental and conservation thinking, which is by no means isolated to the process of marine spatial planning, is the idea that compromise is not only possible, it is perhaps the only way to limit runaway environmental damage. Now a major topic of discussion and debate in the environmental literature (e.g., Holmes et al. 2017; Matulis and Moyer 2017), the idea that combative, oppositional environmentalism is the only way to proceed is perhaps losing favour among at least a component of environmentally focussed organisations. This, of course, does not preclude the right to protest or the more traditional environmentalist perspective of exclusion and protectionism, but it does at least accept the idea that a conversation with 'the enemy' is not only possible, it can also represent the best way to achieve the environmentalist's goals (Chapter 3).

Accepting that these basic tenets of cooperation and collaboration rest at its very foundation, marine spatial planning represents a formal process to unearth optimal trade-offs, provided all of the relevant stakeholders have a seat at the table. Not only does this volume explicitly outline the various, modern methodologies that have been designed to facilitate trade-off analysis and discovery (e.g., Chapters 2 and 9), it also highlights how to get the relevant parties to the table in the first place (Chapter 3) and keep them involved throughout the entire planning process (Chapters 4 and 9). In Chapter 3 in particular, Polsenberg and Kilponen reason eloquently that the conciliatory approach is not only the most practical pathway toward maintained environmental integrity, it also is at heart one of offshore energy's newest 'bottom lines'—the social licence to do business. While accepting that the dominant corporate bottom line will likely remain profit maximisation, social acceptance is increasingly seen not only as a necessary component of social and regulatory licensing, it has also become incorporated into the value system of public-good industries like energy generation in particular (Chapter 3).

In Chapter 9, Yates provides an in-depth assessment of the stakeholder-engagement process, arguing that effective participation is essential to successful planning outcomes and that participation needs to be meaningful to each stakeholder to be effective. In other words, stakeholders need to know clearly how their input will influence decision-making, and so participation processes need to balance power among stakeholders. This is particularly important in marine spatial planning with offshore energy, where energy companies can be seen as large, powerful organisations with national mandates by local stakeholders who believe themselves powerless to influence decisions. Maintaining engagement throughout the planning process is also essential for effective participation. In this light, marine spatial planning requires a good understanding of the psychological and marketing aspects of the stakeholder community to be effective. This is demonstrated clearly in the politically and ethically charged example of protection for a Critically Endangered cetacean (North Atlantic right whale *Eubalaena glacialis*) and offshore wind development on the east coast of the United States. In Chapter 12, Brooks and Jedele demonstrate through acknowledgement that whales in proposed wind-energy areas could be an obstacle for development, environmental advocates promoting both renewable energy and whale conservation effectively used marine spatial planning to achieve reconciliation. The trust built between the parties, while taking years, eventually smoothed the path for developers while winning both the support of once-adversarial environmental organisations, and most importantly, led to greater protection for right whale feeding and resting areas.

The second major evolution in modern environmentalism, at least from the perspective of the environmental sciences, is the idea that environmental integrity is not only difficult to maintain, but it is fraught with ambiguous definition. First, the very terminology 'maintain' insinuates a static baseline to which we are constantly attempting to return, when in reality ecosystems are anything but static. Indeed, the entire discipline of ecology is founded on the notion that ecosystems are dynamic, evolving, and ever-changing entities that defy traditional pigeonholing. Taken to its logical extension and over long-enough timescales, ecological baselines therefore do not exist. The practical implication of this notion for environmental management is that we can find ourselves trapped on a hamster wheel of unattainable environmental targets. If even ecologists struggle with these concepts, imagine how those tasked with enforcing environmental regulation and non-specialist industry representatives can become confused by the constantly shifting goal posts.

Gee and Burkhard (Chapter 7) and Gill *et al.* (Chapter 8) elaborate on how marine spatial planning can evolve to account for this dilemma by focussing not on static environmental baselines, but by placing emphasis instead on the *resilience* of the ecosystem and the *ecosystem services* they provide. Returning to the basic tenet of compromise, an acceptance at least of some change to the marine environment is inevitable if any industry is to proceed; however, marine spatial planning should attempt to implicate the grounding principle of continued or improved delivery of the main ecosystem functions and services in the region of interest. Ideally, this goal requires a more holistic mindset regarding measurement and monitoring of cumulative environmental impacts (Chapter 8), but also includes a socio-economic perspective that assesses whether the entirety of community benefits arising from the use of the sea (e.g., renewable fish stocks to harvest, environmental aesthetics to appreciate, energy to use, and shipping routes to transport goods) will continue in perpetuity. In essence, these are the ecosystem services that need to be maintained. Gill *et al.* also note in Chapter 8 that to date, much of the published evidence for the effects of energy developments on marine organisms focuses on the negative, such that a publication bias might exist given there might be no incentive to report neutral, or even positive benefits.

A novel element to have emerged from the practical implementation of marine spatial planning so far has been the slow dawning on the principal sectors involved that 'co-location' of different sectors—meaning concurrent activities in time and space—is not only possible, but potentially highly beneficial. Described in more conceptual terms in Chapters 10 (Hopper *et al.*) and 11 (Thurstan *et al.*), then elaborated with concrete examples for wind farms by Ashley *et al.* (Chapter 13), the idea generally is that some activities might be compatible in space and time. For instance, offshore wind farms both practically and potentially provide some protection to existing marine ecosystems by virtue of limiting commercial and recreational fishing within their immediately vicinity. These wind farms can therefore act as a type of *de facto* marine protected area that can buffer the greater ecosystem from fishing pressures, even though the marine communities themselves might be altered (i.e., some species will be 'winners', and others will be 'losers'). While more work needs to be done to determine the relative losses from construction *versus* the relative benefits of fisheries exclusion, even the place-ment of artificial infrastructure can potentially enhance marine ecosystems by providing both hard substrata for colonisation (Chapters 10 and 11) and subsequent protection from over-exploitation (Chapter 13). Another prime example of co-location opportunities between offshore energy and other sec-tors is aquaculture on the monopoles of wind farms, which could allow for expansion of aquaculture farther offshore than is currently feasible (Chapter 10). There are regulatory hurdles to be crossed, but if these two industries can work together, infrastructure can be designed to rationalise costs and maximise benefits. Hopefully through marine spatial planning, management of different sectors and associated regulatory requirements can be better in-tegrated, making it easier to realise opportunities like these and optimise benefits across sectors.

These philosophical foundations notwithstanding, this volume also pro-vides some of the more technical 'how to' components of effective marine planning, along with concrete case-study examples that identify both what has worked, and what has not, in the rapidly evolving multi- and transdisci-plinary processes of marine spatial planning with offshore energy. Based on a long and proud history of generic spatial 'conservation planning' (Margules and Pressey 2000, Chan et al. 2006, Pressey et al. 2007, Moilanen et al. 2009), Stevens *et al.* (Chapter 2) describe what they call 'ecosystem service trade-off analysis'. This is a data-driven approach that uses mathematical op-timisation algorithms to help identify the best ways to incorporate sometimes diametrically opposed plans to use the marine environment. Based on the notion of maintaining ecosystem services in the long term as the principal currency to optimise across the seascape, ecosystem service trade-off analysis is one of the most objective methodologies yet constructed to achieve the elu-sive goal of minimising costs and environmental damages while maximising whole-of-community benefits. While certainly a data-hungry approach that requires extensive monitoring and economic data, it represents one of the most powerful and incontestable tools in the marine spatial planner's toolbox.

The human element in marine spatial planning can, however, often usurp or completely displace the mathematical optimisation approaches, especially if stakeholder engagement is done ineffectively. Using robust, transparent processes for gathering and using stakeholder-derived data is therefore es-sential for ensuring meaningful participation and ultimately, fair representa-tion and continued stakeholder buy-in to the planning process (Chapter 9). Tools that allow for the collection of spatial data (quantitative and qualitative) and their incorporation into optimisation methods such as those described in Chapter 9, offer powerful and transparent ways to engage and incorpo-rate stakeholders. Alexander *et al.* (Chapter 15) demonstrate how the process of stakeholder engagement can be a form of data generation and validation that can work in conjunction with ecosystem service trade-off analysis. Us-ing an example from the west coast of Scotland to identify areas of conflict and compromise for a tidal energy installation, the authors demonstrate how opinions and concerns can be turned into useable spatial data through interactive 'geodesign' mapping tools. Indeed, the future of successful marine spatial planning is increasingly likely to depend on the technology that facilitates the transfer of human knowledge, values, and opinions into datasets that can be queried, cross-matched, and optimised formally. The development of such methodological approaches and associated software (and possibly, hardware) interfaces could contribute substantially to effective stakeholder participation and more robust, equitable and ultimately defensible, planning decisions. This is because transparency acts to balance the perception of 'power' in the decision-making process, giving everyone involved an equal voice on how the data are interpreted. This is an especially important element of marine spatial planning with offshore energy.

Of course, even when the data are sufficient to use in optimisation methods to produce ideal planning guidelines, both the legal and governance frameworks must exist within which to implement the optimal pathways. In Chapter 5, van Doorn and Gahlen provide a detailed analysis and discussion of the very real legal impediments and challenges to effective marine spatial planning. In essence, the lack of a dedicated legal framework within which marine spatial planning could ideally operate means that the process is mired within a hodgepodge of other legislations that only touch on elements of the entirety. Depending on the region of implementation, optimised planning can find itself limited by elements of the Law of the Sea Convention that explicitly mandates the free movement of marine vessels on the high seas. Even within the sovereignty of national Exclusive Economic Zones, no existing multistate legal processes yet exist to streamline the planning of potential conflicts among nations. The fact that marine spatial planning has had any room to manoeuvre to date has largely been the result of *post hoc* agreements and the fact that the one of the main underlying components of the Law of the Sea Convention is that environmental integrity (however loosely defined) must be maintained. A clear impediment to the future success of marine spatial planning in Europe and elsewhere is the creation of express legislation purpose-built for maximising the probability of implementing the optimised plans derived for successful multi-sector outcomes.

The other major impediment to effective marine spatial planning is effectively summarised in Chapter 4 by Greenhill. Today, marine spatial planning is housed awkwardly within a fragmented set of governance institutions, from government, to industry, to civil society. Not only is there a major challenge in how decisions are made within and among the different levels of these governance structures, there is an urgent need to standardise even the terminology and concepts of 'sustainability' and 'ecosystem-based management' across them. Akin to the challenges of harmonising the international legal framework within which marine spatial planning must frequently operate, clarifying and integrating governance across the various responsible institutions is a priority for the process. One of the central tenets of marine spatial planning is its 'adaptive' framework, or learning from its past successes and failures (Chapters 2 and 4); without a seamless and supportive governance framework, marine spatial planning is likely to be inefficient and limited in its ability to deliver improved governance of marine activities.

As a clear demonstration of the challenges facing modern marine spatial planning, Mazor *et al.* (Chapter 14) outline today's major obstacles and challenges of reconciling oil and gas development in the Mediterranean with environmental resilience. The two principal problems, foreseen in more conceptual detail in previous chapters, are the difficulties of obtaining adequate monitoring data to assess the long-term effects of development on the marine biota, and the jurisdictional challenges of operating in a seascape governed by more than 20 countries on three different continents. The Mediterranean is the epitome of the problems still facing marine spatial planning in general, and the authors establish conclusively that without realistic means to co-fund data collection, no amount of goodwill and cooperation will adequately buffer marine spatial planning from suboptimal outcomes. Effective collaboration between offshore-energy companies—who are obliged to collect substantial amounts of data as part of their operating licences—and other stakeholders could contribute to resolving this challenge (Chapter 3).

Conclusions and outlook

There are several emerging issues on the horizon that will likely change the face of marine spatial planning with offshore energy as it grows into a mature and universally adopted approach across the world. First, there will likely be mounting pressure to curtail fossil-fuel developments as climate change continues to wreak environmental and social havoc throughout the coming decades (Heard et al. 2017). Although certain governments might resist moves away from carbon-intensive industries because of strong corporate lobbying (Oreskes and Conway 2010), any carbon-intensive developments will inevitably face increasingly intense opposition from a litany of marine users, and not just from the traditional advocates of environmental preservation. Governments around the world are increasingly turning investment toward renewable and low-carbon electricity-generation installations. The flip side, of course, is that increasing interest in marine renewable-energy technologies will likely see expansion into areas traditionally devoid of offshore-energy developments (e.g., floating platforms, tidal energy), and technological advances will also likely enhance co-location possibilities. Thus, marine-based, renewable-energy installations will present different challenges to the traditional procedures associated with fossil-fuel extraction, possibly demanding more space and thus potentially displacing even more fisheries. Aquaculture opportunities might also expand, particularly if co-location opportunities with offshore energy can be realised, potentially providing new challenges to maintaining ecosystem integrity. Thus, with an increasing emphasis on renewable energy, and novel areas of operation, planning will have to evolve in concert. Marine spatial planning will also need to consider new and expanding offshore-energy technologies, which include some uncertainty regarding the type of ecological and social impacts that can be expected. The design of adaptive strategies is essential in this context, whereby desirable development is not hindered, but monitoring refines management appropriate to the environmental and social effects that materialise.

Another consideration is the rising risk of extinction faced by both terrestrial (Ceballos et al. 2017) and marine organisms (McCauley et al. 2015). As awareness of these worrying trends increases, it is likely that legislation safeguarding marine ecosystems will see even more space allocated for conservation in the form of marine protected areas and licencing restrictions for fisheries and offshore-energy developments. For example, the United Nations' Sustainable Development goals explicitly target protection of at least 10% of coastal and marine areas by 2020 (sustainabledevelopment.un. org). Marine spatial planners will therefore be faced with increasing difficulty to promote balanced marine-resource use among different stakeholders as the vulnerabilities of more and more marine species are identified and publicised. From opposition to commercial and recreational fisheries, to general impediments for energy-production infrastructure, environmental concerns are likely to mount, as well as their legal and governance support. As such, it is undoubtedly in the long-term interest of any industry to invest more comprehensively in monitoring and measurement of the ecosystems in which they do business, sharing the data freely with all interested stakeholders. In this regard, the standard measurement protocol insists on pre-development functional baselines (i.e., what constitutes a functional ecosystem, as opposed to static baselines) to contrast with post-development impacts (what is known as before-after-control-impact paired series experimental design) (Thiault et al. 2017).

Within the community of marine spatial planners, there is currently some debate regarding how the entire process should be evaluated, and what metrics and thresholds indicate 'success'. Of course, the definition of 'success' depends on one's perspective, which includes *inter alia* the financial and policy objectives of various sectors, but also that these do not compromise environmental integrity. The latter is currently monitored at different spatial and temporal scales depending on the sector involved—offshore-energy industries tend to monitor at the scale of the specific development project defined by conditions of their operating licence, whereas other organisations such as conservation groups and government environmental agencies tend to examine environmental integrity at species-specific or regional scales. Cooperation across these scales therefore becomes an imperative of marine spatial planning to improve our understanding of ecosystem behaviour in response to marine developments and other human activities on the sea.

But as always, empirical demonstration of impacts—either positive or negative—will be insufficient alone to move marine spatial planning to the next level of sophistication and effectiveness. As discussed throughout this volume, marine spatial planning invokes a complex array of disciplines that must be harmonised—from the scientific, socio-economic, legal, political, to the psychological. Technological innovation is essential in this process, as well as specific training for those would-be marine spatial planners tasked with finding the critical balance between the need for economic development, environmental integrity, energy and food security, livelihoods, and social cohesion. Indeed, the appearance of Marine Spatial Planning Masters programmes in some United Kingdom universities heralds the worldwide recognition of the necessity for custom-designed university degrees in the multidisciplinary field of marine spatial planning. While the challenges are great, marine spatial planning stands to become one of the most important processes required to do business in the sea.

References

Appeltans, W., Shane, T., Ahyong, G., Anderson, M.V., Angel, T., Artois, N., Bailly, R., Bamber, A., Barber, I., Bartsch, A., Berta, M., Błażewicz-Paszkowycz, P., Bock, G., Boxshall, G., Boyko, C.B., Brandão, S.N., Bray, R.A., Bruce, N.L., Cairns, S.D., Chan, T.-Y., Cheng, L., Collins, A.G., Cribb, T., Curini-Galletti, M., Dahdouh-Guebas, F., Davie, P.J.F., Dawson, M.N., De Clerck, O., Decock, W., De Grave, S., de Voogd, N.J., Domning, D.P., Emig, C.C., Erséus, C., Eschmeyer, W., Fauchald, K., Fautin, D.G., Feist, S.W., Fransen, C.H.J.M., Furuya, H., Garcia-Alvarez, O., Gerken, S., Gibson, D., Gittenberger, A., Gofas, S., Gómez-Daglio, L., Gordon, D.P., Guiry, M.D., Hernandez, F., Hoeksema, B.W., Hopcroft, R.R., Jaume, D., Kirk, P., Koedam, N., Koenemann, S., Kolb, J.B., Kristensen, R.M., Kroh, A., Lambert, G., Lazarus, D.B., Lemaitre, R., Longshaw, M., Lowry, J., Macpherson, E., Madin, L.P., Mah, C., Mapstone, G., McLaughlin, P.A., Mees, J., Meland, K., Messing, C.G., Mills, C.E., Molodtsova, T.N., Mooi, R., Neuhaus, B., Peter K., Ng, L., Nielsen, C., Norenburg, J., Opresko, D.M., Osawa, M., Paulay, G., Perrin, W., Pilger, J.F., Poore, G.C.B., Pugh, P., Read, G.B., Reimer, J.D., Rius, M., Rocha, R.M., Saiz-Salinas, J.I., Scarabino, V., Schierwater, B., Schmidt-Rhaesa, A., Schnabel, K.E., Schotte, M., Schuchert, P., Schwabe, E., Segers, H., Self-Sullivan, C., Shenkar, N., Siegel, V., Sterrer, W., Stöhr, S., Swalla, B., Tasker, M.L., Thuesen, E.V., Timm, T., Todaro, M.A., Turon, X., Tyler, S., Uetz, P., van der Land, J., Vanhoorne, B., van Ofwegen, L.P., van Soest, R.W.M., Vanaverbeke, J., Walker-Smith, G., Walter, T.C., Warren, A., Williams, G.C., Wilson, S.P. and Costello, M.J. (2012). The magnitude of global marine species diversity. *Current Biology*, 22(23), pp. 2189–2202.

Bradshaw, C.J.A. and Brook, B.W. (2014). Human population reduction is not a quick fix for environmental problems. *Proceedings of the National Academy of Sciences of the USA*, 111(46), pp. 16610–16615.

Ceballos, G., Ehrlich, P.R. and Dirzo, R. (2017). Biological annihilation via the on-going sixth mass extinction signaled by vertebrate population losses and declines.

Proceedings of the National Academy of Sciences of the USA, 114(30), pp. E6089–E6096. doi:10.1073/pnas.1704949114.

Chan, K.M., Shaw, M.R., Cameron, D.R., Underwood, E.C. and Daily, G.C. (2006). Conservation planning for ecosystem services. *PLoS Biology,* 4(11), p. e379.

Copley, J. (2014). *Just how little do we know about the ocean floor.* The Conversation. [online] Available at https://theconversation.com/just-how-little-do-we-know-about-the-ocean-floor-32751.

Heard, B.P., Brook, B.W., Wigley, T.M.L. and Bradshaw, C.J.A. (2017). Burden of proof: A comprehensive review of the feasibility of 100% renewable-electricity systems. *Renewable and Sustainable Energy Reviews,* 76, pp. 1122–1133.

Holmes, G., Sandbrook, C. and Fisher, J.A. (2017). Understanding conservationists' perspectives on the new-conservation debate. *Conservation Biology,* 31(2), pp. 353–363.

Jay, S. (2010). Planners to the rescue: Spatial planning facilitating the development of offshore wind energy. *Marine Pollution Bulletin,* 60(4), pp. 493–499.

Jones, P., Qiu, W. and Lieberknecht, L. (2013). MESMA Work Package 6 (Governance) Typology of Conflicts in MESMA case studies. Department of Geography, University College London.

Margules, C.R. and Pressey, R.L. (2000). Systematic conservation planning. *Nature,* 405(6783), pp. 243–253.

Matulis, B.S. and Moyer, J.R. (2017). Beyond inclusive conservation: The value of pluralism, the need for agonism, and the case for social instrumentalism. *Conservation Letters,* 10(3), pp. 279–287.

McCauley, D.J., Pinsky, M.L., Palumbi, S.R., Estes, J.A., Joyce, F.H. and Warner, R.R. (2015). Marine defaunation: animal loss in the global ocean. *Science,* 347(6219), pp. 1255641.

Moilanen, A., Wilson, K.A. and Possingham, H. (2009). *Spatial Conservation Prioritization: Quantitative Methods and Computational Tools.* Oxford: Oxford University Press.

Oreskes, N. and Conway, E.M. (2010). *Merchants of Doubt.* New York: Bloomsbury Press.

Pressey, R.L., Cabeza, M., Watts, M.E., Cowling, R.M. and Wilson, K.A. (2007). Conservation planning in a changing world. *Trends in Ecology & Evolution,* 22(11), pp. 583–592.

Thiault, L., Kernaléguen, L., Osenberg, C.W. and Claudet, J. (2017). Progressive-change BACIPS: A flexible approach for environmental impact assessment. *Methods in Ecology and Evolution,* 8(3), pp. 288–296.

Index